Lecture Notes in Computer Science 13217

More information about this series at https://link.springer.com/bookseries/558

Evgeny Burnaev · Dmitry I. Ignatov ·
Sergei Ivanov · Michael Khachay ·
Olessia Koltsova · Andrei Kutuzov ·
Sergei O. Kuznetsov · Natalia Loukachevitch ·
Amedeo Napoli · Alexander Panchenko ·
Panos M. Pardalos · Jari Saramäki ·
Andrey V. Savchenko · Evgenii Tsymbalov ·
Elena Tutubalina (Eds.)

Analysis of Images, Social Networks and Texts

10th International Conference, AIST 2021
Tbilisi, Georgia, December 16–18, 2021
Revised Selected Papers

 Springer

Editors
Evgeny Burnaev (iD)
Skolkovo Institute of Science and Technology
Moscow, Russia

Sergei Ivanov
Skolkovo Institute of Science and Technology
Moscow, Russia

Olessia Koltsova (iD)
National Research University Higher School
of Economics
St. Petersburg, Russia

Sergei O. Kuznetsov (iD)
National Research University Higher School
of Economics
Moscow, Russia

Amedeo Napoli (iD)
LORIA, Campus Scientifique
Vandœuvre lès Nancy, France

Panos M. Pardalos (iD)
University of Florida
Gainesville, USA

Andrey V. Savchenko (iD)
National Research University Higher School
of Economics
Nizhny Novgorod, Russia

Elena Tutubalina (iD)
Kazan Federal University
Kazan, Russia

Dmitry I. Ignatov (iD)
National Research University Higher School
of Economics
Moscow, Russia

Michael Khachay (iD)
Krasovskii Institute of Mathematics
and Mechanics of Russian Academy of Sciences
Yekaterinburg, Russia

Andrei Kutuzov (iD)
University of Oslo
Oslo, Norway

Natalia Loukachevitch (iD)
Lomonosov Moscow State University
Moscow, Russia

Alexander Panchenko (iD)
Skolkovo Institute of Science and Technology
Moscow, Russia

Jari Saramäki (iD)
Aalto University
Espoo, Finland

Evgenii Tsymbalov (iD)
Yandex LLC
Moscow, Russia

ISSN 0302-9743   ISSN 1611-3349 (electronic)
Lecture Notes in Computer Science
ISBN 978-3-031-16499-6   ISBN 978-3-031-16500-9 (eBook)
https://doi.org/10.1007/978-3-031-16500-9

This Springer imprint is published by the registered company Springer Nature Switzerland AG
The registered company address is: Gewerbestrasse 11, 6330 Cham, Switzerland

Preface

This volume contains the refereed proceedings of the 10th International Conference on Analysis of Images, Social Networks, and Texts (AIST 2021)[1]. The previous conferences (during 2012–2020) attracted a significant number of data scientists, students, researchers, academics, and engineers working on interdisciplinary data analysis of images, texts, and social networks. The broad scope of AIST makes it an event where researchers from different domains, such as image and text processing, exploiting various data analysis techniques, can meet and exchange ideas. As the test of time has shown, this leads to the cross-fertilisation of ideas between researchers relying on modern data analysis machinery.

Therefore, AIST 2021 brought together all kinds of applications of data mining and machine learning techniques. The conference allowed specialists from different fields to meet each other, present their work, and discuss both theoretical and practical aspects of their data analysis problems. Another important aim of the conference was to stimulate scientists and people from industry to benefit from knowledge exchange and identify possible grounds for fruitful collaboration.

The conference was held during December 16–18, 2021. The conference was organized in a hybrid mode by Ivane Javakhishvili Tbilisi State University, Georgia (offline on campus)[2] and the Skolkovo Institute of Science and Technology, Russia (online)[3], due to the COVID-19 pandemic constraints.

This year, the key topics of AIST were grouped into five tracks:

1. Data Analysis and Machine Learning chaired by Sergei O. Kuznetsov (HSE University, Russia), Amedeo Napoli (Loria, France), and Evgenii Tsymbalov (Yandex, Russia)
2. Natural Language Processing chaired by Natalia Loukachevitch (Lomonosov Moscow State University, Russia), Andrey Kutuzov (University of Oslo, Norway), and Elena Tutubalina (Kazan Federal University and HSE University, Russia)
3. Social Network Analysis chaired by Sergei Ivanov (Huawei), Olessia Koltsova (HSE University, Russia), and Jari Saramäki (Aalto University, Finland)
4. Computer Vision chaired by Evgeny Burnaev (Skolkovo Institute of Science and Technology, Russia), and Andrey V. Savchenko (HSE University, Russia)
5. Theoretical Machine Learning and Optimization chaired by Panos M. Pardalos (University of Florida, USA) and Michael Khachay (IMM UB RAS and Ural Federal University, Russia)

The Program Committee and the reviewers of the conference included 128 well-known experts in data mining and machine learning, natural language processing, image

---

[1] https://aistconf.org.

[2] https://www.tsu.ge/en.

[3] https://www.skoltech.ru/en.

processing, social network analysis, and related areas from leading institutions of many countries including Australia, Austria, the Czech Republic, France, Germany, Greece, India, Iran, Ireland, Italy, Japan, Lithuania, Norway, Qatar, Romania, Russia, Slovenia, Spain, Taiwan, Ukraine, the UK, and the USA. This year, we received 118 submissions, mostly from Russia but also from Algeria, Brazil, Finland, Germany, India, Norway, Pakistan, Serbia, Spain, Ukraine, the UK, and the USA.

Out of the 118 submissions, 26 were desk rejected. For the remaining 92 papers, only 25 were accepted into this main volume. In order to encourage young practitioners and researchers, we included 17 papers in the companion volume published in Springer's Communications in Computer and Information Science (CCIS) series. Thus, the acceptance rate of this LNCS volume is 27%. Each submission was reviewed by at least three reviewers, experts in their fields, in order to supply detailed and helpful comments.

The conference featured several invited talks and tutorials dedicated to current trends and challenges in the respective areas.

The invited talks from academia were on natural language processing and related problems:

- Magda Tsintsadze, Manana Khachidze, and Maia Archuadze (Tbilisi State University): "On Georgian Text Processing Toolkit Development"
- Jeremy Barnes (University of the Basque Country): "Is it time to move beyond sentence classification?"
- Irina Nikishina (Skolkovo Institute of Science and Technology): "Taxonomy Enrichment with Text and Graph Vector Representations"
- Zulfat Miftahudinov (Kazan Federal University and Insilico Medicine): "Drug and Disease Interpretation Learning with Biomedical Entity Representation Transformer"

The invited industry speakers gave the following talks:

- Iosif Itkin (Exactpro): "Data intensive software testing"
- Aleksandr Semenov (Sber.Games): "Data science in GameDev"
- Alexey Drutsa (Yandex): "Toloka: professional hands-on tools accelerating the data-centric AI"

This year the program also included two round tables. The first one was devoted to the largest AIST track: "NLP on AIST" chaired by Andrey Kutuzov. The other one facilitated discussion on social aspects of data science practices: "Ethical issues and social challenges in data science" chaired by Rostislav Yavorskiy.

We would like to thank the authors for submitting their papers and the members of the Program Committee for their efforts in providing exhaustive reviews.

We would also like to express our special gratitude to all the invited speakers and industry representatives. We deeply thank all the partners and sponsors, especially, the hosting organization: Ivane Javakhishvili Tbilisi State University in Georgia. The local organizing team also included representatives of Exactpro Systems: Iosif Itkin, Natia Sirbiladze, and Janna Zabolotnaya. In addition, we thank the co-organizers from the

Russian side: Alexander Panchenko and Irina Nikishina of the Skolkovo Institute of Science and Technology and the colleagues from various divisions of HSE University. Our special thanks go to Springer for their help, starting from the first conference call to the final version of the proceedings. Last but not least, we are grateful to the volunteers, whose endless energy saved us at the most critical stages of the conference preparation.

Here, we would like to mention that the Russian word "aist" is more than just a simple abbreviation as (in Cyrillic) – it means "stork". Since it is a wonderful free bird, a symbol of happiness and peace, this stork gave us the inspiration to organize the AIST conference series. So we believe that this conference will still likewise bring inspiration to data scientists around the world!

December 2021

Evgeny Burnaev
Dmitry I. Ignatov
Sergei Ivanov
Michael Khachay
Olessia Koltsova
Andrey Kutuzov
Sergei O. Kuznetsov
Natalia Loukachevitch
Amedeo Napoli
Alexander Panchenko
Panos M. Pardalos
Jari Saramäki
Andrey V. Savchenko
Evgenii Tsymbalov
Elena Tutubalina

Organization

The conference was organized by a joint team from Ivane Javakhishvili Tbilisi State University and Exactpro Systems in Georgia, Skolkovo Institute of Science and Technology (Skoltech), the divisions of the National Research University Higher School of Economics (HSE University), and Krasovskii Institute of Mathematics and Mechanics of the Russian Academy of Sciences.

Organizing Institutions

- Ivane Javakhishvili Tbilisi State University (Tbilisi, Georgia)
- Skolkovo Institute of Science and Technology (Moscow, Russia)
- Krasovskii Institute of Mathematics and Mechanics, Ural Branch of the Russian Academy of Sciences (Yekaterinburg, Russia)
- School of Data Analysis and Artificial Intelligence, HSE University (Moscow, Russia)
- Laboratory of Algorithms and Technologies for Networks Analysis, HSE University (Nizhny Novgorod, Russia)
- International Laboratory for Applied Network Research, HSE University (Moscow, Russia)
- Laboratory for Models and Methods of Computational Pragmatics, HSE University (Moscow, Russia)
- Laboratory for Social and Cognitive Informatics, HSE University (St. Petersburg, Russia)
- Research Group "Machine Learning on Graphs", HSE University (Moscow, Russia)

Program Committee Chairs

Evgeny Burnaev	Skolkovo Institute of Science and Technology, Moscow, Russia
Sergei Ivanov	.
Michael Khachay	Krasovskii Institute of Mathematics and Mechanics of Russian Academy of Sciences, Russia & Ural Federal University, Yekaterinburg, Russia
Olessia Koltsova	HSE University, Russia
Andrey Kutuzov	University of Oslo, Norway
Sergei Kuznetsov	HSE University, Moscow, Russia
Amedeo Napoli	LORIA – CNRS, University of Lorraine, and INRIA, Nancy, France
Natalia Loukachevitch	Computing Centre of Lomonosov Moscow State University, Russia

Panos Pardalos	University of Florida, USA
Jari Saramäki	.
Andrey Savchenko	HSE University, Nizhny Novgorod, Russia
Elena Tutubalina	HSE University and Kazan Federal University, Russia
Evgenii Tsymbalov	Yandex, Russia

Proceedings Chair

| Dmitry I. Ignatov | HSE University, Russia |

Steering Committee

Dmitry I. Ignatov	HSE University, Moscow, Russia
Michael Khachay	Krasovskii Institute of Mathematics and Mechanics of Russian Academy of Sciences, Russia & Ural Federal University, Yekaterinburg, Russia
Alexander Panchenko	Skolkovo Institute of Science and Technology, Moscow, Russia
Andrey Savchenko	HSE University, Nizhny Novgorod, Russia
Rostislav Yavorskiy	Tomsk Polytechnic University, Tomsk, Russia

Program Committee

Anton Alekseev	St. Petersburg Department of V.A.Steklov Institute of Mathematics of the Russian Academy of Sciences, Russia
Ilseyar Alimova	Kazan Federal University, Russia
Vladimir Arlazarov	Smart Engines Service LLC, Federal Research Center "Computer Science and Control" of Russian Academy of Sciences, Russia
Aleksey Artamonov	Yandex, Russia
Ekaterina Artemova	HSE University, Russia
Aleksandr Babii	HSE University, Russia
Yulia Badryzlova	HSE University, Russia
Jaume Baixeries	Universitat Politècnica de Catalunya, Spain
Amir Bakarov	HSE University, Russia
Artem Baklanov	International Institute for Applied Systems Analysis, Austria
Nikita Basov	St. Petersburg State University, Russia
Vladimir Batagelj	University of Ljubljana, Slovenia
Tatiana Batura	Ershov Institute of Informatics Systems, Siberian Branch of the Russian Academy of Sciences and Novosibirsk State University, Russia

Malay Bhattacharyya	Indian Statistical Institute, India
Mikhail Bogatyrev	Tula State University, Russia
Elena Bolshakova	Moscow State Lomonosov University, Russia
Evgeny Burnaev	Skolkovo Institute of Science and Technology, Russia
Aleksey Buzmakov	HSE University, Russia
Mikhail Chernoskutov	Krasovskii Institute of Mathematics and Mechanics of Russian Academy of Sciences and Ural Federal University, Russia
Alexey Chernyavskiy	Philips Innovation Labs, Russia
Daniil Chernyshev	Lomonosov Moscow State University, Russia
Vera Davydova	Sber AI, Russia
Boris Dobrov	Lomonosov Moscow State University, Russia
Ivan Drokin	Botkin.ai, Russia
Anton Eremeev	Omsk Branch of Sobolev Institute of Mathematics, Siberian Branch of the Russian Academy of Sciences, Russia
Elena Ericheva	Botkin.ai, Russia
Anna Ermolayeva	Peoples' Friendship University of Russia, Russia
Adil Erzin	Sobolev Institute of Mathematics, Siberian Branch of the Russian Academy of Sciences, Russia
Alena Fenogenova	Sberbank, SberDevices, Russia
Elena Filatova	City University of New York, USA
Yuriy Gapanyuk	Bauman Moscow State Technical University, Russia
Petr Gladilin	ITMO University, Russia
Maksim Glazkov	Neuro.net, USA
Anna Glazkova	University of Tyumen, Russia
Taisia Glushkova	Instituto de Telecomunicações, Portugal
Natalia Grabar	Université de Lille, France
Dmitry Granovsky	Yandex, Russia
Dmitry Ignatov	HSE University, Russia
Dmitry Ilvovsky	HSE University, Russia
Max Ionov	Goethe University Frankfurt, Germany and Moscow State University, Russia
Sergei Ivanov	Skolkovo Institute of Science and Technology, Russia
Vladimir Ivanov	Innopolis University, Russia
Ilia Karpov	HSE University, Russia
Yury Kashnitsky	Elsevier, The Netherlands

Alexander Kazakov   Matrosov Institute for System Dynamics and Control Theory, Siberian Branch of the Russian Academy of Science, Russia

Michael Khachay   Krasovskii Institute of Mathematics and Mechanics of Russian Academy of Sciences, Russia

Javad Khodadoust   Payame Noor University, Iran

Denis Kirjanov   HSE University, Russia

Yury Kochetov   Sobolev Institute of Mathematics, Siberian Branch of the Russian Academy of Sciences, Russia

Sergei Koltcov   HSE University, Russia

Olessia Koltsova   HSE University, Russia

Jan Konecny   Palacký University Olomouc, Czech Republic

Anton Konushin   Smasung and HSE University, Russia

Andrey Kopylov   Tula State University, Russia

Evgeny Kotelnikov   Vyatka State University, Russia

Ekaterina Krekhovets   HSE University, Russia

Angelina Kudriavtseva   Lomonosov Moscow State University, Russia

Sofya Kulikova   HSE University, Russia

Maria Kunilovskaya   University of Wolverhampton, United Kingdom

Anvar Kurmukov   Kharkevich Institute for Information Transmission Problems of the Russian Academy of Sciences, Russia

Andrey Kutuzov   University of Oslo, Norway

Elizaveta Kuzmenko   University of Trento, Italy

Andrey Kuznetsov   Samara National Research University, Russia

Sergei Kuznetsov   HSE University, Russia

Dmitri Kvasov   University of Calabria, Italy

Florence Le Ber   Université de Strasbourg, France

Anna Lempert   Matrosov Institute for System Dynamics and Control Theory, Siberian Branch of Russian Academy of Science, Russia

Alexander Lepskiy   HSE University, Russia

Konstantin Lopukhin   Zyte, Georgia

Natalia Loukachevitch   Lomonosov Moscow State University, Russia

Olga Lyashevskaya   HSE University, Russia

Ilya Makarov   HSE University, Russia

Alexey Malafeev   HSE University, Russia

Yury Malkov   Institute of Applied Physics of the Russian Academy of Sciences, Russia

Valentin Malykh   Institute for Systems Programming of the Russian Academy of Sciences, Russia

Andrey Sozykin	Krasovskii Institute of Mathematics and Mechanics of Russian Academy of Sciences, Russia
Dmitry Stepanov	Program Systems Institute of Russian Academy of Sciences, Russia
Vadim Strijov	Moscow Institute of Physics and Technology, Russia
Tatiana Tchemisova	University of Aveiro, Portugal
Irina Temnikova	Qatar Computing Research Institute, Qatar
Mikhail Tikhomirov	Moscow State University, Russia
Martin Trnecka	Palacký University Olomouc, Czech Republic
Yuliya Trofimova	HSE University, Russia
Christos Tryfonopoulos	University of the Peloponnese, Greece
Evgenii Tsymbalov	Yandex, Russia
Elena Tutubalina	HSE University, Russia
Valery Volokha	ITMO University, Russia
Konstantin Vorontsov	FORECSYS and Moscow Institute of Physics and Technology, Russia
Ekaterina Vylomova	The University of Melbourne, Australia
Dmitry Yashunin	Harman International, USA
Alexey Zaytsev	Skolkovo Institute of Science and Technology, Russia
Nikolai Zolotykh	University of Nizhni Novgorod, Russia

Additional Reviewers

Deligiannis, Kimon
Dokuka, Sofia
Fedyanin, Kirill
Florinsky, Mikhail
Koloveas, Paris
Neznakhina, Ekaterina
Ogorodnikov, Yuri
Pugachev, Alexander
Vytovtov, Petr

Organizing Committee

Dmitry Ignatov	HSE University, Moscow, Russia – AIST Series Head of Organization
Alexander Panchenko	Skolkovo Institute of Science and Technology, Russia – AIST Series Head of Organization
Irina Nikishina	Skolkovo Institute of Science and Technology, Russia – AIST 2020 Secretary

Iosif Itkin	CEO and Co-founder, Exactpro Systems – Local Organizer
Natia Sirbiladze	CEO of Exactpro Systems – Local Organizer
Janna Zabolotnaya	Marketing Manager, Exactpro Systems – Local Organizer
Alexandra Dukhaniana	Skolkovo Institute of Science and Technology, Russia – Administrative Support

Sponsoring Institutions

Boeing
Exactpro Systems
HSE University, Russia
Ivane Javakhishvili Tbilisi State University, Georgia
Skolkovo Institute of Science and Technology, Russia

Abstracts of Invited Talks

Is It Time to Move Beyond Sentence Classification?

Jeremy Barnes🔴

University of the Basque Country, Spain
jeremycb@if.uio.no

Abstract. Many NLP tasks (sentiment analysis, natural language under-standing, etc.) are commonly cast as binary or ternary sentence classifica-tion tasks. This framing allows for quick (often semi-automated) anno-tation, allowing for large amounts of annotated data at sentence-level, which has made these datasets common baselines for deep learning mod-els. Recently, performance on many of these datasets reached human-level performance, which seemed quite promising for NLP. However, it seems that many gains in performance do not lead to models that gener-alize well and often overfit to spurious correlations in the dataset. In this talk, I will detail a set of problems with sentence classification tasks, how they have been affected by BERT-like models, and possible solutions.

Keywords: Natural language processing · Sentence classification · BERT-like models

Drug and Disease Interpretation Learning with Biomedical Entity Representation Transformer

Zulfat Miftahutdinov ⓘ

Kazan Federal University, Russia
ZuSMiftahutdinov@kpfu.ru

Abstract. This talk overviews a medical concept normalization task and its relation to current research in Natural Language Processing (NLP). This task aims to extract medical concepts in real conditions: given a set of documents, a system has to find biomedical entity mentions in a free-form text and map them to a certain medical concept (disease, drug, adverse drug reaction, etc.). In collaboration with my colleagues, we present a simple and effective two-stage neural approach based on fine-tuned BERT architectures. In the first stage, a metric learning model is trained to optimize the relative similarity of mentions and concepts via triplet loss. In the second stage, the closest concept name representation is found in an embedding space to a given clinical mention. Extensive experiments validate the effectiveness of our approach in knowledge transfer from the scientific literature to clinical trials.

Keywords: Biomedical text processing • Knowledge transfer • BERT architectures

Reference

1. Miftahutdinov, Z., Kadurin, A., Kudrin, R., Tutubalina, E.: Medical concept normalization in clinical trials with drug and disease representation learning. Bioinformatics **37**(21), 3856–3864 (2021)

Contents

Computer Vision

Data Analysis and Machine Learning

Social Network Analysis

Theoretical Machine Learning and Optimization

Invited Papers

On Georgian Text Processing Toolkit Development

Magda Tsintsadze$^{(\boxtimes)}$ ⓘ, Manana Khachidze ⓘ, and Maia Archuadze ⓘ

Iv. Javakhishvili Tbilisi State University, Tbilisi, Georgia
{magda.tsintsadze,manana.khachidze,maia.archuadze}@tsu.ge

Abstract. This invited talk presents both recent advances and previous work on text processing and information retrieval for the Georgian Language.

Keywords: Georgian language · Text categorization · Information retrieval

1 Introduction

Information Retrieval represents the classical problem of Informatics. In accordance of giant information flow, the importance of semantic search rises as well [3]. Despite the fact that research in this field has been performed for quite a long time and actively, the search engines are not still perfect, thus the actuality of search engine improvement and new algorithm processing is a quite modern task, especially for those systems that are directly involved in the process and/or performing it firsthand. Search engines are using document keyword base query searching, in fact, this method means to define string (lexical) conformity between the search query and internet document included terms. Information retrieval based on keywords is used for non-structured web documents in general.

One of the most popular internet-searching methods is the Bool retrieval, which is based on keyword different combination discovery using AND, OR, NOT operations. There are also many other types of searching tools like Wildcard Symbol, statistics-based methods (google ranking), context retrieval, saga-oriented searching, searching based on keyword position and etc. Information retrieval based on the above-mentioned techniques requires indexation of billions (and the number is growing each second) of web pages, thus the web-based document content analysis problem is getting more and more important. To dig a little deeper into information retrieval history we will find out that the importance of internet-document content analysis rises in accordance of a huge increase in web-page numbers. In 2001 the semantic web technology was presented by W3C (World Wide Web Consortium), according to their definition

E. Burnaev et al. (Eds.): AIST 2021, LNCS 13217, pp. 3–8, 2022.
https://doi.org/10.1007/978-3-031-16500-9_1

the main target of these technologies is to transform non-structured or semi-structured web documents into the "web of data". The semantic web is based on the Resource Description Framework (RDF) which represents the standard model of information exchange; it has the peculiarity to unite different schemas based on data. RDF is generalizing the link-based structure of the document and using URI for relationships and metadata for websites, it allows representation and sharing of structured and semi-structured data among different applications.

The main purpose of the semantic web is to generalize existing www by inserting metadata and making it easier for machines (search engines or any other automatic agents) to "understand" and respond accordingly to human requests. To accomplish this objective and "understand" the human requested query, the appropriate information source should be semantically structured. Thus, using tools specially designed for data like Resource Description Framework (RDF), Web Ontology Language (OWL), and Extensible Markup Language (XML) it is possible to describe things like animals, cars, people, and even parts of the construction in contrast to HTML, that is describing documents and links between them. The terms metadata, ontologies, semantics, and the semantic web are inseparable.

Web Ontology Language (OWL) is a specially designed language for Semantic Web and is used for things or thing groups' characteristics complex knowledge representation, also to describe the relations between them. Based on computation logic, the knowledge provided by OWL might be compiled by different program applications. In fact, OWL documents are the ontologies themselves. Along with RFD, RDF1, and SPARQL, the OWL is an inseparable part of Semantic web technologies [4]. The semantic web search technologies (Onto Search, Semantic Portals, Semantic Wikis, Multiagent P2P) proposed by the experts of semantic retrieval are using ontologies to form the knowledge base. The semantic web document might be defined as a document that has ontology as its content or as an ordinary web document "labeled" with specific tags taken from the so-called Domain Ontology. According to ontology types used for labeling (make annotations) the annotated web documents can be divided into different types.

Keyword-based search technology might also be used in SWD retrieval by matching query terms to terms that lexicalize ontology elements in a document. Swoogle search engine is using this method of searching where the semantics of the SWD is not used, instead, the term similarity is defined using the lexical methods, but for the semantic search algorithm term lexical and syntactic similarity is less important, main role here plays their "meaning" similarity.

A document's semantic concept knowledge is necessary for semantic matching. In the case of formal query each term semantics might be defined in an explicit way, thus if we have to deal with ontology'query, then each term concept is defined by semantic relation between this term and other term ontologies. Such a relation is not only is-a, "but also, part-of", meronym, "synonym" end etc. In the case of informal query, for example, query based on natural language, each query containing term semantics should be somehow defined. The problem is how the machine will handle the problem of query "real meaning" understanding to provide a request-appropriate document.

Intelligent search engines like AskJeeves5 are trying to solve this problem by analyzing the terms and their relationships using natural language processing techniques, or by re-defining the query with users. Other methods are mapping each query term to its "guessed" meaning, for it they are using the LSI (Latent Semantic Indexing) and WorldNet For semantic reference, (on additional request) the document should also contain its own semantics. In the case of Semantic web documents, the semantics are defined in ontology in a formal and disambiguate way. It is necessary to use modern ontology technologies for non-structured documents.

2 Problems

Problems posed by natural language are very hard and almost impossible for the solution without getting the real meaning of the words in a search query. Uncertainty is becoming more crucial when the system fails to gain knowledge and can't perform the real word "human way" perception. Semantic search systems can solve some of the most common problems associated with information retrieval:

2.1 Too Many Synonyms

All that we like so much in natural language is representing a big problem in information retrieval, for instance, according to writers skills same meaning might be represented in several variations of text, thus search engine is facing a problem of understanding what user was really meaning while performing this or that query search. Also, the existing synonyms of the same word for different professionals of different countries may vary. All the natural languages have this problem and Georgian especially for different corners of Georgia we have different words for the same meaning, so we are in need of a retrieval system that is able to catch the idea and not the word itself. The semantic search system should handle this problem by the generalization of the searched query with synonyms.

2.2 Polysemy

In all-natural languages, one word might have several meanings according to the context it presents. Semantic search systems are able to perform query compilation in accordance with their context and may serve as a tool for solving polysemy.

Despite the fact that the Semantic web provides a better solution for information retrieval, we have to mention some challenges that it poses, and even this minute the active research/work by WC3 is in the process to handle them:

The already existed a huge amount of unstructured internet documents requiring labeling in order to use the semantic search system. Partially this problem is solved as already there have been presented automatic "labeling" systems that can on fly transform structured queries to RDF. Free form processed query (request provided on natural language or the set of key-words) automatic transformation method development and etc.

In the Georgian language performed retrieval problems should be stated separately [5]. Simple experiments using search engines will show that the same information retrieval, based on Georgian language query and English language query will derive quite different results. The reasons are several: very low Georgian language web documents presented in semantic web document form, the morphological-syntactic complexity of Georgian language, no or very little SEO and least but not last very important problem of Georgian language corps: there are couple Georgian Language corps, but they are not available online, and can't be used by search engines to form dictionaries and ontologies not only for retrieval purposes but for automatic translation also.

NLP is a form of artificial intelligence that helps the machine to "read" and "perceive" the text provided in the natural language. NLP technology is an integral and fundamental part of interactive applications such as Apple's Siri, Online Banking, Automatic Translation, etc. The application of NLP is the best way to automatically get the desired content and/or context from big data sources based on Natural Language.

NLP technologies include a variety of methods, including linguistic, semantic, statistical or machine learning, to identify textual content. Statistical algorithms, automated analytical models, as well as neural networks can be used to make decisions in machine learning [5]. Inductive transfer training has greatly simplified the solution of computer vision tasks [7], however, specific interventions/modifications are still required for complete adjustment in NLP tasks [6]. Despite the abundance of approaches and algorithms, selecting the right method for a particular task is a constant challenge in this area.

3   Main Tasks for Toolkit Development

Keeping in mind that the semantic web was developed as an extension of the ordinary web and it is oriented on the data (not document) web, the information search process might be presented as the sequence of the following stages:

- The required information analysis and appropriate query formation
- The definition of information array source(s)
- The information selection process from defined arrays
- The retrieved information and retrieval result analysis

We have to note that the recall of the process is fully dependent on each level's success. Due to the peculiarities and complexity of the Georgian language, in order to get an adequate result, it is necessary to have a large database of words. The existence of Georgian "text bank" will be necessary for knowledge base and Georgian word/word-formation database development.

We have implemented algorithms that collect nouns, pronouns and numerical names in texts. Accordingly, we currently have a 30 000 words database. In addition to this database, kartu-verbs database [1] containing approximately 16,000 verbs. It has to be mentioned that this amount of verbs have over 5 million conjugated forms in 11 different tenses. Each form has 11 characteristics. There are more than 80 million links in the database. Like a database of verbs, we will develop word-formation forms for different parts of speech. We can extract information from the Clarino database created by Paul Meurer and convert it to RDF format to get triplets (<subject> <predicate> <object>.) The obtained knowledge graphs are required to be loaded into the sparql database. The Sparklis web system works on end-points loaded in the sparql database and can remove all attributes associated with columns, thus allowing us to extract the lemma from any attribute of a word [2].

This process involves the development of a data structure that reflects document information. The natural language processing toolkit should be developed in the following modules:

- Stemming – Text segmentation is not a hard task as in Georgian language words are separated with space, thus braking sentence in words will be quite easy, but the second part - stemming and term selection is quite hard for Georgian language–Georgian language stemmer might be used for this task;
- Outlining the word form in the text (can be natural language-based query as well) - identification might be useful in order to correctly assign part of speech for common root having words presented in different forms;
- Separation of syntactic-morphological parameters (parts of speech part-of-speech tagging) - is the process of attaching appropriate part of speech tag to member words of the sentence. There are several tagging mechanisms known, but stochastic and rule-based tagging are the most common. The first one using the principle: "use all information you have and guess", but the second one says: "do not guess, eliminate the impossible", we are going to use the second one which is known to be the Constrain Grammar (CG) tagging. To process with this version we have to develop rules and have the vocabulary (the database we are going to develop will serve for this task);
- Parsing – top-down parsing model based on formal grammar rules might be applied.

To optimize the search BERT – the modern machine learning model, used for NLP tasks and trained in Georgian was chosen. BERT is an encoder only. There are two pre-trained variations of BERT – the base model with 110 million parameters and the large model with 340 million parameters. One of the characteristics of BERT is that all layers maintain context in both left and right directions. Consequently, this model can create a new so-called. state-of-the-art models with only one additional layer for a wide range of tasks. We put an additional layer on top of the transformer which makes it a classifier. A very small amount of data is needed to train the obtained neural network further. Simulations were performed in the Google collab through the nvidia tesla t4 GPU using the Python framework pytorch. Gained results looks promising and further simulations are on the way to provide detailed analysis.

4   Conclusion

In the presented article we introduced the semantic web and information retrieval peculiarities. The problems associated with the retrieval of Georgian texts and the necessity of a large collection of words are due to the complexity of the Georgian language, especially the verb. Steps for Georgian text processing toolkit development were presented and the possibility of transformer application possibility is noted. The work is in progress to end up with a powerful instrument for Georgian text processing and retrieval.

References

1. Ducassé, M., et al.: Kartu-verbs: un système d'informations logiques de formes verbales fléchies pour contourner les problèmes de lemmatisation des verbes géorgiens. Extraction et Gestion des Connaissances: EGC'2022 38 (2022)
2. Ferré, S.: Sparklis: an expressive query builder for sparql endpoints with guidance in natural language. Semant. Web **8**(3), 405–418 (2017)
3. Finin, T., Mayfield, J., Joshi, A., Cost, R.S., Fink, C.: Information retrieval and the semantic web. In: Proceedings of the 38th Annual Hawaii International Conference on System Sciences, pp. 113a–113a. IEEE (2005)
4. Horridge, M., Patel-Schneider, P.F.: Owl 2 web ontology language: manchester syntax. World Wide Web Consortium, Working Draft WD-owl2-manchester-syntax-20081202 (2008)
5. Khachidze, M., Tsintsadze, M., Archuadze, M., Besiashvili, G.: Concept pattern based text classification system development for georgian text based information retrieval. Baltic J. Mod. Comput. **3**(4), 307 (2015)
6. Lashkarashvili, N., Tsintsadze, M.: Toxicity detection in online georgian discussions. Int. J. Inf. Manag. Data Insights, **2**(1), 100062 (2022)
7. Soselia, D., Tsintsadze, M., Shugliashvili, L., Koberidze, I., Amashukeli, S., Jijavadze, S.: On georgian handwritten character recognition. IFAC-PapersOnLine **51**(30), 161–165 (2018)

Taxonomy Enrichment with Text and Graph Vector Representation

Irina Nikishina[✉] [ID]

Skolkovo Institute of Science and Technology, Moscow, Russia
irina.nikishina@skoltech.ru

Abstract. Knowledge graphs such as DBpedia, Freebase or Wikidata always contain a taxonomic backbone that allows the arrangement and structuring of various concepts in accordance with hypo-hypernym ("class-subclass") relationship. With the rapid growth of lexical resources for specific domains, the problem of automatic extension of the existing knowledge bases with new words is becoming more and more widespread. In this talk, she addresses the problem of taxonomy enrichment which aims at adding new words to the existing taxonomy.

The author presents a new method which allows achieving high results on this task with little effort, described in [16]. It uses the resources which exist for the majority of languages, making the method universal. The method is extended by incorporating deep representations of graph structures like node2vec, Poincaré embeddings, GCN etc. that have recently demonstrated promising results on various NLP tasks. Furthermore, combining these representations with word embeddings allows them to beat the state of the art.

Keywords: Knowledge graphs · Taxonomy enrichment · Graph vector representation

1 Introduction

Taxonomy is a graph structure, where words or word phrases are nodes (synsets) and the relations between them are the edges between them. There exist several relations between nodes, however, in the current work we focus on hypo-hypernym relations. According to [13], hypernymy relation exists between objects X and Y if native speakers accept sentences constructed using such patterns as "An X is a (kind of) Y". "Y" is hypernym for the word "X" and "X" is hyponym for the word "Y". Therefore, in order to add a word to a taxonomy, we need to find its hypernym among the entities (synsets in case of wordnets) of this taxonomy. Here, we refer to a word absent from the taxonomy (a word which we would like to add) as a *query word*. Our task is to attach query words to an existing taxonomic tree.

The state-of-the-art taxonomy enrichment methods have two main drawbacks. First of all, they often use unrealistic formulations of the task. For example, SemEval-2016 task 14 [8] which was the first effort to evaluate this task

© The Author(s), under exclusive license to Springer Nature Switzerland AG 2022
E. Burnaev et al. (Eds.): AIST 2021, LNCS 13217, pp. 9–19, 2022.
https://doi.org/10.1007/978-3-031-16500-9_2

in a controlled environment, provided definitions of the query words (words to be added to a taxonomy). This is very informative resource, so the majority of the presented methods heavily depended on those definitions [5, 20]. However, in the real-world scenarios, such information is usually unavailable, which makes the developed methods inapplicable. We tackle this problem by testing our new methods and the state-of-the-art methods in a realistic setting.

Another gap in the existing research is that the majority of methods use the information from only one source. Namely, some researchers use the information from distributional word embeddings, whereas others consider graph-based models which represent a word based on its position in a taxonomy. Our intuition is that the information from these two sources is complementary, so combining them can improve the performance of taxonomy enrichment models. Therefore, we propose a number of ways to incorporate various sources of information.

First, we propose the new **DWRank** method which uses only distributional information from pre-trained word embeddings and is similar to other existing methods. We then enable this method to incorporate the different sources of graph information. We compare the various ways of getting the information from a knowledge graph. Finally, another modification of our method successfully combines the information from different sources, beating the current state of the art. To place our models in the context of the research on taxonomy enrichment, we compare them with a number of state-of-the-art models. To the best of our knowledge, this is the first large-scale evaluation of taxonomy enrichment methods. We are also the first to evaluate the methods on datasets of different sizes and in different languages. The code and the dataset are published in the Github repository[1].

2   Related Work

Until recently, the only dataset for the taxonomy enrichment task was created under the scope of SemEval-2016. It contained definitions for the new words, so the majority of models solving this task used the definitions. For instance, *Deftor* team [20] computed definition vector for the input word, comparing it with the vector of the candidate definitions from WordNet using cosine similarity. Another example is *TALN* team [5] which also makes use of the definition by extracting noun and verb phrases for candidates generation.

This scenario may be unrealistic for manual annotation because annotators are writing a definition for a new word and adding new words to the taxonomy simultaneously. Having a list of candidates would not only speed up the annotation process but also identify the range of possible senses. Moreover, it is possible that not yet included words may have no definition in any other sources: they could be very rare ("apparatchik", "falanga"), relatively new ("selfie", "hashtag") or come from a narrow domain ("vermiculite").

Thus, following RUSSE-2020 shared task [14], we stick to a more realistic scenario when we have no definitions of new words, but only examples of their

---

[1] https://github.com/skoltech-nlp/diachronic-wordnets.

usage. For this shared task we provide a baseline as well as training and evaluation datasets based on RuWordNet [11] which will be discussed in the next section. The task exploited words which were recently added to the latest release of RuWordNet and for which the hypernym synsets for the words were already identified by qualified annotators. The participants of the competition were asked to find synsets which could be used as hypernyms.

The participated systems mainly relied on vector representations of words and the intuition that words used in similar contexts have close meanings. They cast the task as a classification problem where words need to be assigned one or more hypernyms [9] or ranked all hypernyms by suitability for a particular word [4]. They also used a range of additional resources, such as Wiktionary, dictionaries, additional corpora [1]. Interestingly, only one of the well-performing models [21] used context-informed embeddings (BERT) or external tools such as online Machine Translation (MT) and search engines (the best-performing model denoted as *Yuriy* in the workshop description paper).

3 Base Methods

Here we first describe our baseline model which is a method of synset ranking based on distributional embeddings and hand-crafted features (the method was proposed as a baseline for RUSSE-2020 shared task [15]). We then propose extending it with new features extracted from Wiktionary and use the alternative sources of information about words (e.g. graph representations) and their combinations. The extended list of methods is presented in the full paper [16].

3.1 Baseline

We consider the approach by Nikishina et al. [15] as our baseline. There, we first create a vector representation for each synset in the taxonomy by averaging vectors (pretrained embeddings) of all words from this synset. Then, we retrieve top 10 synsets whose vectors are the closest to that of the *query word* (we refer to these synsets as *synset associates*). For each of these *associates*, we extract their immediate hypernyms and hypernyms of all hypernyms (second-order hypernyms). This list of the first- and second-order hypernyms forms our *candidate set*. We need to rank the candidates by their relevance for the query word. Note that the lists of candidates for different associates can have intersections. When forming the overall candidate set, we make sure that each candidate occurs in it only once.

The intuition behind the method is the following. We propose that if a synset of a taxonomy is a *parent* of a word which is similar to our query word, it can also be a parent of this query word.

To rank the candidate set of synsets we train a Linear Regression model with L2-regularisation on the training dataset formed of the words and synsets of WordNet. Candidate hypernyms are ranked by their model output score. We limit the output to the $k = 10$ best candidates.

We rank the candidate set using the following features:

- $n \times sim(v_i, v_{h_j})$, where v_x is a vector representation of a word or a synset x, h_j is a hypernym, n is the number of occurrences of this hypernym in the merged list, $sim(v_i, v_{h_j})$ is the cosine similarity of the vector of the input word i and hypernym vector h_j;
- the candidate presence in the Wiktionary hypernyms list for the input word (binary feature);
- the candidate presence in the Wiktionary synonyms list (binary feature);
- the candidate presence in the Wiktionary definition (binary feature);
- the average cosine similarity between the candidate and the Wiktionary hypernyms of the input word.

3.2 DWRank

We present a new method of taxonomy enrichment—Distributional Wiktionary-based synset Ranking (**DWRank**). It combines distributional features with features from Wiktionary. DWRank builds up on the baseline described in Sect. 3.1. We extend the baseline Logistic Regression model with the new features which mainly account for the number of occurrences of a synset in the candidate lists of different synset associates (nearest neighbours) of the query word. We introduce the following new features:

- the number of occurrences (n) of the synset in the merged candidate list and the quantity $log_2(2 + n)$ which serves for smoothing,
- the minimum, average, and maximum proximity level of the synset in the merged candidate list:
  - the level is 0 if the synset was added based on similarity to the query word,
  - the level of 1 is for the immediate hypernyms of the query word,
  - the level of 2 is for the hypernyms of the hypernyms,
- the minimum, average, and maximum similarities of the query word to all words of the synset,
- the features based on hyponyms of a candidate synset ("children-of-parents"):
  - we extract all hyponyms ("children") of the candidate synset,
  - for each word/phrase in each hyponym synset we compute their similarity to the query word,
  - we compute the minimum, average, and maximum similarity for each hyponym synset,
  - we form three vectors: a vector of minimums of similarities, average similarities, and maximum similarities of hyponym synsets,
  - for each of these vectors we compute minimum, average, and maximum. We use these resulting 9 numbers as features.

These features account for different aspects of similarity of the candidate's children to the query word and help defining if these children can be the query word's co-hyponyms ("siblings").

Moreover, in this approach we use cross-validation and feature scaling when training the Logistic Regression model.

This methods could be easily extended to other languages that possess a taxonomy, a wiki-based open content dictionary (Wiktionary) and text embeddings like fastText or/and word2vec and GloVe.

3.3 Word Representations for DWRank

We test our baseline approach and DWRank with different types of embeddings: fastText [2], word2vec [12] embeddings for English and Russian datasets and also GloVe embeddings [17] for the English dataset.

We use the fastText embeddings from the official website[2] for both English and Russian, trained on Common Crawl from 2019 and Wikipedia CC including lexicon from the previous periods as well. For word2vec we use models from [6, 10] for both English[3] and Russian.[4] We lemmatise words and synsets for both languages with the same UDPipe [19] model which was used while training the representations. For the out-of-vocabulary (OOV) words we find all words in the vocabulary with the longest prefix matching this word and average their embeddings like in [4]. As for the GloVe embeddings, we also use them from the official website[5] trained on Common Crawl, the vocabulary size is 840 billion tokens.

4 DWRank-Meta

In DWRank we employed only distributional information, i.e. pre-trained word embeddings, whereas in DWRank-Graph we represented words using the information from the graph structure of the taxonomy and usually ignoring their distributional properties. Meanwhile, taxonomy enrichment models may benefit from combining these two types of information. Therefore, we present **DWRank-Meta**—an extension of DWRank which combines multiple types of input word representations. The description of this approach is out of scope of the paper. Its description could be found in the full paper [16].

5 Evaluation

In the following Section we present the datasets used for the task, the evaluation metrics and present the results.

5.1 Dataset

Each dataset consists of a taxonomy and a set of novel words to be added to this resource. The detailed description of the datasets is presented in [16]. The statistics are provided in Table 1.

---

[2] https://fasttext.cc/docs/en/crawl-vectors.html.
[3] http://vectors.nlpl.eu/repository/20/29.zip.
[4] http://vectors.nlpl.eu/repository/20/185.zip.
[5] https://nlp.stanford.edu/projects/glove/.

Table 1. Statistics of two diachronic WordNet datasets used in this study.

Dataset	Nouns	Verbs
WordNet1.6 - WordNet3.0	17 043	755
WordNet1.7 - WordNet3.0	6 161	362
WordNet2.0 - WordNet3.0	2 620	193
RuWordNet1.0 - RuWordNet2.0	14 660	2 154
RUSSE'2020	2 288	525

5.2   Evaluation Metrics

The goal of diachronic taxonomy enrichment is to build a newer version of a wordnet by attaching the new given terms to the older wordnet version. We cast this task as a soft ranking problem and use Mean Average Precision (MAP) score for the quality assessment:

$$MAP = \frac{1}{N} \sum_{i=1}^{N} AP_i,$$

$$AP_i = \frac{1}{M} \sum_{i}^{n} prec_i \times I[y_i = 1],$$

(1)

where N and M are the number of predicted and ground truth values, respectively, $prec_i$ is the fraction of ground truth values in the predictions from 1 to i, y_i is the label of the i-th answer in the ranked list of predictions, and I is the indicator function.

This metric is widely acknowledged in the Hypernym Discovery shared tasks, where systems are also evaluated over the top candidate hypernyms [3]. The MAP score takes into account the whole range of possible hypernyms and their rank in the candidate list.

5.3   Results

DWRank-Meta. Figures 1 and 2 show that the leaderboard for both English and Russian nouns is dominated by DWRank-Meta models. While English benefits from the union of distributional and graph embeddings, for Russian word embeddings alone perform on par with their combinations with graph embeddings. Besides that, high-performing variants of DWRank-Meta for English feature TADW, node2vec, and GraphSAGE, whereas for Russian TADW is the only graph embedding model which does not decrease the scores of DWRank-Meta.

DWRank-Graph. On the other hand, DWRank-Graph fails in the task of taxonomy extension for all datasets. TADW model is the only graph embedding model which can compete with DWRank-Meta models. This can be explained by the fact that TADW is an extended version of DeepWalk and applies the skipgram model with the pre-trained fastText representations. In contrast to that, the other graph models suffer from the noisy representations of OOV query

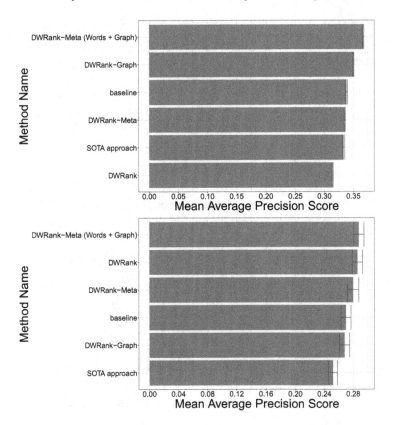

Fig. 1. Performance of different models on the English WordNet1.6-3.0 dataset: nouns (left) and verbs (right).

words. At the same time, despite the success of TADW, it does not outperform models based solely on distributional embeddings, showing that graph representations apparently do not contribute any information which is not already contained in distributional word vectors.

Baselines. We also notice that for both languages the baselines are quite competitive. They are substantially worse than the best-performing models, but they are much simpler to implement and is easy to train. Therefore, we suggest that it should be preferred in the situation of limited resources and time.

However, the choice of embedding models is crucial for the baselines (as well as for the vanilla DWRank which performs closely). We see that fastText outperforms word2vec and GloVe embeddings for almost all languages and datasets. The low scores of GloVe and word2vec embeddings on baseline and DWRank methods can be explained by data coverage issues. Fixed vocabularies of word2vec and Glove do not allow generating any representation for missing query words, whereas fastText can handle them.

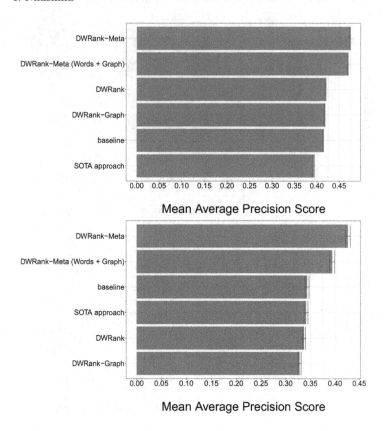

Fig. 2. Performance of different models on the Russian non-restricted dataset: nouns (left) and verbs (right).

SOTA Models. Neither of SOTA models managed to outperform the fastText baseline or approach the best DWRank-Meta variants. Web-based synset ranking (WBSR) model shows that the information from online search engines and Machine Translation models is beneficial for the task – its performance without this information drops dramatically. However, this information is not enough to outperform the word embedding-based models.

The performance of hypo2path model is even lower than that of WBSR. Being an autoregressive generative model, it is very sensitive to its own mistakes. Generating one senseless hypernym can ruin all the following chain. Conversely, when starting with the root hypernym "entity.n.01", it often takes a wrong path. Finally, TaxoExpan model relies on definitions of words which we did not provide in this task. Therefore, its results are close to zero. We do not consider them credible and provide them in italics.

6 Conclusions

In this work, we performed a large-scale computational study of various methods for taxonomy enrichment on English and Russian datasets. We presented a new taxonomy enrichment method called DWRank, which combines distributional information and the information extracted from Wiktionary outperforming the baseline method from [15] on the English datasets. We also presented its extensions: DWRank-Graph and DWRank-Meta which use graph and meta-embeddings via a common interface.

According to our experiments, word vector representations are simple, powerful, and extremely effective instrument for taxonomy enrichment, as the contexts (in a broad sense) extracted from the pre-trained word embeddings (fastText, word2vec, GloVe) and their combination are sufficient to attach new words to the taxonomy. TADW embeddings are also useful and efficient for the taxonomy enrichment task and in combination with the fastText, word2vec and GloVe approaches demonstrate SOTA results for the English language and compatible results for the Russian language.

Despite the mixed results of the application of graph-based methods, we propose further exploration of the graph-based features as the existing resource contains principally different and complementary information to the distributional signal contained in text corpora. One way to improve their performance, may be to use more sophisticated non-linear projection transformations from word to graph embeddings. Another promising way in our opinion is to explore other types of meta-embeddings to mix word and graph signals, e.g. GraphGlove [18]. Moreover, we find it promising to experiment with temporal embeddings such of those of [7] for the taxonomy enrichment task.

References

1. Arefyev, N., Fedoseev, M., Kabanov, A., Zizov, V.: Word2vec not dead: predicting hypernyms of co-hyponyms is better than reading definitions. In: Computational Linguistics and Intellectual Technologies: Papers from the Annual conference "Dialogue" (2020)
2. Bojanowski, P., Grave, E., Joulin, A., Mikolov, T.: Enriching word vectors with subword information. Trans. Assoc. Comput. Linguist. **5**, 135–146 (2017)
3. Camacho-Collados, J., et al.: SemEval-2018 task 9: hypernym discovery. In: Proceedings of The 12th International Workshop on Semantic Evaluation, pp. 712–724. Association for Computational Linguistics, New Orleans, Louisiana (2018). www. aclweb.org/anthology/S18-1115, https://doi.org/10.18653/v1/S18-1115
4. Dale, D.: A simple solution for the taxonomy enrichment task: discovering hypernyms using nearest neighbor search. In: Computational Linguistics and Intellectual Technologies: Papers from the Annual Conference "Dialogue" (2020)
5. Espinosa-Anke, L., Ronzano, F., Saggion, H.: TALN at SemEval-2016 task 14: semantic taxonomy enrichment via sense-based embeddings. In: Proceedings of the 10th International Workshop on Semantic Evaluation (SemEval-2016), pp. 1332–1336. Association for Computational Linguistics, San Diego, California (2016). www.aclweb.org/anthology/S16-1208, https://doi.org/10.18653/v1/S16-1208

6. Fares, M., Kutuzov, A., Oepen, S., Velldal, E.: Word vectors, reuse, and replicability: Towards a community repository of large-text resources. In: Proceedings of the 21st Nordic Conference on Computational Linguistics, pp. 271–276. Association for Computational Linguistics, Gothenburg, Sweden (2017). www.aclweb.org/anthology/W17-0237
7. Goel, R., Kazemi, S.M., Brubaker, M., Poupart, P.: Diachronic embedding for temporal knowledge graph completion. In: Proceedings of the AAAI Conference on Artificial Intelligence, vol. 34. no. 04, pp. 3988–3995 (2020). http://ojs.aaai.org/index.php/AAAI/article/view/5815, https://doi.org/10.1609/aaai.v34i04.5815
8. Jurgens, D., Pilehvar, M.T.: SemEval-2016 task 14: semantic taxonomy enrichment. In: Proceedings of the 10th International Workshop on Semantic Evaluation (SemEval-2016), pp. 1092–1102. Association for Computational Linguistics, San Diego, California (2016). www.aclweb.org/anthology/S16-1169, https://doi.org/10.18653/v1/S16-1169
9. Kunilovskaya, M., Kutuzov, A., Plum, A.: Taxonomy enrichment: linear hyponym-hypernym projection vs synset ID classification. In: Computational Linguistics and Intellectual Technologies: Papers from the Annual Conference 'Dialogue' (2020)
10. Kutuzov, A., Kuzmenko, E.: WebVectors: a toolkit for building web interfaces for vector semantic models. In: Ignatov, D.I., et al. (eds.) AIST 2016. CCIS, vol. 661, pp. 155–161. Springer, Cham (2017). https://doi.org/10.1007/978-3-319-52920-2_15
11. Loukachevitch, N.V., Lashevich, G., Gerasimova, A.A., Ivanov, V.V., Dobrov, B.V.: Creating Russian wordnet by conversion. In: Computational Linguistics and Intellectual Technologies: Papers from the Annual Conference "Dialogue", pp. 405–415 (2016)
12. Mikolov, T., Sutskever, I., Chen, K., Corrado, G.S., Dean, J.: Distributed representations of words and phrases and their compositionality. In: Burges, C.J.C., Bottou, L., Welling, M., Ghahramani, Z., Weinberger, K.Q. (eds.) Advances in Neural Information Processing Systems, vol. 26, pp. 3111–3119. Curran Associates, Inc (2013)
13. Miller, G.A.: WordNet: An electronic lexical database. MIT press, Cambridge (1998)
14. Nikishina, I., Logacheva, V., Panchenko, A., Loukachevitch, N.: RUSSE'2020: findings of the first taxonomy enrichment task for the russian language. In: Computational Linguistics and Intellectual Technologies: Papers from the Annual Conference "Dialogue" (2020)
15. Nikishina, I., Panchenko, A., Logacheva, V., Loukachevitch, N.: Studying taxonomy enrichment on diachronic wordnet versions. In: Proceedings of the 28th International Conference on Computational Linguistics. Association for Computational Linguistics, Barcelona, Spain (2020)
16. Nikishina, I., Tikhomirov, M., Logacheva, V., Nazarov, Y., Panchenko, A., Loukachevitch, N.: Taxonomy enrichment with text and graph vector representations. Semantic Web, pp. 1–35 (2022). https://doi.org/10.3233/SW-212955
17. Pennington, J., Socher, R., Manning, C.: GloVe: Global vectors for word representation. In: Proceedings of the 2014 Conference on Empirical Methods in Natural Language Processing (EMNLP), pp. 1532–1543. Association for Computational Linguistics, Doha, Qatar (2014). www.aclweb.org/anthology/D14-1162, https://doi.org/10.3115/v1/D14-1162
18. Ryabinin, M., Popov, S., Prokhorenkova, L., Voita, E.: Embedding words in non-vector space with unsupervised graph learning. arXiv preprint arXiv:2010.02598 (2020)

19. Straka, M., Straková, J.: Tokenizing, POS tagging, lemmatizing and parsing UD 2.0 with UDPipe. In: Proceedings of the CoNLL 2017 Shared Task: Multilingual Parsing from Raw Text to Universal Dependencies, pp. 88–99. Association for Computational Linguistics, Vancouver, Canada (2017). www.aclweb.org/anthology/K/K17/K17-3009.pdf

20. Tanev, H., Rotondi, A.: Deftor at SemEval-2016 task 14: Taxonomy enrichment using definition vectors. In: Proceedings of the 10th International Workshop on Semantic Evaluation (SemEval-2016), pp. 1342–1345. Association for Computational Linguistics, San Diego, California (2016). www.aclweb.org/anthology/S16-1210, https://doi.org/10.18653/v1/S16-1210

21. Tikhomirov, M., Loukachevitch, N., Parkhomenko, E.: Combined approach to hypernym detection for thesaurus enrichment. In: Computational Linguistics and Intellectual Technologies: Papers from the Annual conference 'Dialogue' (2020)

Natural Language Processing

Natural Language Processing

Near-Zero-Shot Suggestion Mining
with a Little Help from WordNet

Anton Alekseev[1,2,3(✉)] [iD], Elena Tutubalina[3,4,5] [iD], Sejeong Kwon[6],
and Sergey Nikolenko[1,7] [iD]

[1] Steklov Mathematical Institute, St. Petersburg, Russia
anton.m.alexeyev@gmail.com
[2] St. Petersburg University, St. Petersburg, Russia
[3] Kazan (Volga Region) Federal University, Kazan, Russia
[4] HSE University, Moscow, Russia
[5] Sber AI, Kazan, Russia
[6] Samsung Research, Seoul, Korea
[7] Neuromation OU, Tallinn, Estonia

Abstract. In this work we explore the *constructive* side of online reviews: advice, tips, requests, and suggestions that users provide about goods, venues and other items of interest. To reduce training costs and annotation efforts needed to build a classifier for a specific label set, we present and evaluate several entailment-based zero-shot approaches to suggestion classification in a *label-fully-unseen* fashion. In particular, we introduce the strategy of assigning target class labels to sentences with user intentions, which significantly improves prediction quality. The proposed strategies are evaluated with a comprehensive experimental study that validated our results both quantitatively and qualitatively.

Keywords: Text classification · Suggestion mining · Zero-shot learning

1 Introduction

Online user reviews often provide feedback that extends much further than just the overall or aspect-specific sentiment. Users can describe their experience in detail and, in particular, provide advice tips or suggestions that can be useful both to other users and service providers. One of the most valuable entities both for peer users and the reviewed object's owners/manufacturers/developers/sellers/etc. are reviews with user-generated suggestions that help other users make informed decisions and choices, while providers responsible for reviewed items get more specific advice on which modifications to make or on the selling strategy.

Therefore, with growing volumes of opinions and reviews posted online, it is important to develop an effective way to extract advice/tips/suggestions for highlighting or aggregation. The task of automatic identification of suggestions

© The Author(s), under exclusive license to Springer Nature Switzerland AG 2022
E. Burnaev et al. (Eds.): AIST 2021, LNCS 13217, pp. 23–36, 2022.
https://doi.org/10.1007/978-3-031-16500-9_3

in a given text is known as *suggestion mining* [2]. It is usually defined as a sentence classification task: each sentence of a review is assigned a class of either "suggestion" or "non-suggestion" [23]. Here are three sample sentences from the dataset [24]:

- "Having read through the cumbersome process, I am more incline to build FREE apps with subscription rather than getting any money through the Marketplace";
- "Why can't Microsoft simplify it?";
- "Even something as simple as ctrl+S would be a godsend for me".

In this case, the first two sentences do not propose any specific changes or improvements to the systems the users are writing about, while the third one does. Hence, the latter can be called a *suggestion* and the first two cannot. Also note that this suggestion example is not formulated as a direct request, which demonstrates that the task is harder than simply learning to recognize a fixed set of lexical patterns.

Suggestion mining finds many applications in a number of industries, from consumer electronics to realty. In each of these domains, however, suggestions and tips are proposed in different ways, so NLP approaches to processing online reviews differ across domains as well. That is why training domain-independent embeddings, e.g., for the task of cross-domain suggestion mining is a hard challenge; see, e.g., *SemEval-2019 Task 9B* [24]. Since most state-of-the-art models use machine learning to tune their parameters, their performance—and relevance to real-world implementations—is highly dependent on the dataset on which they are trained [5,34].

In this work, we attempt to avoid this problem by focusing on the *(near-) zero-shot learning* approach, which does not require any training data. We extend the methodology recently proposed by [35] to this new task. Our approach is based on natural language inference that includes a sentence to be classified as premise and a sentence describing a class as hypothesis.

In order to make the proposed methods more universally applicable across different domains, and following the methodology proposed in [35], we explore the extent to which suggestion classification can be solved with zero-shot approaches. For that purpose, we use the BART model [16] pretrained for natural language inference [33]. The main contributions of this work include:

(1) new WordNet-based modifications to the basic approach to zero-shot *label-fully-unseen* classification of suggestions (*near*-zero-shot classification) and
(2) established benchmarks in zero-shot learning for suggestion classification in English.

We also test whether the best method is easily transferable to other domains and report negative results.

The paper is structured as follows. Section 2 surveys related work in opinion mining and natural language inference. In Sect. 3 we describe the datasets used for this work. Section 4 introduces the proposed approaches to zero-shot and near-zero-shot learning, Sect. 5 presents and discusses the numerical results of our experiments, and Sect. 6 concludes the paper.

2   Related Work

The two main topics related to this work are

(1) opinion mining, especially in regard to the datasets collected and used here,
(2) natural language inference, where most of our methods come from.

In this section, we briefly survey the state of the art in both topics.

In opinion mining, existing research has been focused on expressions of opinion that convey positive or negative sentiments [3,7,17,18,26,31,34]. For example, two widely studied topics in text classification are whether a given text is

(i) positive or negative, a problem known as *sentiment analysis*, and
(ii) subjective or objective, which distinguishes subjective and fact-based opinions; see comprehensive surveys of both topics in [3,18,26,29].

A fact-based opinion is a regular or comparative opinion implied in an objective or factual statement [17]. In a pioneering work, Pang and Lee [25] presented a preprocessing filter that labels text sentences as either subjective or objective. It removes the latter and uses subjective sentences for sentiment classification of movie reviews. Raychev and Nakov [28] proposed a novel approach that assigns position-dependent weights to words or word bigrams based on their likelihood of being subjective; their evaluation shows a naïve Bayes classifier on a standard dataset of movie reviews on par with state-of-the-art results based on language modeling. Recent works recorded eye movements of readers for the tasks of sentiment classification and sarcasm detection in sentiment text [21,22]. The results show that words pertaining to a text's subjective summary attract a lot more attention. Mishra et al. [22] proposed a novel multi-task learning approach for document-level sentiment analysis by considering the modality of human cognition along with text.

In contrast with movie reviews, Chen and Sokolova [4] investigated subjective (positive or negative) and objective terms in medical and scientific texts, showing that the majority of words for each dataset are objective (89.24%), i.e., sentiment is not directly stated in the texts of these domains.

Little work has been done on mining suggestions and tips until recently. *SemEval-2019 Task 9* (see [24]) was devoted to suggestion mining from user-generated texts, defining suggestions as sentences expressing advice, tips, and recommendations relating to a target entity. This task was formalized as sentence classification. Crowdsourced annotations were provided with sentences from software suggestion forums and hotel reviews from *TripAdvisor*. Each text had been annotated as a *suggestion* or *non-suggestion*. The authors presented two subtasks:

- *Task A* for training and testing models on the dataset of software comments;
- *Task B* for testing on hotel reviews without an in-domain training set.

As expected, the best systems used ensemble classifiers based on BERT and other neural networks. In particular, the winning team *OleNet@Baidu* [19] achieved F-scores of 0.7812 and 0.8579 for subtasks A and B respectively. However, a rule-based classifier based on lexical patterns obtained an F-score of 0.858 in subtask B [27]. The *SemEval-2019 Task 9* dataset shows that many suggestions from software forums include user requests, which is less frequent in other domains.

Several studies in opinion mining have focused on the detection of complaints or technical problems in electronic products and mobile applications [6,10,11, 13,30]. Khan et al. [13] described the "No Fault Found" phenomenon related to software usability problems. Iacob and Harrison [10] applied linguistic rules to classify feature requests, e.g., "(the only thing) missing *request*", while Ivanov and Tutubalina [11] proposed a clause-based approach to detect electronic product failures. To sum up, detection of subjective and prescriptive texts (i.e., tips and suggestions) has not been extensively studied even though sentiment analysis is a very well researched problem with many different techniques.

As for natural language inference (NLI), there exist three main sources of NLI datasets presented by Williams et al. [33], Bowman et al. [1], and Khot et al. [14]. In these datasets, the basic NLI task is to classify the relation between two sentences (e.g., entailment). NLI has become popular in the last decade, especially rising in popularity with recently developed approaches that utilize Transformer-based models such as BERT [12], where it was adopted for few-shot and zero-shot learning along with text classification [35] and natural language understanding [15]. The success of this line of work based on NLI models has motivated our choice of the approach, and in this work we extend this methodology to suggestion detection.

3 Data

The dataset for suggestion mining used in this work was presented by Negi et al. [24] and used in *SemEval2019, Task 9*[1]; we refer to [24] for a description of data collection and annotation procedures, which include a two-phase labeling procedure by crowd workers and experts respectively.

In our setup we do not use the original validation (development) set but rather the so-called "trial test set", also released during the shared task run. Table 1 shows the statistics for this version of the dataset, including the training, validation, and testing parts of each dataset. We also show samples from the dataset in Table 2.

The data covers two independent domains.

1. *Software suggestion forum.* Sentences for this dataset were scraped from the Uservoice platform that provides customer engagement tools to brands and hosts dedicated suggestion forums for certain products. Posts were scraped and split into sentences using the Stanford CoreNLP toolkit. Many suggestions are in the form of requests, which is less frequent in other domains. The text contains highly technical vocabulary related to the software being discussed.

---

[1] https://github.com/Semeval2019Task9.

Table 1. Dataset statistics. The Train/validation/test split is shown for suggestions and non-suggestions for the two subtasks of [24].

Task/Domain		Suggestions		Non-Suggestions	
A	Software development forums (Uservoice)	Training	2085	Training	6415
		Validation	296	Validation	296
		Testing	87	Testing	746
B	Hotel reviews (TripAdvisor)	Training	2085	Training	6415
		Validation	404	Validation	404
		Testing	348	Testing	476

Table 2. Samples from the SemEval2019-Task9 suggestion mining dataset: software development (SDE) forums and hotel reviews.

	Suggestion	Non-suggestion
SDE forums	The proposal is to add something like: // Something happened update your UI or run your business logic	I write a lot support ticket on this, but no one really cares on this issue
Hotel reviews	For a lovely breakfast, turn left out of the front entrance - on the next corner is a cafe with fresh baked breads and cooked meals	A great choice!!

2. *Hotel reviews.* Wachsmuth et al. [32] provide a large dataset of *TripAdvisor* hotel reviews, split into statements so that each statement has only one manually assigned sentiment label. Statements are equivalent to sentences, and consist of one or more clauses. [24] annotated these segments as suggestions and non-suggestions.

Following [24], we use two datasets, called Subtask A and B. In Subtask A the training, validation, and test parts are from the software forum domain, while Subtask B uses the same training part as Subtask A but with different test and validation parts. Since we consider the zero-shot setting, we do not use the training part.

4   Approach

Following the work on zero-shot learning for text classification benchmarking [35] and using `facebook/bart-large-mnli` (BART [16] trained on [33]) as a foundation model, we have tried several different procedures for preparing labels in the *label-fully-unseen* setting.

Approach 1. In this approach we directly test whether the premise is suggesting something. The following statements are used as hypotheses: "This text is a

Table 3. Hyponyms of WordNet synset *message.n.02*. The possible candidates for direct mapping as a suggestion are checkmarked in column "?".

?	Synset name	Lemma #1	Definition
	acknowledgment.n.03	acknowledgment	A statement acknowledging something/someone
	approval.n.04	Approval	A message expressing a favorable opinion
	body.n.08	Body	The central message of a communication
	commitment.n.04	commitment	A message that makes a pledge
	corker.n.01	Corker	(Dated slang) a remarkable or excellent thing
	digression.n.01	Digression	A message that departs from the main subject
✓	*direction.n.06*	Direction	A message describing how something is
	disapproval.n.02	Disapproval	The expression of disapproval
	disrespect.n.01	Disrespect	An expression of lack of respect
	drivel.n.01	Drivel	A worthless message
✓	*guidance.n.01*	Guidance	Something that provides direction or advice
	information.n.01	Information	A message received and understood
	interpolation.n.01	interpolation	A message (spoken or written) that is introduced or inserted
	latent_content.n.01	latent_content	(Psychoanalysis) hidden meaning of a fantasy
	meaning.n.01	Meaning	The message that is intended or expressed or
	narrative.n.01	Narrative	A message that tells the particulars of an act or occurrence or course of events
	nonsense.n.01	Nonsense	A message that seems to convey no meaning
✓	*offer.n.02*	Offer	Something offered (as a proposal or bid)
	opinion.n.02	Opinion	A message expressing a belief about something
✓	*promotion.n.01*	Promotion	A message issued in behalf of some product or
✓	*proposal.n.01*	Proposal	Something proposed (such as a plan or
	refusal.n.02	Refusal	A message refusing to accept something that
✓	*reminder.n.01*	Reminder	A message that helps you remember something
✓	*request.n.01*	Request	A formal message requesting something that is
	respects.n.01	Respects	(Often used with 'pay') a formal expression of
	sensationalism.n.01	sensationalism	Subject matter that is calculated to excite and
	shocker.n.02	Shocker	A sensational message (in a film or play or novel)
	statement.n.01	Statement	A message that is stated or declared
	statement.n.04	Statement	A nonverbal message
	subject.n.01	Subject	The subject matter of a conversation
✓	*submission.n.01*	Submission	Something (manuscripts or...) submitted for the judgment of others
	wit.n.01	wit	A message whose ingenuity or verbal skill or

suggestion.", "This text is not a suggestion." We have also experimented with direct reformulations, e.g. "This text suggests/is suggesting" as well, with similar results both on test and development sets.

Table 4. Results of Approaches 1 & 2 and the best label subset from Approach 3 (Subtask A).

Premise	Task	Dev. set		Test set	
		F1	Acc.	F1	Acc.
"This text is [not] a suggestion." (A1)	A	0.6727	0.5152	0.1961	0.1536
	B	0.6616	0.5	0.5806	0.4163
"This text is [not] suggesting."(A1)	A	0.6712	0.5118	0.1898	0.1393
	B	0.6617	0.4988	0.5840	0.4175
3 definitions VS "This text is **not** a suggestion." (A2)	A	0.6689	0.5051	0.1925	0.1237
	B	0.6656	0.5025	0.5876	0.4175
"The best subset" (A3)	A	0.7517	0.7568	0.4479	0.8283
	B	0.4635	0.6361	0.4841	0.6699

Approach 2. For labels, we use the following definitions of "suggestion" from WordNet [8, 20]:

- *suggestion.n.01* ("This text is an idea that is suggested"),
- *suggestion.n.02* ("This text is a proposal offered for acceptance or rejection"),
- *suggestion.n.04* ("This text is persuasion formulated as a suggestion").

Other definitions were discarded as irrelevant (*trace.n.01*: a just detectable amount, *suggestion.n.05*: the sequential mental process in which one thought leads to another by association, *hypnotism.n.01*: the act of inducing hypnosis). The non-suggestion label text used was as follows: "This text is not a suggestion."

Results of these simple approaches were arguably unsatisfactory. Each of the two classes ("suggestion" and "non-suggestion") varies in the type of the possible message conveyed by the authors. Suggestion can be a plea, a question, a request etc., and non-suggestions are even more diverse: questions, comments, jokes, complaints and so on.

Approach 3. Given a wide variety of possible "message types", a natural idea would be to classify sentences into all of them and them map some labels to "suggestions" and others to "non-suggestions".

Thus, we have analyzed the place of "suggestion" in WordNet, exploring the hyperonyms of "suggestion". The synset whose direct hyponyms in *Word-Net* [8, 20] described the types of the users' "message" and which we had found suitable for the task was the *message.n.02* WordNet synset: "what a communication that is about something is about". A list of the direct hyponyms is presented in Table 3[2]. Further, we have selected a list of candidates from those hyponyms to be later mapped to suggestions: *direction.n.06, guidance.n.01, offer.n.02, promotion.n.01, proposal.n.01, reminder.n.01, request.n.01, submission.n.01*. We have

---

[2] See also WordNet Web: wordnetweb.princeton.edu.

Table 5. Development set, Subtask A. Top-3 label subsets results for each subset size from 4 to 8 selected by F1 measure.

Size	Subset of labels	F1	Accuracy
4	Guidance, proposal, reminder, request	0.7509	0.7568
	Guidance, promotion, proposal, request	0.7486	**0.7652**
	Guidance, offer, proposal, request	0.7455	0.7382
5	Guidance, promotion, proposal, reminder, request	**0.7517**	0.7568
	Guidance, offer, proposal, reminder, request	0.7484	0.7297
	Guidance, offer, promotion, proposal, request	0.7463	0.7382
6	Guidance, offer, promotion, proposal, reminder, request	0.7492	0.7297
	Direction, guidance, promotion, proposal, reminder, request	0.7462	0.7449
	Guidance, promotion, proposal, reminder, request, submission	0.7445	0.7449
7	Direction, guidance, offer, promotion, proposal, reminder, request	0.7443	0.7179
	Guidance, offer, promotion, proposal, reminder, request, submission	0.7427	0.7179
	Direction, guidance, promotion, proposal, reminder, request, submission	0.7393	0.7331
8	All 8 labels	0.7380	0.7061

formulated the labels as "This text is a [LEMMA]", where [LEMMA] is the first lemma in the WordNet lemma synset list[3]. We have used the development set of the *SemEval2019 Task 9*, Subtask A to find the best subset of candidate labels in terms of F1-measure of the "suggestion" class.

5   Results

Approaches 1 & 2. The results of "asking" entailment models whether the text is a suggestion or not are shown in Table 4. The difference in performance of all models on test sets of Subtasks A and B can be explained by the differences in target classes distributions; we also note that Transformer-based models often struggle with negations [9]. Similar to [35], using definitions seemingly does not guarantee better results for entailment-based zero-shot text (suggestion) classification.

---

[3] We have also tried to enrich the labels by adding all lemmas joined by "or", but with no improvement; results of these experiments are given in Table 6.

Table 6. Development set. Top-3 label (all lemmas concatenated with "or") extended subsets results for each subset size from 5 to 8 selected by F1 measure.

Size	Labels subset	F1	Acc.
4	Direction or instruction, proposal, reminder, request or petition or postulation	0.6590	0.6976
	direction or instruction, promotion or publicity or promotional_material or packaging, proposal, request or petition or postulation	0.6589	0.7027
	Direction or instruction, guidance or counsel or counseling or counselling or direction, promotion or publicity or promotional_material or packaging, proposal	0.6562	0.7027
5	Direction or instruction, guidance or counsel or counseling or counselling or direction, proposal, reminder, request or petition or postulation	0.6654	0.7010
	Direction or instruction, guidance or counsel or counseling or counselling or direction, promotion or publicity or promotional_material or packaging, proposal, request or petition or postulation	0.6654	0.7061
	Direction or instruction, promotion or publicity or promotional_material or packaging, proposal, reminder, request or petition or postulation	0.6642	0.6993
6	Direction or instruction, guidance or counsel or counseling or counselling or direction, promotion or publicity or promotional_material or packaging, proposal, reminder, request or petition or postulation	0.6704	0.7027
	Direction or instruction, guidance or counsel or counseling or counselling or direction, promotion or publicity or promotional_material or packaging, proposal, request or petition or postulation, submission or entry	0.6679	0.7078
	Direction or instruction, guidance or counsel or counseling or counselling or direction, proposal, reminder, request or petition or postulation, submission or entry	0.6679	0.7027
7	Direction or instruction, guidance or counsel or counseling or counselling or direction, promotion or publicity or promotional_material or packaging, proposal, reminder, request or petition or postulation, submission or entry	0.6729	0.7044
	Direction or instruction, guidance or counsel or counseling or counselling or direction, offer or offering, promotion or publicity or promotional_material or packaging, proposal, reminder, request or petition or postulation	0.6704	0.6993
	Direction or instruction, guidance or counsel or counseling or counselling or direction, offer or offering, promotion or publicity or promotional_material or packaging, proposal, request or petition or postulation, submission or entry	0.6679	0.7044
8	Direction or instruction, guidance or counsel or counseling or counselling or direction, offer or offering, promotion or publicity or promotional_material or packaging, proposal, reminder, request or petition or postulation, submission or entry	0.6728	0.7010

Table 7. Development set, Subtask B. Top-3 label subsets results for each subset size from 4 to 8 selected by F1 measure.

Size	Labels subset	F1	Accuracy
4	Guidance,offer,proposal,reminder	0.6934	0.7351
	Offer,proposal,reminder,request	0.6897	0.7327
	Direction,offer,proposal,reminder	0.6802	0.7277
5	Guidance,offer,proposal,reminder,request	0.7097	0.7438
	Direction,guidance,offer,proposal,reminder	0.7007	0.7389
	Direction,offer,proposal,reminder,request	0.6970	0.7364
6	Direction,guidance,offer,proposal,reminder,request	0.7167	0.7475
	Guidance,offer,proposal,reminder,request,submission	0.7123	0.7450
	Guidance,offer,promotion,proposal,reminder,request	0.7067	0.7401
7	Direction,guidance,offer,proposal,reminder,request,submission	**0.7192**	0.7488
	Direction,guidance,offer,promotion,proposal,reminder,request	0.7137	0.7438
	Guidance,offer,promotion,proposal,reminder,request,submission	0.7093	0.7413
8	(all 8 labels)	0.7163	0.7450

Table 8. Test set, Subtask B. Top-3 label subsets results for each subset size from 4 to 8 selected by F1 measure.

Size	Labels subset	F1	Accuracy
4	Offer,proposal,reminder,request	0.5963	0.6845
	Guidance,offer,proposal,reminder	0.5833	0.6723
	Offer,promotion,proposal,reminder	0.5804	0.6772
5	Guidance,offer,proposal,reminder,request	0.6030	0.6820
	Offer,promotion,proposal,reminder,request	0.6006	0.6869
	Offer,proposal,reminder,request,submission	0.5997	0.6857
6	Guidance,offer,promotion,proposal,reminder,request	0.6073	0.6845
	Guidance,offer,proposal,reminder,request,submission	0.6063	0.6833
	Direction,guidance,offer,proposal,reminder,request	0.6054	0.6820
7	Guidance,offer,promotion,proposal,reminder,request,submission	0.6105	0.6857
	Direction,guidance,offer,promotion,proposal,reminder,request	0.6096	0.6845
	Direction,guidance,offer,proposal,reminder,request,submission	0.6087	0.6833
8	(all 8 labels)	**0.6129**	0.6857

Approach 3. Results of the best subsets of labels (we evaluated all subsets of size 4 to 8) using the development set of Subtask A are shown in Table 5. The best subset is {guidance, promotion, proposal, reminder, request}, with F1-measure 0.7517 and accuracy 0.7568. Half of the development set data points are annotated with "suggestions", while the test set is imbalanced (for details see Sect. 3), which may explain the difference in performance.

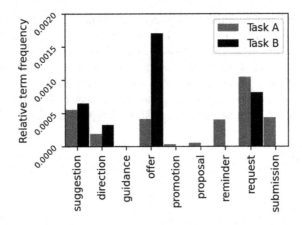

Fig. 1. Relative word frequency in suggestions in Subtask A (software, train+dev+test) and Subtask B (hotels, dev+test).

We apply the same entailment-based zero-shot prediction procedure to the test sets of the two subtasks of *SemEval2019, Task 9*. For the best combination of "suggestion-related" labels we achieve the following results.

A. F1-measure: 0.4479, accuracy: 0.8283. For comparison, the best F1-measure in the supervised settings is 0.7812 [19], while random uniform sampling yields mean F1-measure of 0.1734.

B. F1-measure: 0.4841, accuracy: 0.6699. Here, the best F1-measure in the supervised setting is 0.858 [27], and random sampling of labels yields 0.4566.

The results are also reported in Table 4. Clearly, F1-measures achieved for Subtask B are not far from random prediction results for the label set tuned for Subtask A. We have also carried out the same entailment-based near-zero-shot classification using all possible subsets of labels from size 4 to size 8 on Subtask B development and test sets as well. The 3 best results for each label subset size are reported in Tables 7 and 8, respectively. The best results for Subtask B are now much better: F1-measure on the development set is 0.7192 (all labels except *promotion*) and 0.6129 on the test set (all 8 labels). Thus, the proposed method can achieve much better results with specially tuned label sets; this means that this approach is not entirely domain-independent. Figure 1 supports this claim showing that the frequencies of words commonly used for suggestions are very different in hotel and software domains.

6 Conclusion

In this work, we have evaluated several approaches to *label-fully-unseen* zero-shot suggestion mining. We have proposed an approach based on hyponyms of the word *message* from *WordNet* that outperforms direct labeling "This text is

[not] a suggestion" and definitions of *suggestion* from WordNet. However, choosing the best-performing subset of the hyponyms for one domain (softwarerelated discussions) does not perform well on another (hotel reviews), which suggests that the subset of hyponyms is a new hyperparameter for the proposed approach. Either this hyperparameter needs to be tuned for each new domain, or new automated approaches for finding the best subset of hyponyms have to be developed, which we suggest as an interesting direction for further work.

Acknowledgements. This work was supported by the Russian Science Foundation grant # 18-11-00284.

References

1. Bowman, S.R., Angeli, G., Potts, C., Manning, C.D.: A large annotated corpus for learning natural language inference. In: EMNLP (2015)
2. Brun, C., Hagege, C.: Suggestion mining: Detecting suggestions for improvement in users' comments. Res. Comput. Sci. **70**(79), 171–181 (2013)
3. Chaturvedi, I., Cambria, E., Welsch, R.E., Herrera, F.: Distinguishing between facts and opinions for sentiment analysis: survey and challenges. Inf. Fusion **44**, 65–77 (2018)
4. Chen, Q., Sokolova, M.: Unsupervised sentiment analysis of objective texts. In: Meurs, M.-J., Rudzicz, F. (eds.) Canadian AI 2019. LNCS (LNAI), vol. 11489, pp. 460–465. Springer, Cham (2019). https://doi.org/10.1007/978-3-030-18305-9_45
5. Chen, X., Cardie, C.: Multinomial adversarial networks for multi-domain text classification. arXiv preprint arXiv:1802.05694 (2018)
6. Dong, L., Wei, F., Duan, Y., Liu, X., Zhou, M., Xu, K.: The automated acquisition of suggestions from tweets. In: Proceedings of the Twenty-Seventh AAAI Conference on Artificial Intelligence, pp. 239–245 (2013)
7. Feldman, R.: Techniques and applications for sentiment analysis. Commun. ACM **56**(4), 82–89 (2013)
8. Fellbaum, C.: WordNet: An Electronic Lexical Database. MIT Press, Cambridge (1998)
9. Hossain, M.M., Kovatchev, V., Dutta, P., Kao, T., Wei, E., Blanco, E.: An analysis of natural language inference benchmarks through the lens of negation. In: Proceedings of the 2020 Conference on Empirical Methods in Natural Language Processing (EMNLP), pp. 9106–9118. Association for Computational Linguistics, Online (Nov 2020). www.aclweb.org/anthology/2020.emnlp-main.732
10. Iacob, C., Harrison, R.: Retrieving and analyzing mobile apps feature requests from online reviews. In: Mining Software Repositories (MSR), 2013 10th IEEE Working Conference on, pp. 41–44. IEEE (2013)
11. Ivanov, V., Tutubalina, E.: Clause-based approach to extracting problem phrases from user reviews of products. In: Ignatov, D.I., Khachay, M.Y., Panchenko, A., Konstantinova, N., Yavorskiy, R.E. (eds.) AIST 2014. CCIS, vol. 436, pp. 229–236. Springer, Cham (2014). https://doi.org/10.1007/978-3-319-12580-0_24
12. Kenton, J.D.M.W.C., Toutanova, L.K.: Bert: pre-training of deep bidirectional transformers for language understanding. In: Proceedings of NAACL-HLT, pp. 4171–4186 (2019)

13. Khan, S., Phillips, P., Jennions, I., Hockley, C.: No fault found events in mainte-
nance engineering part 1: current trends, implications and organizational practices.
Reliab. Eng. Syst. Saf. **123**, 183–195 (2014)

14. Khot, T., Sabharwal, A., Clark, P.: Scitail: a textual entailment dataset from science question answering. In: AAAI, vol. 17, pp. 41–42 (2018)

15. Kumar, A., Muddireddy, P.R., Dreyer, M., Hoffmeister, B.: Zero-shot learning
across heterogeneous overlapping domains. In: INTERSPEECH, pp. 2914–2918
(2017)

16. Lewis, M., et al.: Bart: denoising sequence-to-sequence pre-training for natural language generation, translation, and comprehension. arXiv preprint
arXiv:1910.13461 (2019)

17. Liu, B.: Many facets of sentiment analysis. In: Cambria, E., Das, D., Bandyopad-
hyay, S., Feraco, A. (eds.) A Practical Guide to Sentiment Analysis. Socio-Affective
Computing, vol. 5, pp. 11–39. Springer, Cham (2017). https://doi.org/10.1007/
978-3-319-55394-8_2

18. Liu, B., et al.: Sentiment analysis and subjectivity. Handb. Nat. Lang. Process.
2(2010), 627–666 (2010)

19. Liu, J., Wang, S., Sun, Y.: Olenet at semeval-2019 task 9: Bert based multi-
perspective models for suggestion mining. In: Proceedings of the 13th International
Workshop on Semantic Evaluation, pp. 1231–1236 (2019)

20. Miller, G.A.: Wordnet: a lexical database for English. Commun. ACM **38**(11),
39–41 (1995)

21. Mishra, A., Kanojia, D., Nagar, S., Dey, K., Bhattacharyya, P.: Leveraging cognitive features for sentiment analysis. In: Proceedings of The 20th SIGNLL Conference on Computational Natural Language Learning, pp. 156–166 (2016)

22. Mishra, A., Tamilselvam, S., Dasgupta, R., Nagar, S., Dey, K.: Cognition-cognizant
sentiment analysis with multitask subjectivity summarization based on annotators'
gaze behavior. In: Thirty-Second AAAI Conference on Artificial Intelligence (2018)

23. Negi, S., Asooja, K., Mehrotra, S., Buitelaar, P.: A study of suggestions in opin-
ionated texts and their automatic detection. In: Proceedings of the Fifth Joint
Conference on Lexical and Computational Semantics, pp. 170–178 (2016)

24. Negi, S., Daudert, T., Buitelaar, P.: Semeval-2019 task 9: Suggestion mining from
online reviews and forums. In: Proceedings of the 13th International Workshop on
Semantic Evaluation, pp. 877–887 (2019)

25. Pang, B., Lee, L.: A sentimental education: sentiment analysis using subjectivity summarization based on minimum cuts. In: Proceedings of the 42nd Annual
Meeting on Association for Computational Linguistics, pp. 271-284 (2004)

26. Pang, B., Lee, L.: Opinion mining and sentiment analysis. Found. Trends Inf.
Retrieval **2**(1–2), 1–135 (2008)

27. Potamias, R.A., Neofytou, A., Siolas, G.: NTUA-ISLAB at SemEval-2019 task 9:
mining suggestions in the wild. In: Proceedings of the 13th International Workshop
on Semantic Evaluation, pp. 1224–1230 (2019)

28. Raychev, V., Nakov, P.: Language-independent sentiment analysis using subjectivity and positional information. In: Proceedings of the International Conference
RANLP-2009, pp. 360–364 (2009)

29. Tsytsarau, M., Palpanas, T.: Survey on mining subjective data on the web. Data
Mining Knowl. Discov. **24**(3), 478–514 (2012)

30. Tutubalina, E.: Target-based topic model for problem phrase extraction. In: Hanbury, A., Kazai, G., Rauber, A., Fuhr, N. (eds.) ECIR 2015. LNCS, vol. 9022, pp.
271–277. Springer, Cham (2015). https://doi.org/10.1007/978-3-319-16354-3_29

31. Tutubalina, E., Nikolenko, S.: Inferring sentiment-based priors in topic models. In: Lagunas, O.P., Alcántara, O.H., Figueroa, G.A. (eds.) MICAI 2015. LNCS (LNAI), vol. 9414, pp. 92–104. Springer, Cham (2015). https://doi.org/10.1007/978-3-319-27101-9_7

32. Wachsmuth, H., Trenkmann, M., Stein, B., Engels, G., Palakarska, T.: A review corpus for argumentation analysis. In: Gelbukh, A. (ed.) CICLing 2014. LNCS, vol. 8404, pp. 115–127. Springer, Heidelberg (2014). https://doi.org/10.1007/978-3-642-54903-8_10

33. Williams, A., Nangia, N., Bowman, S.: A broad-coverage challenge corpus for sentence understanding through inference. In: Proceedings of the 2018 Conference of the North American Chapter of the Association for Computational Linguistics: Human Language Technologies, volume 1 (Long Papers), pp. 1112–1122. Association for Computational Linguistics (2018). http://aclweb.org/anthology/N18-1101

34. Wu, F., Huang, Y.: Collaborative multi-domain sentiment classification. In: 2015 IEEE International Conference on Data Mining, pp. 459–468. IEEE (2015)

35. Yin, W., Hay, J., Roth, D.: Benchmarking zero-shot text classification: Datasets, evaluation and entailment approach. In: Proceedings of the 2019 Conference on Empirical Methods in Natural Language Processing and the 9th International Joint Conference on Natural Language Processing (EMNLP-IJCNLP), pp. 3914–3923 (2019)

Selection of Pseudo-Annotated Data for Adverse Drug Reaction Classification Across Drug Groups

Ilseyar Alimova[1] and Elena Tutubalina[1,2,3(✉)]

[1] Kazan (Volga Region) Federal University, Kazan, Russia
tlenusik@gmail.com
[2] HSE University, Moscow, Russia
[3] Sber AI, Kazan, Russia

Abstract. Automatic monitoring of adverse drug events (ADEs) or reactions (ADRs) is currently receiving significant attention from the biomedical community. In recent years, user-generated data on social media has become a valuable resource for this task. Neural models have achieved impressive performance on automatic text classification for ADR detection. Yet, training and evaluation of these methods are carried out on user-generated texts about a targeted drug. In this paper, we assess the robustness of state-of-the-art neural architectures across different drug groups. We investigate several strategies to use pseudo-labeled data in addition to a manually annotated train set. Out-of-dataset experiments diagnose the bottleneck of supervised models in terms of breakdown performance, while additional pseudo-labeled data improves overall results regardless of the text selection strategy.

Keywords: Biomedical text mining · Text classification · Neural models

1 Introduction

Pharmacovigilance from social media data that focuses on discovering adverse drug effects (ADEs) from user-generated texts (UGTs). ADEs[1] are unwanted negative effects of a drug, in other words, harmful and undesired reactions due to its intake.

In recent years, researchers have increasingly applied neural networks, including Bidirectional Encoder Representations from Transformers (BERT) [4], to various biomedical tasks, including as text-level or entity-level ADR classification of user-generated texts [2,8,13,16]. The text-level ADR classification task aims to detect whether a given short text contains a mention of an adverse drug effect. The entity-level task focuses on the classification of a biomedical entity

---

[1] The terms ADEs and adverse drug reactions (ADRs) are often used interchangeably.

© The Author(s), under exclusive license to Springer Nature Switzerland AG 2022
E. Burnaev et al. (Eds.): AIST 2021, LNCS 13217, pp. 37–44, 2022.
https://doi.org/10.1007/978-3-031-16500-9_4

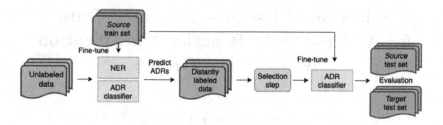

Fig. 1. An overview of our pipeline.

or a phrase within a text. The first type of classification is needed to filter irrelevant texts from a data collection. In the second case, classification models process results of named entity recognition (NER) tools. However, most recent studies mostly share the same limitations regarding their training strategy: classification systems rely only on existing manually annotated training data for supervised machine learning. These annotated corpora include texts about a small number of drugs (at best about dozens) while there are over 20,000 prescription drug products approved for marketing[2]. Moreover, the model performance is frequently evaluated under the implicit hypothesis that the training data (source) and the test data (target) come from the same underlying distribution (i.e., both sets include drugs from a particular Anatomical Therapeutic Chemical (ATC) group). This hypothesis can cause overestimated results on ADR classification since the same drug classes shared similar patterns of ADR presence [17].

In this paper, we take the task a step further from existing research by exploring how well a BERT-based classification model trained on texts about drugs from one ATC group (*source*) performs on texts about other drugs from the second ATC group (*target*). First, we train a base model on some amount of *source* labeled data and use this model to pseudo-annotate unlabeled data. Second, the original labeled data is augmented with the pseudo-labeled data and used to train a new model. Third, both models are evaluated on source data (*in-dataset*) and target data (*out-of-dataset*) (Fig. 1). In particular, we focus on the automatic expansion of training data. We explore several strategies to select a subset to train our ADR classification model for target data. In this work, we seek to answer the following research questions:

RQ1: Do in-dataset evaluation with training and testing on each benchmark separately lead to a significant overestimation of performance?

RQ2: How can we utilize raw reviews without manual annotations to improve model performance on a particular drug group?

2   Related Work

ADR classification is usually formulated as a classification or ranking problem. A number of supervised neural models have been proposed; please see recent

---

[2] https://www.fda.gov/about-fda/fda-basics/fact-sheet-fda-glance.

state-of-the-art models in Social Media Mining for Health shared tasks [8,13]. Further, we describe several studies that leveraged different types of unlabeled data to deal with data bias or a small number of labeled samples.

The work that is the closest to ours and considers unlabeled data is [10]. The authors proposed utilized semi-supervised convolutional neural network models for ADR classification in tweets. The method works in two phases: (1) unsupervised phrase embedding learning on unlabeled data, and (2) integrating the learned embeddings into the supervised training that uses labeled data. The authors used several types of unlabeled data: (i) collection of random tweets; (ii) corpus of health-related texts; (iii) tweets with drug names; (iv) tweets that mention a health condition. Health-related texts include sentences from the clinical medicine category Wikipedia pages, PubMed articles, and UMLS Medical Concept Definitions. These types of unlabeled data were used to learn phrase embeddings for the semi-supervised classification. The experiments show that the model trained on the tweets with health conditions outperformed in average models trained on other unlabeled datasets by 3.3% of F-score. This model also outperformed trained only on annotated dataset model by 8.9% of F-score. Gupta et al. present a semi-supervised Bi-directional LSTM based model for ADR mention extraction from tweets [6]. The large corpus of unlabeled tweets was collected for unsupervised learning. The semi-supervised approach improved the F1-score metric by 2.2%. [5] proposed a semi-supervised co-training based learning method to tackle the problem of labeled data scarcity for adverse drug reaction mention extraction task. The methods were evaluated on two corpora of tweets. The results show that increasing the dataset leads to improving results. Thus, the model co-trained on 100K tweets outperformed baselines on F-Score by 6.52% and 4.8% on Twitter ADR and TwiMed corpora, respectively. Perez et al. investigated the semi-supervised approach for detecting ADR mentions on Spanish and Swedish clinical corpora [14]. The authors applied unlabeled data of EHR texts to obtain Brown trees and semantic space cluster features. The features were enhanced to maximum probability, CRF, Perceptron, and SVM classifiers. The results showed that the semi-supervised approaches significantly improved standard supervised techniques for both languages on average by 12.26% and 4.56% of F-score for Spanish and Swedish languages, respectively.

Our work differs from the studies discussed above in the following important aspects. First, we annotate unlabeled data with noise labels and use it to train for a classification model instead of language pre-training. Second, we utilize state-of-the-art BERT-based models. Finally, we evaluate models across different drug groups with in-dataset and out-of-dataset setup.

3   Datasets and Models

We perform all experiments in four steps (Fig. 1). First, we train two models on source data: (i) a classifier for ADR identification, (i) a NER model for detection of biomedical disease-related entities. Second, we apply both models to extract medical entities from an unlabeled corpus and classify these entities as ADRs.

Third, we select a set of texts with ADR mentions. Finally, we train the ADR classifier from Step 1 on the pseudo-annotated data and evaluate this model on the target corpus. Further, we describe two manually annotated datasets, a raw corpus of reviews, and neural models.

CADEC. CSIRO Adverse Drug Event Corpus (CADEC) [7] consists of annotated user reviews from https://www.askapatient.com/ about 12 drugs, divided into two groups: 1. Diclofenac; 2. Lipitor.

Diclofenac is linked with the ATC group *Nervous system* (N). Lipitor is included into three ATC main groups: *Sensory organs* (S), *Musculoskeletal system* (M), *Dermatologicals* (D). The corpus contains 6,320 entities, 5,770 of them marked as 'ADR'. Following [1,11,15], we group non-ADR types (diseases, symptoms and findings) in a single opposite class for binary classification.

PsyTAR. (Psychiatric Treatment Adverse Reactions (PsyTAR) corpus [18] is the first open-source corpus of user-generated posts about psychiatric drugs taken from https://www.askapatient.com/. This dataset includes 887 posts about four psychiatric medications of two classes: 1. Zoloft and Lexapro; 2. Effexor and Cymbalta.

Zoloft, Lexapro, Effexor, and Cymbalta are included in the ATC group *Nervous system* (N). All posts were annotated manually for 4 types of entities:(i) adverse drug reactions (ADR); (ii) withdrawal symptoms (WD); (iii) drug indications (DI); (iv) sign/symptoms/illness (SSI). The total number of entities is 7,415. We joined WD, DI and SSI entities in a single class.

Unlabeled Data. We collect 113,836 reviews about 1,593 drugs from https://www.askapatient.com/. There are 35,712 reviews with a rating 1. The total number of reviews about drugs from the CADEC corpus is 173 and from the PsyTAR corpus is 6,590. We create several subsets for our experiments:

1. The full set of reviews. This set is abbreviated as AskaPatient$_{full}$;
2. The set of reviews about target drugs (AskaPatient$_{target}$);
3. The set of reviews with the lowest rating (AskaPatient$_1$). Each review is associated with an overall rating, which is a numeric score between 1 (dissatisfied) and 5 (very satisfied). The lowest rating indicates that a user would not recommend taking this medicine.

Models. In particular, we utilize two models:

1. LSTM-based Interactive Attention Network (IAN) [12];
2. BERT [4,9].

[3] showed the superiority of IAN over other neural models on four of the five corpora for entity-level ADR text classification. We used 15 epochs to train IAN on each dataset, the batch size of 128, the number of hidden units for LSTM layer 300, the learning rate of 0.01, L2 regularization of 0.001, dropout 0.5. We applied the implementation of the model from https://github.com/songyouwei/ABSA-PyTorch. We trained BioBERT model [9] for 15 epochs.

Table 1. Classification results of IAN and BERT evaluated on CADEC.

Train set	Model	ADR-class			Macro-averaged		
		P	R	F	P	R	F
In-dataset performance							
CADEC	IAN	.966	.972	.969	.832	.805	.815
CADEC	BERT	.947	.983	.965	.824	.702	.746
Out-of-dataset performance							
PsyTAR	IAN	.898	.885	.891	.615	.624	.619
PsyTAR	BERT	.909	.914	.911	.673	.669	.671
PsyTAR + AskaPatient$_{full}$	IAN	.903	.895	.899	.636	.642	.638
PsyTAR + AskaPatient$_1$	IAN	.896	.951	.922	.693	.626	.647
PsyTAR + AskaPatient$_{target}$	IAN	.950	.968	.959	.861	.823	.841
PsyTAR + AskaPatient$_{full}$	BERT	.912	.918	.915	.681	.675	.677
PsyTAR + AskaPatient$_1$	BERT	.959	.839	.895	.695	.806	.724
PsyTAR + AskaPatient$_{target}$	BERT	.952	.929	.941	.782	.817	.798

Table 2. Classification results of IAN and BERT evaluated on PsyTAR.

Train set	Model	ADR-class			Macro-averaged		
In-dataset performance							
PsyTAR	IAN	.902	.913	.909	.868	.868	.868
PsyTAR	BERT	.904	.917	.910	.884	.882	.881
Out-of-dataset performance							
CADEC	IAN	.685	.982	.807	.765	.582	.553
CADEC	BERT	.711	.984	.825	.808	.629	.623
CADEC + AskaPatient$_{full}$	IAN	.693	.836	.751	.591	.577	.563
CADEC + AskaPatient$_1$	IAN	.766	.931	.841	.784	.703	.721
CADEC + AskaPatient$_{target}$	IAN	.695	.964	.807	.739	.598	.582
CADEC + AskaPatient$_{full}$	BERT	.718	.986	.831	.810	.627	.630
CADEC + AskaPatient$_1$	BERT	.764	.973	.856	.834	.714	.731
CADEC + AskaPatient$_{target}$	BERT	.681	.978	.803	.746	.573	.633

4 Experiments

All models were evaluated by 5-fold cross-validation. We computed averaged
recall (R), precision (P), and F_1-measures (F) for ADR and non-ADR classes
separately and then macro-average of these values for both classes. The average
F-score results for IAN and BERT models are presented in Tables 1 and 2.

To answer **RQ1**, we compare in-dataset and out-of-dataset results in Tables 1
and 2. The in-dataset performance of both models is significantly higher than

out-of-dataset results. In particular, BERT trained on PsyTAR and CADEC achieved macro-averaged F_1 of 0.881 and 0.623 on the PsyTAR sets, respectively. This indicates the impact of different contexts and entity mentions on cross-dataset performance.

To answer **RQ2**, we compare models trained on (i) manually annotated data and (ii) extended corpora with additional pseudo-annotated data. The results show that in cross-dataset experiments training the model on additional data improves results regardless of the selected noisy data in comparison to models trained only on labeled data. IAN and BERT trained on PsyTAR + Askapatient$_{target}$ showed the highest results on the CADEC corpus (0.841 and 0.798 F-score, respectively). These models outperformed models trained only on the CADEC corpus by 0.26 of F-score. On the PsyTAR corpus, the models trained on the CADEC + Askapatient$_1$ achieved the best results among models trained on additional noisy data (0.72). However, in this case, the F-score of the model trained on PsyTAR corpus is higher than the results of models trained on CADEC corpus and additional noisy data. Such a difference in results is due to reviews with a rating of 1 include 1,307 reviews about drugs from the PsyTAR corpus and do not contain reviews about drugs from the CADEC corpus.

Among the models additionally trained on noisy data the models trained on the full noisy corpus gave the smallest improvement in F-score metric on both corpora in comparison to models trained on annotated data. For the IAN model, the increase of F-score was 0.19 and 0.07, for the BERT model the improvement was 0.06 and 0.07 on the CADEC and PsyTAR cases, respectively.

Considering results in terms of 'ADR' class metrics the same models achieved the best F-score metrics: trained on PsyTAR + AskaPatient$_{target}$ (IAN - 0.959 and BERT - 0.941) for CADEC corpus and trained on CADEC + AskaPatient$_1$ (IAN - 0.84 and BERT - 0.856) for PsyTAR corpus. However, in this case, none of the models outperformed the results of the model evaluated within a single corpus. Moreover, several models did not improve the F-score metrics in comparison to the same models trained only on annotated corpus: IAN trained on CADEC + AskaPatient$_{full}$ (0.75), BERT trained on PsyTAR + AskaPatient$_1$ (0.895) and BERT trained on CADEC + AskaPatient$_{target}$ (0.803).

5   Conclusion

In this paper, we studied the task of discovering the presence of adverse drug effects in user reviews about anti-inflammatory and psychiatric drugs. We perform an extensive evaluation of LSTM-based and BERT-based models on two datasets in the cross-dataset setup. Our evaluation shows the great divergence in performance between splits of two corpora: BERT-based models trained on one corpus and evaluated on another show a substantial decrease in performance (-7.5% and -25.8% macro-averaged F_1 for the CADEC and PsyTAR sets, respectively) compared to in-dataset models. Using in-dataset models, we automatically annotate raw user posts and extend a train set with texts about a wide range of drugs. Retraining of models shows the increase in performance (e.g., up to $+12.7\%$ macro-averaged F_1 for CADEC) compared to out-of-dataset models. We foresee three directions for future work. First, promising research directions include adaptive self-training and meta-learning techniques for training neural models with few labels. Second, future research may focus on advanced labeled data acquisition strategies such as uncertainty-based methods. Third, knowledge transfer from the clinical domain to the social media domain across different drug groups remains to be explored.

Acknowledgments. This work was supported by the Russian Science Foundation grant # 18-11-00284.

References

1. Alimova, I., Tutubalina, E.: Automated detection of adverse drug reactions from social media posts with machine learning. In: van der Aalst, W.M.P., et al. (eds.) AIST 2017. LNCS, vol. 10716, pp. 3–15. Springer, Cham (2018). https://doi.org/10.1007/978-3-319-73013-4_1

2. Alimova, I., Tutubalina, E.: Multiple features for clinical relation extraction: a machine learning approach. J. Biomed. Inform. **103**, 103382 (2020)

3. Alimova, I., Tutubalina, E.: Entity-level classification of adverse drug reaction: a comparative analysis of neural network models. Program. Comput. Softw. **45**(8), 439–447 (2019)

4. Devlin, J., Chang, M.W., Lee, K., Toutanova, K.: Bert: pre-training of deep bidirectional transformers for language understanding. In: Proceedings of the 2019 Conference of the North American Chapter of the Association for Computational Linguistics: Human Language Technologies, Volume 1 (Long and Short Papers), pp. 4171–4186 (2019)

5. Gupta, S., Gupta, M., Varma, V., Pawar, S., Ramrakhiyani, N., Palshikar, G.K.: Co-training for extraction of adverse drug reaction mentions from tweets. In: Pasi, G., Piwowarski, B., Azzopardi, L., Hanbury, A. (eds.) ECIR 2018. LNCS, vol. 10772, pp. 556–562. Springer, Cham (2018). https://doi.org/10.1007/978-3-319-76941-7_44

6. Gupta, S., Pawar, S., Ramrakhiyani, N., Palshikar, G.K., Varma, V.: Semi-supervised recurrent neural network for adverse drug reaction mention extraction. BMC Bioinform. **19**(8), 212 (2018). https://doi.org/10.1186/s12859-018-2192-4

7. Karimi, S., Metke-Jimenez, A., Kemp, M., Wang, C.: CADEC: a corpus of adverse drug event annotations. J. Biomed. Inform. **55**, 73–81 (2015)

8. Klein, A., et al.: overview of the fifth social media mining for health applications (# smm4h) shared tasks at COLING 2020. In: Proceedings of the Fifth Social Media Mining for Health Applications Workshop and Shared Task, pp. 27–36 (2020)

9. Lee, J., Yoon, W., Kim, S., Kim, D., So, C., Kang, J.: BioBERT: a pre-trained biomedical language representation model for biomedical text mining. Bioinformatics **36**(4), 1234–1240 (2020)

10. Lee, K., et al.: Adverse drug event detection in tweets with semi-supervised convolutional neural networks. In: Proceedings of the 26th International Conference on World Wide Web, pp. 705–714 (2017)

11. Li, Z., Yang, Z., Luo, L., Xiang, Y., Lin, H.: Exploiting adversarial transfer learning for adverse drug reaction detection from texts. J. Biomed. Inform. **106**, 103431 (2020)

12. Ma, D., Li, S., Zhang, X., Wang, H.: Interactive attention networks for aspect-level sentiment classification. In: Proceedings of the 26th International Joint Conference on Artificial Intelligence, pp. 4068–4074 (2017)
13. Magge, A., et al.: Overview of the sixth social media mining for health applications (# smm4h) shared tasks at NAACL 2021. In: Proceedings of the Sixth Social Media Mining for Health (# SMM4H) Workshop and Shared Task, pp. 21–32 (2021)
14. Perez, A., Weegar, R., Casillas, A., Gojenola, K., Oronoz, M., Dalianis, H.: Semi-supervised medical entity recognition: a study on Spanish and Swedish clinical corpora. J. Biomed. Inform. **71**, 16–30 (2017)
15. Rakhsha, M., Keyvanpour, M.R., Shojaedini, S.V.: Detecting adverse drug reactions from social media based on multichannel convolutional neural networks modified by support vector machine. In: 2021 7th International Conference on Web Research (ICWR), pp. 48–52. IEEE (2021)
16. Tutubalina, E., Nikolenko, S.: Exploring convolutional neural networks and topic models for user profiling from drug reviews. Multimed. Tools Appl. **77**(4), 4791–4809 (2018)
17. Wu, L., et al.: Study of serious adverse drug reactions using FDA-approved drug labeling and MedDRA. BMC Bioinform. **20**(2), 129–139 (2019)
18. Zolnoori, M., et al.: A systematic approach for developing a corpus of patient reported adverse drug events: a case study for SSRI and SNRI medications. J. Biomed. Inform. **90**, 103091 (2019)

Building a Combined Morphological Model for Russian Word Forms

Elena I. Bolshakova[1,2]([envelope]) [ORCID] and Alexander S. Sapin[1] [ORCID]

[1] Lomonosov Moscow State University, Moscow, Russia
eibolshakova@gmail.com
[2] National Research University Higher School of Economics, Moscow, Russia

Abstract. In recent years, high-precision machine learning models for traditional inflectional morphological analysis, as well as models for morpheme segmentation of words were built for Russian. Two these morphological tasks are evidently related, and some NLP applications may require to perform both of them, so development and evaluation of combined morphological model is of research interest. Such a model is supposedly useful for processing texts in languages with rich morphology (e.g., Russian), in particular, for deriving meaning of new words rarely encountered in texts. The paper presents a neural model implementing both inflectional analysis of Russian word forms (with morphological disambiguation) and their segmentation into constituent morphs with their classification. To train the model, a relevant dataset was built, by morphemic labeling of SynTagRus corpus, and transfer learning techniques were applied. Experimental evaluation of the model has shown its sufficiently high quality: 94.2% of precision for morphological tags disambiguation and 88–91% of word-level classification accuracy for segmentation.

Keywords: Automatic morphological analysis · Morpheme segmentation with classification · Neural network models for morphology

1   Introduction

Morphological analysis is the basic stage of text processing, which is used in many natural language processing (NLP) tasks. Traditional and well-studied problems of morphological analysis involve lemmatization of a given word form and its inflectional analysis resulted in morphological characteristics, or tags (case, tense, number, etc.), among them the most important is part of speech (POS). Inflectional analysis also implies disambiguating morphological homonymy (a rather frequent phenomenon in natural languages), arising when several variants of tags are possible for the processed word form. Morphological disambiguation is the most difficult problem of inflectional analysis, especially for highly-inflecting languages (e.g., Russian or Finnish), and to resolve the problem, machine learning methods are usually applied, giving up to 93–97% accuracy, depending on particular language and texts for training.

E. Burnaev et al. (Eds.): AIST 2021, LNCS 13217, pp. 45–55, 2022.
https://doi.org/10.1007/978-3-031-16500-9_5

Another task of morphological analysis is morpheme segmentation, two kinds of the task were studied [10,11,22]:

- pure morpheme segmentation, i.e. splitting a given word into constituent morphs, e.g., *impossible* → *im-poss-ible*, Rus.: *не-воз-мож-н-ый*.
- morpheme segmentation with classification of segmented morphs (morphemes), the main types of morphemes are Prefix, Root, Suffix, Ending, Postfix, e.g., *impossible* → *im*:PREF/*poss*:ROOT/*ible*:SUFF, Rus: *невоз-можный* → *не*:PREF/*воз*:PREF/*мож*:ROOT/*н*:SUFF/*ый*:END.

Morphemes (roots and affixes) are the smallest meaningful language units, therefore information about morphemic structure of words is helpful for various NLP auxiliary tasks and applications, including machine translation [2], creating derivational trees of words [15], recognition of semantically related words (cognates, paronyms, etc.), constructing word embeddings [9] for handling rare and out-of-vocabulary words, and improving contextualized embeddings [12].

Inflectional morphological analysis, as well as morpheme segmentation, are especially topical and at the same time more difficult for languages with rich inflection and complex word formation (such as Russian having many affixes of various types and meanings).

In recent years, several high-precision machine learning models have been developed for Russian, implementing traditional inflectional morphological analysis [1,13,16,23] and morpheme segmentation with classification [4,5,22]. The models for morpheme segmentation with classification were developed only for lemmas (normalized forms of words) and are not suitable for processing Russian texts with significantly varying grammatical forms of words. The recent work [6] presents a high-precision neural model for morpheme segmentation with classification for Russian word forms. However, to implement segmentation, this model needs information about POS of word form being segmented, which requires its preliminary inflectional morphological analysis performed by an appropriate processor capable to resolve morphological disambiguation.

Having in mind that morphology is the unified language mechanism, and some NLP applications may require results of both above considered tasks of morphological analysis, we have developed a combined morphological model for Russian word forms, which implements inflectional analysis and morphemic segmentation with classification. To perform the first task, the developed model exploits dictionary data from open-source morphological processor and disambiguates homonymy for the given word form (for its morphological tags), if any. Convolutional neural networks (CNN) were chosen as the core of our model, and to train the model, we have developed a dataset with morphological and morpheme labeling, by automatic morpheme segmentation of SynTagRus[1] text corpus.

Experimental evaluation has shown that the combined model gives 94.2% of precision for morphological tags disambiguation (which is sufficient for many applied tasks), and for morpheme segmentation, it demonstrates 88–91% of

---

[1] https://github.com/UniversalDependencies/UD_Russian-SynTagRus.

word-level classification accuracy, which is comparable with the state-of-the art results (89–91%) achieved in the previous works [5,6,22].

The paper starts with a brief overview of main results obtained for traditional inflectional analysis of Russian word forms, and also basic results in morpheme segmentation task. Then architecture of our combined morphological model is described, followed by key issues of its training (including development of the dataset needed for training). The results of experiments with the model are also reported and discussed, and some conclusions are finally presented.

2 Related Work

The most widely used morphological processors for Russian (including open-code Pymorphy [14] and freely available Mystem [18]) are based on dictionaries of word forms, and analysis of a given word form relies on search in such a dictionary, generally resulting in several possible variants of morphological tags (POS, case, tense, number, etc.). To disambiguate morphological homonymy, an additional procedure is required, which is usually built with various supervised machine learning techniques (e.g., [19,24]).

In recent years, another approach to morphological analysis is evolving: machine learning models simultaneously perform inflectional analysis and disambiguation, producing a single variant of morphological tags for each word form [1,13,23]. Such models make use of word vector representations (embeddings) taken from modern language models of various types: FastText [3], ELmO [17], BERT [8], which significantly improves the quality of disambiguation, but the models have low performance, compared to dictionary approach. The model proposed in [1] and based on contextualized embeddings from BERT, along with multiclass logistic regression has achieved the best quality for the particular diverse Russian text collection: about 95% accuracy for morphological tags. The morphological analysis methods tested in [21] showed accuracy of 93–97%, for other Russian text collections.

For automatic morpheme segmentation, the first purely statistical methods (dictionary-based [11] and corpus-based [10,20]) performed only segmentation without classification and showed rather low quality. The most known solution implemented in Morfessor system [20] achieves about 70–80% F-measure for detected morpheme boundaries, for English, Finnish, and Turkish.

Significant progress in automatic morpheme segmentation is related to supervised machine learning and creation of necessary datasets with segmented and labeled morphemes. For Russian language, such datasets have recently appeared, the most representative RuMorphs-Lemmas[2] dataset with about 96 thou. words was obtained from Tikhonov's derivation dictionary [25]. This dataset contains lemmas (normalized word forms) split into morphs, the morphs are classified according to main morpheme types (prefix, root, suffix, ending, postfix), and successive prefixes and suffixes (if any) are also labeled, for example:

---

[2] https://github.com/cmc-msu-ai/NLPDatasets/releases/download/v0.2/RuMorphs
-Lemmas.txt.

c:PREF/*маз*:ROOT/*ыва*:SUFF/*ть*:SUFF/*ся*:POSTFIX,
o:PREF/*сложн*:ROOT/*ени*:SUFF/*e*:END.

Based on RuMorphs-Lemmas, several high-precision methods (models) of morpheme segmentation with classification for Russian lemmas have been developed [4,5,22], with various powerful ML techniques: convolutional neural network (CNN), Gradient boosted decision trees (GBDT), Long short-term memory neural network (Bi-LSTM). These developed models perform the task by classifying letters of words according to main types of morphs and differ in additional techniques:

- The CNN model [22] uses labeling scheme with 22 classes of letters, an auxiliary rule-based procedure for post editing of predicted letter classes, as well as ensemble of three identical CNN models, for increasing the model quality.
- GBDT model [4] exploits a reduced set of 10 letter classes (but sufficient for recognizing successive affixes and roots) and information about morphological tags (POS, gender, number, case, time) of the word being segmented.
- Bi-LSTM model [5] also uses the set of 10 classes and additional morphological information, but only POS of the word, and the model does not require any correcting procedure.

All these morpheme segmentation models give comparable scores evaluated on RuMorphs-Lemmas dataset: 97–98% of F-measure on morpheme boundaries, about 98% of classification accuracy for letters, and up to 89% of accuracy on whole words, which is the best results for Russian lemmas. Nevertheless, for significantly varying Russian word forms they work poorly: only about 34% of word-level classification accuracy. Recent paper [6] presents both a new neural model built specifically for Russian word forms and RuMorphs-Words[3] dataset developed for training the model (by construction, it contains segmented and labeled word forms of a diverse lexicon). This segmentation model for word forms is applicable for processing Russian texts and achieves 91% of word-level classification accuracy (that is more than the score for lemmas), but it makes use of POS for the word form under processing, which requires to preliminarily apply morphological analyzer. Therefore, it seems reasonable to develop and evaluate a combined morphological neural model producing both morphological tags and segmented word forms, without any additional techniques. Such a model is expected to be more convenient and faster[4] than successive application of two different models (for inflective analysis and segmentation, respectively).

3   Architecture of Combined Model

To develop a combined morphological model, we have chosen one-dimensional convolutional neural networks (CNN) as the core of architecture, because CNNs

---

[3] https://github.com/cmc-msu-ai/NLPDatasets/releases/download/v0.2/RuMorphs -Words.txt.

[4] Further experiments have shown performance gain about 20%.

are faster to train and take less memory, and at the same time, they do not lose in quality for the morpheme segmentation task [4]. CNNs work with sequences of fixed length, so our model analyzes the input text sentence by sentence, sentences are considered to be sequences of 9 words. Each word form of the input sentence is accordingly considered to be of 20 letters (most Russian words contain fewer letters), shorter words are padded with blanks, and longer ones are split into parts.

In order to facilitate the neural model, morphological tags for input word forms are taken from the available dictionary-based morphological processor[5] for Russian (moreover, the dictionary data is more complete than any training corpus, and this can presumably improve the quality of the model). In general case, several variants of main morphological tags (POS, case, number, gender, time) are provided by morphological processor, and our combined model implements their morphological disambiguation.

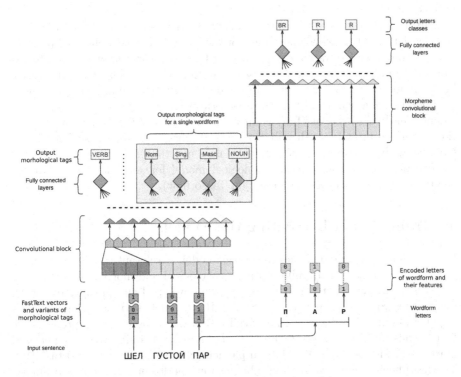

Fig. 1. Combined morphological model

The model architecture is shown in Fig. 1, the submodel responsible for the disambiguation is presented to the left, and the submodel implementing morpheme segmentation is given to the right. Each submodel contains three con-

---

[5] https://github.com/alesapin/XMorphy.

nected "convolution blocks" followed by fully connected network layers with softmax activation function acting as classifier.

Since vector representations of words significantly improve the quality of morphological analysis, as input for the disambiguation submodel, we use FastText embeddings concatenated with encoded morphological tags for each word form (all variants of tags received from the morphological processor). FastText was chosen as computationally simple and at the same time it can produce vectors even for unknown words.

Each convolutional block of the disambiguation submodel consists of 1D convolution layer, max pooling, and dropout layers. Max pooling layers significantly speed up training and inference of the model, while dropout layers help to avoid overfitting. For activating convolution layers, ReLU function is used, as simple and well-proven in practice. The output of the last convolution block is passed to fully connected layers classifying values of morphological tags for the word forms (POS, case, gender, number, time). For each word form, its own set of fully connected layers is applied.

Output POS tag from the fully connected layer is concatenated with one-hot encoded letters of the corresponding word form and also with encoded features of the letters (are letters vowels or not), and the resulted vector is given to the submodel implementing morpheme segmentation with classification. Its convolutional blocks are similar to those in the disambiguation submodel but without max pooling. Segmentation of each word form from input sequence is performed independently, and the last fully connected layers with softmax activation function outputs a probability distribution over all possible letter classes. Similar to [5,6] we apply labeling scheme with 10 classes of letters.

The described combined morphological model for word forms was implemented with Keras library [8] (based on Tensorflow).

4 Training and Evaluating the Model

For training the combined morphological model, a relevant labeled dataset is needed, but datasets containing both morphological and morpheme labeling of Russian word forms were absent. So we have developed an appropriate one by segmenting words of the text corpus SynTagRus[6], which has morphological labeling (for about 1.1 million words). SynTagRus was chosen as a representative text corpus, and at the same time, it was used for training many supervised morphological models for Russian, in particular, models of the shared task [16]. The morpheme segmentation and labeling were performed in a semi-automatic mode, with the aid of the open-source morpheme segmentation model for Russian word forms [6] (while processing, numbers and non-cyrillic words were marked as unknown and further were not taken into account for training and evaluation). Automatic segmentation was followed by manual validation of some fragments of resulted dataset and also rule-based correction of certain errors.

---

[6] https://github.com/UniversalDependencies/UD_Russian-SynTagRus.

For our experiments, the developed dataset, hereafter, RuMorphs-SynTagRus[7], was randomly divided in proportion 70:10:20 for training, validation, and testing. Various hyperparameters of the combined model were experimentally tested, the best ones turned to be the following: the number of units in layers of convolution block is accordingly 512, 256, 192, respectively; the gradient descent algorithm is Adam, with learning rate of 0.001 and dropout of 0.3, max pooling size equals three. Experiments have also shown that additional convolution blocks insignificantly improve the quality of the model, but it became too heavy both for training and for evaluation.

Evaluation of the resulted model showed its disambiguation accuracy 97.9 for POS tag and 94.2% for all tags, the latter is only slightly lower than in [5], while word-level classification accuracy for morpheme segmentation turned 96.5%, which is significantly higher than for the best morpheme segmentation models (89–91%). It should be noted that to avoid overestimation while evaluating quality of morpheme segmentation, all words from the test subset that are shorter than three letters were not taken into account, since their segmentation is trivial.

However, testing our model on representative and lexically diverse RuMorphs-Words dataset [6] has shown only 47.3% word-level classification accuracy, which signals about overfitting to RuMorphs-SynTagRus corpus. In our opinion, the main reason is very low "morpheme diversity" of words in RuMorphs-SynTagRus: it contains the limited number of different words, many short words and words similar in structure, and so few different affixes and their combinations.

To overcome the detected overfitting, we have applied transfer learning techniques often used for creating NLP neural models (in particular, for classification models based on BERT [7]). Training of the combined model was carried out in three stages. At the first stage, the submodel for morpheme segmentation was trained separately on RuMorphs-Words dataset (it includes POS tag for all segmented word forms). At the second stage, the weights in this trained submodel were frozen (i.e., excluded from training), and all the combined model was trained on RuMorphs-SynTagRus corpus. At the third stage, the weights of the morpheme submodel were defrosted, learning rate was decreased to 0.00001 (in order to keep the knowledge gained at the first stage), and the whole combined model was trained once again, with the reduced maximum number of iterations (by the same reason). Except the learning rate, the other hyperparameters of the model did not change at all stages.

Thereby, the described model has saved the knowledge about morpheme segmentation obtained after the first training stage and at the same time has learned to disambiguate word forms from RuMorphs-SynTagRus. As a result, for morpheme segmentation, the combined model achieves 88.3% word-level classification accuracy for word forms from RuMorphs-SynTagRus and 91.7% for RuMorphs-Words. All evaluation scores are given in Table 1: the quality of

---

[7] https://github.com/cmc-msu-ai/NLPDatasets/releases/download/v0.2/RuMorphs-SynTagRus.txt.

segmentation is measured in precision, recall, and F-measure for detecting morpheme boundaries, while classification accuracy is computed for letters and for whole words, respectively. The former is the ratio of correctly recognized classes of letters to the number of all letters, the latter estimates the ratio of completely correctly segmented words with true classes of all their letters (thus, the score for words is usually lower than for letters).

One can notice, that all scores for morpheme segmentation turned out to be highly close to the maximum possible, while classification accuracy achieves the known best results (98% for letters and 89–91% for whole words) only on RuMorphs-Words data set. Although the combined model shows worse (but not significantly) quality on RuMorphs-SynTagRus dataset (compared to RuMorphs-Words), nevertheless, final results can depend on the training process. At the third stage of training, with increasing the number of iterations, the morpheme segmentation accuracy for word forms from RuMorphs-SynTagRus increases, while for RuMorphs-Words it falls. Thus, by tuning the number of iterations at the third stage, the model can be adjusted to the particular dataset.

Table 1. Evaluation of morpheme segmentation for the combined model (%).

DataSet for evaluation	Segmentation			Classification	
	Precision	Recall	F-measure	Letter Accuracy	Word Accuracy
RuMorphs-SynTagRus	97.1	97.4	97.3	95.7	88.3
RuMorphs-Words	98.6	99.1	98.8	98.0	91.7

Additionally, we have evaluated ratio of various errors in morpheme segmentation, depending on wrong boundaries between morphemes of various types, the results are presented in Table 2. The most frequent errors are expectedly related to wrong boundaries between roots and suffixes, almost half of the errors (column ROOT-SUFF in Table 2). Here is an example of such error: incorrect segmentation of verb form *пришла* (*came* for feminine, singular, third person): *при*:PREF/*шл*:ROOT/*a*:END instead of *при*:PREF/*ш*:ROOT/*л*:SUFF/*a*:END (because of suppletive verb). Other types of frequent errors are ROOT-END, for example, instead of correct segmentation *сн*:ROOT/*a*:END (the word form of *сон* – *dream*) erroneous result is *сна*:/ROOT (the ending was not detected), and PREF-ROOT errors, that is erroneous segmentation of prefix, for example, word *уткин* (*duck's*) is split as *у*:PREF/*ткин*:ROOT instead of correct *утк*:ROOT/*ин*:SUFF (since prefix *у*- is rather frequent in Russian).

Table 2. Types of errors in morpheme segmentation (%).

PREF-PREF	PREF-ROOT	ROOT-ROOT	ROOT-SUFF	ROOT-END	SUFF-SUFF	SUFF-END
0.1	14.25	4.79	47.66	19.92	4.34	8.94

5   Conclusions

We have presented the combined morphological model based on convolutional neural networks and intended for processing Russian texts by performing twofold task: morphological inflectional analysis of word forms and their morpheme segmentation with classification. Such twofold task seems to be useful in NLP tasks not only for Russian, but also for other morphologically rich and highly inflective languages.

The inflectional analysis is carried out with the help of morphological dictionary providing several variants of morphological tags for the given word form, and the neural model itself resolves this morphological homonymy.

To train the developed model, the necessary dataset was built, based on SynTagRus corpus and the available tool for morpheme segmentation of separate Russian word forms. Transfer learning techniques applied for training the combined model make it possible to achieve sufficiently good quality of the model: 94.2% for accuracy of morphological tags, F-measure up to 98% on morpheme boundaries, and 88–91% of word-level classification accuracy. For a particular training data set, quality of morpheme segmentation can be improved by changing parameters of training.

The developed combined morphological model[8] and the dataset (RuMorphs-SynTagRus) are freely available and can be used in various NLP experiments with Russian text. The helpful feature of the model is its rather high performance and relatively small size (due to usage of CNNs): about 1.9 thou. words per second (in the single core mode of Intel i7-10850H) and size of 5.3 MB.

Evidently, our combined model needs further investigations: first of all, correction of errors remaining in RuMorphs-SynTagRus dataset and its enlargement are important tasks, but changes in neural architecture may also improve quality of the model.

Acknowledgements. The research has been partially supported by the Interdisciplinary Scientific and Educational School of Lomonosov Moscow State University «Brain, Cognitive Systems, Artificial Intelligence».

References

1. Anastasyev, D.G.: Exploring pretrained models for joint morpho-syntactic parsing of Russian. In: Computational Linguistics and Intellectual Technologies: Proceedings of the International Conference "Dialogue 2020", Moscow, pp. 1–12 (2020)

---

[8] https://github.com/alesapin/XMorphy/blob/master/scripts/joined_model.py.

2. Botha, J., Blunsom, P.: Compositional morphology for word representations and language modelling. In: International Conference on Machine Learning, pp. 1899–1907 (2014)
3. Bojanowski, P., Grave, E., Joulin, A., Mikolov, T.: Enriching word vectors with subword information. Trans. Assoc. Comput. Linguist. **5**, 135–146 (2017)
4. Bolshakova, E., Sapin, A.: Comparing models of morpheme analysis for Russian words based on machine learning: In: Computational Linguistics and Intellectual Technologies: Proceedings of the International Conference "Dialogue 2019", Moscow, pp. 104–113 (2019)
5. Bolshakova, E., Sapin, A.: Bi-LSTM model for morpheme segmentation of Russian words. In: Ustalov, D., Filchenkov, A., Pivovarova, L. (eds.) AINL 2019. CCIS, vol. 1119, pp. 151–160. Springer, Cham (2019). https://doi.org/10.1007/978-3-030-34518-1_11
6. Bolshakova, E., Sapin, A.: Building dataset and morpheme segmentation model for Russian word forms. In: Computational Linguistics and Intellectual Technologies: Proceedings of the International Conference "Dialogue 2021", Moscow, pp. 154–161 (2021)
7. Devlin, J., et al.: BERT: pre-training of deep bidirectional transformers for language understanding. In: Proceedings of the 2019 Conference of the NAACL: Human Language Technologies, vol. 1, pp. 4171–4186. ACL (2019)
8. Chollet, F.: Keras: Deep learning library for Theano and TensorFlow. https://keras.io/. Accessed 17 Nov 2021
9. Cotterell, R., Schutze, H.: Morphological word-embeddings. In: Proceedings of the 2015 Conference of the NAACL: Human Language Technologies, pp. 1287–1292. ACL (2015)
10. Creutz, M., Lagus, K.: Unsupervised models for morpheme segmentation and morphology learning. ACM Trans. Speech Lang. Process. **4**(1), Article no. 3 (2007)
11. Harris, Z.S.: Morpheme boundaries within words: report on a computer test. Transformations Discourse Anal. Pap. **73**, 68–77 (1967)
12. Hofmann, V., Pierrehumbert, J.B., Schutze, H.: Superbizarre is not superb: improving BERT's interpretations of complex words with derivational morphology. In: Proceedings of the 59th Annual Meeting of the ACL, pp. 3594–3608 (2021)
13. Kanerva, J., et al.: Turku neural parser pipeline: an end-to-end system for the CoNLL 2018 shared task. In: Proceedings of the CoNLL 2018 Shared Task: Multilingual Parsing from Raw Text to Universal Dependencies, pp. 133–142 (2018)
14. Korobov, M.: Morphological analyzer and generator for Russian and Ukrainian languages. In: Khachay, M.Y., Konstantinova, N., Panchenko, A., Ignatov, D.I., Labunets, V.G. (eds.) AIST 2015. CCIS, vol. 542, pp. 320–332. Springer, Cham (2015). https://doi.org/10.1007/978-3-319-26123-2_31
15. Lango, M., Žabokrtský, Z., Ševčíková, M.: Semi-automatic construction of word-formation networks. In: Proceedings of the 11th International Conference on Language Resources & Evaluation (2018)
16. Lyashevskaya, O.N., et al.: GRAMEVAL 2020 shared task: Russian full morphology and universal dependencies parsing. In: Computational Linguistics and Intellectual Technologies: Proceedings of the International Conference "Dialogue 2020", Moscow, pp. 553–569 (2020)
17. Peters, M.E., et al.: Deep contextualized word representations. In: Proceedings of North American Association for Computational Linguistics (NAACL), pp. 2227–2237 (2018)

18. Segalovich, I.: A fast morphological algorithm with unknown word guessing induced by a dictionary for a web search engine. In: MLMTA 2003, pp. 273–280. CSREA Press (2003)

19. Schmid, H.: Probabilistic part-of-speech tagging using decision trees. In: Proceedings of the International Conference on New Methods in Language Processing, pp. 44–49 (1994)

20. Smit, P., Virpioja S., Gronroos S., Kurimo M.: Morfessor 2.0: toolkit for statistical morphological segmentation. In: Proceedings of the Demonstrations at the 14th Conference of the European Chapter of the ACL, Gothenburg, pp. 21–24 (2014)

21. Sorokin, A., et al.: MorphoRuEval-2017: an evaluation track for the automatic morphological analysis methods for Russian. In: Computational Linguistics and Intellectual Technologies: Proceedings of the International Conference "Dialogue 2017", Moscow, vol. 1, pp. 297–314 (2017)

22. Sorokin, A., Kravtsova, A.: Deep convolutional networks for supervised morpheme segmentation of Russian language. In: Ustalov, D., Filchenkov, A., Pivovarova, L., Žižka, J. (eds.) AINL 2018. CCIS, vol. 930, pp. 3–10. Springer, Cham (2018). https://doi.org/10.1007/978-3-030-01204-5_1

23. Sorokin, A., Smurov, I., Kirianov, P.: Tagging and parsing of multidomain collections. In: Computational Linguistics and Intellectual Technologies: Proceedings of the International Conference "Dialogue 2020", Moscow, pp. 670–683 (2020)

24. Straka, M., Straková, J., Hajic, J.: Prague at EPE 2017: the UDPipe system. In: Proceedings of the 2017 Shared Task on Extrinsic Parser Evaluation at the Fourth International Conference on Dependency Linguistics and the 15th International Conference on Parsing Technologies, pp. 65–74 (2017)

25. Tikhonov, A.N.: Word Formation Dictionary of Russian Language. Russkiy yazyk Publication, Moscow (1990)

SocialBERT – Transformers for Online Social Network Language Modelling

Ilia Karpov$^{(\boxtimes)}$ (ID) and Nick Kartashev (ID)

National Research University Higher School of Economics,
Moscow, Russian Federation
karpovilia@gmail.com

Abstract. The ubiquity of the contemporary language understanding tasks gives relevance to the development of generalized, yet highly efficient models that utilize all knowledge, provided by the data source. In this work, we present SocialBERT - the first model that uses knowledge about the author's position in the network during text analysis. We investigate possible models for learning social network information and successfully inject it into the baseline BERT model. The evaluation shows that embedding this information maintains a good generalization, with an increase in the quality of the probabilistic model for the given author up to 7.5%. The proposed model has been trained on the majority of groups for the chosen social network, and still able to work with previously unknown groups. The obtained model is available for download and use in applied tasks (https://github.com/karpovilia/SocialBert).

Keywords: Language modelling · Natural language processing · Social network analysis · Graph embeddings · Knowledge injection

1 Introduction

Online Social Networks (OSN) texts corpora size is comparable with the largest journalism, fiction, and scientific corpora. Evaluations within computational linguistics conferences and Kaggle competitions prove the feasibility of their automatic analysis. Traditional text processing tasks like morphological analysis, sentiment analysis, spelling correction are highly challenging in such texts. As a rule, by analyzing OSN texts we can observe a decrease in most quality metrics by 2–7%. For instance, the best result on the sentiment detection on Twitter dataset at SemEval 2017 [1] has an accuracy of 65.15, while a year earlier on the track SemEval 2016 [2] the best result for the general English language has an accuracy of 88.13. Regarding the Russian language, comparison within the competition of morphological analysis tools MorphoRuEval-2017 [3] shows that the same tools work worse on the texts of online social networks than on fiction and news corpus - the lemmatization accuracy best result for OSN dataset is 92.29, while best result for literature dataset is 94.16.

E. Burnaev et al. (Eds.): AIST 2021, LNCS 13217, pp. 56–70, 2022.
https://doi.org/10.1007/978-3-031-16500-9_6

Text processing quality decreasement is usually caused by a great amount of slang, spelling errors, region- and theme-specific features of such texts. This can be explained by the specifics of these texts being written by non-professionals, i.e., by the authors without the journalist education who have no opportunity or need for professional editing of their texts. Existing research also indicates the specifics of the social network communication itself, such as the tendency to transform oral speech to written text (orality), tendency to express emotions in written texts (compensation), and tendency to reduce typing time (language economy) [4].

BERT [5], and its improvements to natural language modeling, which apply to extremely large datasets and sophisticated training schemes, solves the problems above to a great extent, by taking into consideration the corpora vocabulary at the pre-training step, and providing knowledge transfer from other resources at the fine-tuning step. For instance, Nguyen reports successful application of the RoBERTa model to OSN texts of Twitter users [6]. Nevertheless, there is no consistent approach for analyzing social network users' texts and no effective generalized language models that utilize the structure of OSN.

Unlike many other text sources, any social network text has an explicitly identifiable and publicly accessible author. This leads to a model that processes such a text, taking into consideration the characteristic features of its author. Such a model would make text analysis depending on the author's profile, significantly simplifying such tasks like correction of typos or word disambiguation by taking into account thematic interests and author's speech characteristics.

Thus, our objective is to define author latent language characteristics that capture language homophily. The homophily principle stipulates that authors with similar interests are more likely to be connected by social ties. The principle was introduced in the paper "Homogeneity in confiding relations" [7], by Peter Marsden. An interest's homophily analysis can be found in the Lada A. Adamic paper on U.S. Elections [8]. It shows that people with similar political views tend to make friends with each other. Usually, online social network users simultaneously have several interests and tend toward network-based homophily only with respect to some projection, such as political views, as shown by Lada Adamic. At the same time, online social network groups do not require any projections and are preferable for language structure modeling due to the following characteristics:

- Groups and public pages (hereinafter "groups") have their own pages. Texts, posted at group pages are mostly monothematic since group users are sharing the same interest or discussing news, important for a certain geographical region. In both cases, it is possible to identify group's specific vocabulary and speech patterns.
- The number of groups is two orders less than that of the users. This enables us to train a language model suitable for the entire online social network, without significant node filtering and computation costs.
- Groups generate a major part of text content, whereas many social network users do not write a single word for years because they act only as content consumers. At the same time, users' interests are rather easily expressed through the groups they are subscribed to.

Due to the reasons above, in this paper we focus on generating a group language model and keep the user language model outside the scope of this paper. Hereinafter we interpret a group as an author of texts written in this group's account. In the absence of explicit group attributes like age and gender, we focus on group homophilous relationships. We model groups social homophily through common users intersection to encourage groups with shared social neighborhoods to have similar language models.

In this paper we focused on the masked language modelling (MLM) task because, as a result of training such a model, one can create a better basic model for the analysis of social network texts which may be further adapted to the applied tasks listed above. We will discuss the effect of the basic model on the applied tasks in the Results section.

Our key contributions are as follows:

- We have generated a network embedding model, describing each group containing 5, 000 and more members. Observed groups have no topic limitations, so we can say that our training has covered all currently existing themes in the OSN, assuming that if some topic does not have at least one group with 5, 000 subscribers, then it is not important enough for language modeling. This can potentially lead to the model's inability to take into account highly specialized communities and tiny regional agglomerations, but does not affect the main hypothesis that author dependent language modeling can be more effective.
- We have proposed several new BERT-based models that can be simultaneously trained with respect to the group embedding, and performed training of an MLM-task using the group texts. Our best model achieved 7.5% perplexity increasement in comparison to the basic BERT model training for the same text corpora. This proves the appropriateness of the chosen approach.

The rest of this paper proceeds as follows. Section 2 summarizes the related work on the modelling of OSN authors as network nodes and the existing approaches to language modelling and knowledge injection. Section 3 presents our proposed approach to continuous MLM with respect to network embedding. Section 4 presents details of the experimental setup, including the description of data collection and model training hyperparameters. We present the experimental results in Section 5 before making conclusions.

2   Related Work

In this section, we discuss related work on network node description and language modelling.

2.1   Author as Network Node

The idea to use the author's demographic features in order to improve the analysis quality had been offered before transformer based models were applied.

The existing research e.g. of political preferences on Twitter [8] or comments on Facebook [9], proves the users' inclination to establish relations with users with similar interests. In this work we want to model online social network group structure and language. Given users and groups simultaneously interact in OSN, we can use bipartite graphs to describe groups by their users and vice versa. For *user → community* bipartite graph, the affinity of groups may be described by number of common users: the more common subscribers they have, the greater their similarity is. Therein, various metrics can be applied such as correlation, for instance the Jaccard coefficient, cosine similarity etc. After calculating pairwise distances, one can obtain an adjacency matrix between all groups of the network. In order to reduce its dimensionality, methods based on random walk [10] and autoencoder models such as Deep Walk [11], Node2Vec [12] or a matrix factorization algorithm like NetMF [13] may be used. Attention models may also be used for social representations. They are GraphBERT [14] or Graph Attention Networks [15], but they are much more computationally expensive, and for this reason their use is limited for graphs of over 10^5 node degree.

2.2 Language Modelling

To the best of our knowledge, at the time of writing this paper, there is no published approach to the injection of online social network structure inside transformer-based deep learning language models. Various themes in the network may be considered standalone domains. Thus, the researches applying domain-specific adaptation become relevant [16,17]. These studies show that domain-relevant data is useful for training with both excessive and low resource problems. Since we want to develop single continuous model for all possible topics inside OSN, those approaches are significantly leveraged by a large (about 100–300) number of topics, depending on the granularity degree and absence of distinct borders between the social network communities.

Multi-domain adaptation [18] mechanisms applied in computer-based translation [19] are also of interest. Use of knowledge distillation [20] in training produces a positive result when there are split domains and their number is rather small. It makes the development of the continuous domain adaptation model relevant.

2.3 Knowledge Injection

One of the existing way of enhancing existing deep learning architectures is based on the knowledge injection approach. An example of graph data injection in BERT is the work of VGCN-BERT [21], which adds graph information as a null token. This approach is similar to the first of our two proposed methods. The difference is that Lu proposes the addition of an ontological graph rather than a social network graph. Another approach, based on inserting additional layers to BERT model, is provided by Lauscher [22]. Authors show accuracy increasement up to 3% on some datasets, simultaneously having the same or

$$(X_{v_i}, X_{v_k}) = \sum_{j \in A \cap B} \frac{(1 - \frac{c_j}{M})^2}{c_j - \frac{c_j^2}{M}} + \sum_{j \in \Omega \setminus (A \cup B)} \frac{(\frac{c_j}{M})^2}{c_j - \frac{c_j^2}{M}} + \sum_{j \in (A \oplus B)} \frac{(\frac{c_j}{M} - \frac{c_j^2}{M^2})}{c_j - \frac{c_j^2}{M}} \quad (1)$$

lower accuracy on other datasets. Our second model also modifies one of the BERT layers, but the proposed injection architecture is quite different.

3 Proposed Approach

The proposed model takes into consideration the characteristics of a domain using a pre-computed social vector for the analysis of each token of incoming text. The general training process is as follows:

- Generating adjacency matrices on the basis of network data - matrices preparation to evaluate the adjacency of two groups, based on mutual group members.
- Learning social vectors - obtaining the author's vectors using factorization and random walk algorithms.
- BERT training, given pre-trained social vectors

3.1 Adjacency Matrix Generation

When computing the social vector, we intended to have the opportunity to use the information on the community's local environment as well as a description of its global position relative to all groups. To simulate local context we have chosen the DeepWalk algorithm. To capture the structure of our social graphs on a more global level, we used factorization of different kinds of pairwise distance matrices between the groups.

To calculate pairwise intersection sizes for our set of groups, we created a multithreaded C++ library, which yields an intersection matrix which, as shown later, is transformed into one of the various adjacency metrics.

3.2 Correlation Coefficient

Our first algorithm was based on factorizing a pairwise correlation matrix of our set of groups. Given a group as a vector of zeros and ones, having a length equal to the total number of users in the social network N, and containing a 1 for users who subscribed to our group, and a 0 otherwise. So, in this model, we represent the set of groups of size M as a set of vectors X_{v_i}, each containing a sampling of a Bernoulli distribution. Without physical uploading of all vectors in our RAM, due to huge size of the resulting matrix, we calculated the sample correlation of our vectors based only on these easily computable variables: For set A of subscribers of the group a, and the set B of subscribers of the group b, we, as described in a previous section, calculated the intersection size $|A \cap B|$. Then, using this equation we obtain the correlation coefficient that will be used as one of the possible group affinity variables:

$$cor(a, b) = \frac{|A \cap B| \cdot N - |A| \cdot |B|}{\sqrt{|A| \cdot |B| \cdot (N - |A|) \cdot (N - |B|)}} \quad (2)$$

3.3 Cosine Coefficient

The most important difference of using cosine similarity instead of correlation as a distance metric between our groups, is that we normalized each user's subscription string, therefore lowering the effect that users with a higher subscription count have on the resulting matrix.

Supposing that user j is the member of c_j groups. First of all, we need to subtract the mean from the respective row in our matrix of vectors X_{v_i}. After this transformation we will have $1 - \frac{c_j}{M}$ for positive subscription position and $-\frac{c_j}{M}$ for negative. We then need to divide each value by the standard deviation of a row. Dispersion equals $\frac{c_j^2}{M} + c_j - 2\frac{c_j^2}{M} = c_j - \frac{c_j^2}{M}$.

Therefore, after this transformation we will have value $(1 - \frac{c_j}{M})/\sqrt{c_j - \frac{c_j^2}{M}}$ for positive subscription indicator and $-\frac{c_j}{M}/\sqrt{c_j - \frac{c_j^2}{M}}$ for negative.

So, the final expression will be as shown in Eq. 1, where c_j denotes the count of subscriptions from user j. M denotes the total number of groups. A denotes set of subscribers of group v_k, B denotes set of subscribers of group v_i. X_{v_i} is a normalized vector for group v_i, and X_{v_k} is a normalized vector for group v_k. Ω denotes set of users of the social network, and $\bigoplus$ denotes the symmetric difference between two sets. This formula depicts one of the metrics we used to calculate the similarity between two groups on a scale from -1 to 1.

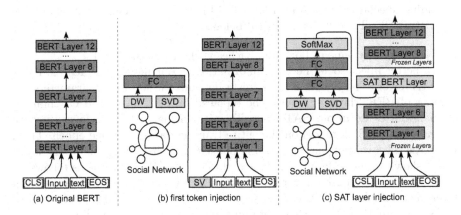

Fig. 1. Social vector injection methods

3.4 Matrix Factorization

We use a truncated SVD algorithm to closely estimate the distance between our groups with a pairwise scalar product of our embedded vectors.

For pairwise distance matrix A, we compute $U, \Sigma, V = SVD(A)$, and then obtain our vectors as rows of matrix $U \cdot \sqrt{\Sigma}$.

3.5   Random Walk

Given the group membership data, we can describe the measure of their closeness based on the Jaccard coefficient, which normalizes the number of common members of two groups by their size.

$$Jac(a,b) = \frac{|A \cap B|}{|A| + |B| - |A \cap B|} \tag{3}$$

Such a metric can be efficiently used for a random walk since it describes the probability of a transition from group A to group B with the "common user" edge. The resulting walks were used to train the DeepWalk model with the parameters recommended by the authors: $\gamma = 80$, $t = 80$, $w = 10$

3.6   BERT Training

The vectors obtained independently as a result of random walk and SVD were integrated into the existing BERT Base model. The main purpose of the training was to teach the model to pay attention to the network vector. Here we used several different ways of embedding:

- adding of a special social vector which concatenates both characteristics at the beginning of each sequence (**Zero token** injection).
- adding special **Social ATtention** (SAT) layer at various positions of the existing BERT model as described below.

The general scheme of both approaches is shown at the Fig. 1.

To better inject social network information in our model, we created a special SAT layer. The injection mechanism depends on two hyperparameters: i - number of BERT layer, chosen to be replaced by SAT layer, and C - number of channels to use in our SAT layer. To inject Social Attention layer, first we pretrain basic BERT model on the entire training dataset for one epoch. Then, we freeze all layers of our model, and substitute i-th layer by our SAT layer, which shown in more detail at 2. The architecture of SAT layer is as follows:

First, we build a 2-layer perceptron with GELU activation function between layers and SoftMax activation after second layer. We pass social network embeddings through that MLP thus obtaining new vectors W with dimensionality reduced to C. Then, we create C parallel BERT Layers, each initialised as substituted i-th layer of the original BERT. To compute output of SAT layer, we multiply each of the parallel bert layers with corresponding element of our resulting social vector W, and then summarise resulting vector sequences. The idea behind this method is to train each of our C BERT layers to be responsive for a superset of social network topics, and then represent each author as a composition of this supersets.

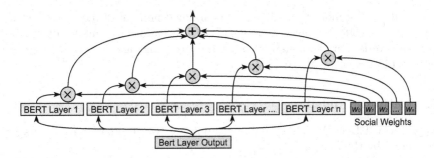

Fig. 2. Social Attention layer architecture

4   Experimental Setup

This section describes the parameters of the approach we have proposed, and varying hyperparameters of the trained models.

4.1   Data Collection

We have used the social network VKontakte, which comprises 600 million users and 2.9 million groups. The majority of its users are Russian-speaking Internet users. The social network has a rich API that automatically provides a significant amount of data related to texts and network characteristics of the network nodes (community members, users' friends).

The VKontakte social network makes it possible to receive messages from groups, public pages, and event pages, through the program interface for non-commercial use[1]. Storage and transmission of users' personal data, including the user's primary key (identifier), are restricted.

We performed sha3 hashing of all user identifiers during the data collection step. This operation makes it impossible to calculate precisely exact user's membership in a community, while at the same time, preserving the bipartite graph structure. We did hashing with python-sha3 library[2].

First we collected information on the size of common ties, then we selected the communities comprising 5,000 or more members. We have established this threshold because small groups are often closed or updated irregularly. It lengthens the stage of the matrix preprocessing and does not improve the quality of the model training. Our final network included 309,710 communities with given sizes.

We collected 1,000 messages from 2019 for each selected group. The above period was chosen because we intended to obtain a thematic structure of the network which was not influenced by recent epidemiological issues. If a group wrote fewer than 1,000 messages, we used the actual number of texts. If the

---

[1] https://vk.com/dev/rules.
[2] https://github.com/bjornedstrom/python-sha3.

community wrote more than the abovementioned number of messages, we randomly chose 1, 000 messages. The length of the majority of messages was less than 500 words. For the construction of the text language model, we used only the first 128 tokens of the text.

4.2   Network Modelling

We performed DeepWalk [11] training, applying the standard parameters recommended by the authors.

Independently, we managed to obtain our social vectors using DeepWalk and SVD of correlation matrices and cosine group-wise distance. To compute SVD part of social embedding computing, we were using fbpca[3] framework by Facebook, with n_iters = 300 and others parameters set by default, and which was running for 10 h on 20 CPU cores (40 threads) Intel Xeon(R) Gold 5118 CPU with 1TB RAM.

4.3   Language Model Training

As a base for our language modelling experiment, we used RuBERT: Multilingual BERT, adapted and pre-trained for the Russian language by DeepPavlov [23]. We ran a series of experiments to compare the ways in which social embeddings were integrated into BERT and the ways in which they were obtained.

For the purpose of the training and the evaluation of our model, we created 3 different datasets, as to better illustrate the performance of our model in different situations. First dataset, containing posts from 278, 739 randomly chosen groups, is used by our model on the training stage, so it will be referenced as training set below. Second dataset, also containing posts from the same 278, 739 groups as in training, contains new text data, previously unseen by our model on the training stage. This dataset is called validation-known *(val-k)*. The final datasets, contains posts from remaining 30, 971 groups, so when validating on this data the only information our model knows about the source is the social embeddings we pass to our model, which makes the task a little more challenging, because unlike in the validation-known dataset our model hasn't seen different posts from the same author on the training stage. This dataset is called validation-unknown *(val-u)*.

From our initial set of 309, 710 groups we selected 189, 496 groups which contained at least 5 texts of sufficient length, with mean number of texts per group of 174.48, and standard deviation of 125.5.

All our experiments were conducted on a machine with Tesla V100 GPU with 32 GB of video memory for BERT training and Intel Xeon(R) Gold 5118 CPU with 1TB RAM for random walk and matrix factorization.

There was a total count of 43, 232, 000 training sequences, 5, 404, 000 val-k sequences, and 5, 404, 000 *val-u* sequences in our data.

---

[3] https://fbpca.readthedocs.io.

Each experiment was trained for a total of one to two weeks on Tesla V100, in each experiment $1,351,000 - 2,702,000$ training steps were made (training was stopped in case of overfitting).

Each experiment used a learning rate of 1e-5, and an Adam optimizer with a warmup of $20,000$ steps. Random seed was fixed for each series of experiments, and a total of 5 series of experiments with different random seeds were conducted.

- Social embedding vector added to zero token embedding, uses concatenation of vectors from SVD of correlation matrix and vectors obtained by DeepWalk.
- Social embedding vector added to zero token embedding, uses concatenation of vectors from SVD of cosine similarity matrix and vectors obtained by DeepWalk.
- Social embedding vector added to zero token embedding, uses just vectors obtained by DeepWalk.
- Baseline BERT, no social network embeddings used.
- BERT with SAT layer. It uses concatenation of vectors of SVD of the correlation matrix and the vectors obtained by DeepWalk as a social embedding.

Throughout our experiments we found out that layer number hyperparameter i has no significant meaning on our model performance, so we chose $i = 11$, as the best value for i with insignificant lead. We found no improvement when increasing C past 32, however, times and memory costs were very high, so we stopped with value $C = 32$.

5   Results

We evaluated the obtained model using the quality of predicting the missing token in the sentence and the perplexity measure, used in the original works such as BERT and RoBERTa. The absolute value of perplexity for the given model depends on many parameters such as the size of the model vocabulary, tokenization parameters, and fine-tuning dataset. Thus, it is rather difficult to evaluate the direct perplexity influence on the solution of any applied problem. Our case is further complicated by the necessity to prepare our own benchmark, since, as far as we know, none of the existing datasets contain the information about the author's social ties used by our model.

On the other hand, perplexity difference for the same basic model, trained on exactly the same corpus with the same preprocessing, must affect the quality of the applied tasks, as shown by the RoBERTa and original BERT paper authors: as perplexity decreases, the quality of classification on the SST-2 (for RoBERTa and BERT), MNLI-m and MRPC (for BERT) dataset increases. Thus, the perplexity difference for two initially identical BERT models, trained on the same texts, indicates better trainability and further effectiveness for the model with lower perplexity.

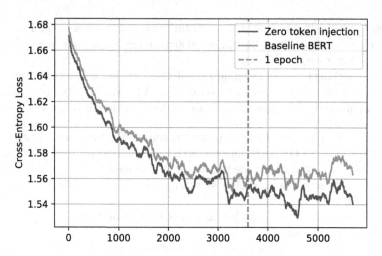

Fig. 3. Zero injection test evaluation, *val-u* dataset

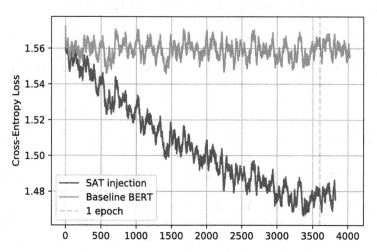

Fig. 4. SAT layer test evaluation on *val-u* dataset

The original BERT paper [5] reports a perplexity of 3.23 for the 24 layer model with 1024 token input. The BERT Base model, trained on the same corpora has a perplexity of at least 3.99 both for English and Russian language. This can be explained by the significant variation in topics, and even languages, covered by those models. Since online social network (OSN) texts are a subset of the entire text array, training only on OSN reduces perplexity to 2.83 for the multilanguage BERT Base model (RuBert OSN). Further improvement is possible through the use of additional information regarding social vectors, allowing the evaluation measure to be reduced to 2.72, as shown in Table 2.

Table 1. Comparison of various network vectors and strategies of BERT Pretraining

SNA Model	LM Model	Data	Loss
—	Baseline BERT		1.568
DW only	Zero token inj		1.563
Cos. & DW	Zero token inj	val-u	1.551
Corr. & DW	Zero token inj		1.542
Corr. & DW	SAT injection		**1.473**
—	Baseline BERT		1.500
Corr. & DW	Zero token inj	val-k	1.486
Corr. & DW	SAT injection		**1.393**

Table 1 shows the averaged loss function for last 50 iterations before the model stops training. The best result is achieved when using the concatenation of the Deep Walk (DW) embedding and the correlation coefficient (Corr.) as the network vector. Concatenation of cosine similarity and Deep Walk (DW) shows a bit worse results. Validation of model on *val-u* dataset (Fig. 3) shows that injecting (Corr. & DW) network vector into a zero token improves the base BERT model by no more than 0.03 points of loss function. Further training doesn't lead to any improvements. Replacing the eleventh layer of the BERT model with the SAT layer improves the model by 0.21 points compared to the baseline BERT loss results (Fig. 4).

Figures 3 and 4 are built for the unknown texts of earlier unknown groups (dataset *val-u*). Evaluation on the unknown texts of the known groups (dataset *val-k*) shows more significant increase up to 0.11 points of loss function. Most of the groups we have selected do not change subscribers and topic significantly over time, which will allow either, to use pretrained group embedding groups for analysis, or to search for the most similar community, based on social network characteristics.

Table 2. BERT base models perplexity

Model	Perplexity	Loss
BERT 12L	3.54	1.82
BERT Large	3.23	1.69
RuBert	4.0	2.00
RuBert OSN	2.83	1.50
SocialBERT	**2.62**	**1.39**

Table 2 shows relative difference of perplexity and loss function for different forks of initial BERT Base model. We can observe that a two times increase in number of BERT layers can reduce perplexity by 8.8% from 3.54 to 3.23. At the same time, the use of network vectors can reduce perplexity by 7.5% from 2.83 to 2.62 for the *val-k* dataset. This result is comparable to the 8% perplexity improvement within the RoBERTa model.

We consider the proposed model useful for all language understanding tasks that implicitly use probabilistic language modeling, first of all, entity linking, spell-checking, and fact extraction. The model shows very promising results on short messages and texts with poor context:

(1) The obtained examples demonstrate that the model successfully learns regional specifics. For example, for the *"[MASK] embankment"* pattern, the basic BERT model recommendation is *"Autumn embankment"*, while the model initialized with the Saint Petersburg regional groups offers *"**Nevskaya** embankment"* based on the Neva River in the Saint Petersburg.

(2) The model can be useful for Link Prediction tasks on short texts. For example, for the pattern *"we read Alexander [MASK] today"* baseline BERT model returns *"we read Alexander Korolev today"* (actor and producer) while model with poetry group vector initialization returns *"we read Alexander **Blok** today"* (well known poet) .

(3) It is also useful in tasks of professional slang detection. For example, given the pattern "Big [MASK]", basic BERT model returns *"Big **bro**"* while model with Data Science group vector returns *"Big **data**"*.

6   Conclusion

In this paper we present the SocialBERT model for the author aware language modelling. Our model injects the author social network profile to BERT, thus turning BERT to be author- or domain- aware.

We select 310 thousand groups and 43 million texts that describe nearly all topics being discussed in the entire network. The proposed model demonstrates its effectiveness by improving the value of perplexity for the Masked Language Modelling task by up to 7.5%. The model has the best results for new texts of already seen groups, still showing good transfer learning for texts of earlier unseen groups. We believe that the proposed model can be useful as a basic model for text analysis of Online Social Network texts and lead to author-aware generative models.

Acknowledgements. The article was prepared within the framework of the HSE University Basic Research Program and through computational resources of HPC facilities provided by NRU HSE.

References

1. Rosenthal, S., Farra, N., Nakov, P.: SemEval-2017 task 4: sentiment analysis in Twitter. In Proceedings of the 11th International Workshop on Semantic Evaluation (SemEval-2017), pp. 502–518. Association for Computational Linguistics, Vancouver, Canada (2017)

2. Pontiki, M., et al.: SemEval-2016 task 5: aspect based sentiment analysis. In Proceedings of the 10th International Workshop on Semantic Evaluation (SemEval-2016), pp. 19–30. Association for Computational Linguistics, San Diego, California (2016)
3. Sorokin, A., et al.: MorphoRuEval-2017: an evaluation track for the automatic morphological analysis methods for Russian. Komp'yuternaya lingvistika i intellektual'nyye tekhnologii **1**, 297–313 (2017)
4. Crystal, D.: Language and the Internet. Cambridge University Press, Cambridge (2006)
5. Devlin, J., Chang, M.W., Lee, K., Toutanova, K.: Pre-training of deep bidirectional transformers for language understanding, Bert (2019)
6. Nguyen, D.Q., Vu, T., Nguyen, A.T.: BERTweet: a pre-trained language model for English tweets. In: Proceedings of the 2020 Conference on Empirical Methods in Natural Language Processing: System Demonstrations, pp. 9–14. Association for Computational Linguistics (2020)
7. Marsden, P.V.: Homogeneity in confiding relations. Soc. Netw. **10**(1), 57–76 (1988)
8. Adamic, L.A., Glance, N.: The political blogosphere and the 2004 U.S. election: Divided they blog. In: Proceedings of the 3rd International Workshop on Link Discovery, LinkKDD'05, pp. 36–43. Association for Computing Machinery, New York (2005)
9. Rieder, B.: Studying facebook via data extraction: the Netvizz application. In: Proceedings of the 5th Annual ACM Web Science Conference, WebSci'13, pp. 346–355. Association for Computing Machinery, New York(2013)
10. Keikha, M.M., Rahgozar, M., Asadpour, M.: Community aware random walk for network embedding. Knowl. Based Syst. **148**, 47–54 (2018)
11. Perozzi, B., Al-Rfou, R., Skiena, S.: Deepwalk. In: Proceedings of the 20th ACM SIGKDD International Conference on Knowledge Discovery and Data Mining (2014)
12. Grover, A., Leskovec, J.: node2vec: scalable feature learning for networks (2016)
13. Zhang, Z., Cui, P., Wang, X., Pei, J., Yao, X., Zhu, W.: Arbitrary-order proximity preserved network embedding, pp. 2778–2786. Association for Computing Machinery, New York (2018)
14. Zhang, J., Zhang, H., Xia, C., Sun, L.: Graph-bert: only attention is needed for learning graph representations (2020)
15. Veličković, P., Cucurull, G., Casanova, A., Romero, A., Lio, P., Bengio, Y.: Graph attention networks (2018)
16. Gururangan, S., et al.: Don't stop pretraining: adapt language models to domains and tasks (2020)
17. Han, X., Eisenstein, J.: Unsupervised domain adaptation of contextualized embeddings for sequence labeling (2019)
18. Yogatama, D., et al.: Learning and evaluating general linguistic intelligence (2019)
19. Mghabbar, I., Ratnamogan, P.: Building a multi-domain neural machine translation model using knowledge distillation (2020)
20. Hinton, G., Vinyals, O., Dean, J.: Distilling the knowledge in a neural network (2015)
21. Lu, Z., Du, P., Nie, J.-Y.: VGCN-BERT: augmenting BERT with graph embedding for text classification. In: Jose, J.M., et al. (eds.) ECIR 2020. LNCS, vol. 12035, pp. 369–382. Springer, Cham (2020). https://doi.org/10.1007/978-3-030-45439-5_25

22. Lauscher, A., Majewska, O., Ribeiro, L.F., Gurevych, I., Rozanov, N., Glavaš, G.: Common sense or world knowledge? investigating adapter-based knowledge injection into pretrained transformers. In: Proceedings of Deep Learning Inside Out (DeeLIO): The First Workshop on Knowledge Extraction and Integration for Deep Learning Architectures, pp. 43–49. Association for Computational Linguistics (2020)
23. Kuratov, Y., Arkhipov, M.: Adaptation of deep bidirectional multilingual transformers for Russian language (2019)

Lexicon-Based Methods vs. BERT
for Text Sentiment Analysis

Anastasia Kotelnikova[1] , Danil Paschenko[1] , Klavdiya Bochenina[2] ,
and Evgeny Kotelnikov[1,2(✉)]

[1] Vyatka State University, Kirov, Russia
[2] ITMO University, Saint-Petersburg, Russia
kotelnikova.av@gmail.com

Abstract. The performance of sentiment analysis methods has greatly
increased in recent years. This is due to the use of various models based
on the Transformer architecture, in particular BERT. However, deep
neural network models are difficult to train and poorly interpretable.
An alternative approach is rule-based methods using sentiment lexicons.
They are fast, require no training, and are well interpreted. But recently,
due to the widespread use of deep learning, lexicon-based methods have
receded into the background. The purpose of the article is to study the
performance of the SO-CAL and SentiStrength lexicon-based methods,
adapted for the Russian language. We have tested these methods, as
well as the RuBERT neural network model, on 16 text corpora and
have analyzed their results. RuBERT outperforms both lexicon-based
methods on average, but SO-CAL surpasses RuBERT for four corpora
out of 16.

Keywords: Sentiment analysis · Sentiment lexicons · SO-CAL ·
SentiStrength · BERT

1 Introduction

The performance of sentiment analysis has improved dramatically[1] over the past
few years, for example:

- the English-language corpus SST-5 (Stanford Sentiment Treebank – 5 classes),
  the accuracy increased from 45.70 (RNTN model [28]) to 59.10 (RoBERTa-
  large+Self-Explaining [29]);
- for the English-language corpus Yelp Reviews (5 classes), the error decreased
  from 37.95 (Char-level CNN model [36]) to 27.05 (XLNet model [35]);
- for the Russian-language news corpus ROMIP-2012 (3 classes) [6] F1-score
  increased from 62.10 (lexicon-based Polyarnik system) [17] to 72.69 [9].

---

[1] See for example: https://paperswithcode.com/task/sentiment-analysis.

E. Burnaev et al. (Eds.): AIST 2021, LNCS 13217, pp. 71–83, 2022.
https://doi.org/10.1007/978-3-031-16500-9_7

This improvement in performance is mainly associated with the development of deep learning methods, especially various models based on the Transformer architecture [34], in particular, BERT [8].

However, deep neural network models are difficult to train: they need large amounts of data; training requires powerful expensive video cards with a large memory size; the learning process is time consuming and energy intensive [18]. Another issue is the complexity of interpreting the results of the models [1].

An alternative are rule-based (or lexicon-based) methods using sentiment lexicons [30]. They are fast, do not require training and are well interpreted [2]. But recently, due to the widespread use of deep learning, lexicon-based methods have receded into the background.

For the Russian language, there are several recent studies of deep learning models for sentiment analysis [9,27]. However, there are currently no studies devoted to comparing lexicon-based methods and deep learning models.

We strive to close this gap and compare the fine-tuned deep neural network model RuBERT [16] with two lexicon-based methods adapted for the Russian language - SO-CAL [31] and SentiStrength [32]. For testing we have used 16 Russian-language text corpora, labelled by sentiment into 3 classes.

The contribution of this article is as follows:

- lexicon-based methods SO-CAL and SentiStrength have been adapted for the Russian language;
- performance evaluation of lexicon-based methods and RuBERT has been carried out for 16 Russian-language text corpora;
- the performance of SO-CAL and SentiStrength for 17 sentiment lexicons have been estimated: 9 publicly available Russian sentiment lexicons and a set of 8 combined lexicons;
- the classification results of lexicon-based methods and RuBERT have been analyzed.

2 Lexicon-Based Methods and Tools

There are several tools for sentiment analysis on the base of sentiment lexicons:

- open source: SO-CAL [31], VADER [10], Pattern, TextBlob;
- proprietary: SentiStrength [32], SentText [25].

Taboada et al. developed SO-CAL[2] (Semantic Orientation CALculator) - a method and tool for determining the sentiment of texts in English and Spanish [31]. The sentiment is recognized on the basis of counting the weights of the sentiment words included in the text (only nouns, adjectives, verbs and adverbs are taken into account). A system of rules is also involved to account for the influence of lexical markers such as modifiers, negations, and irrealis markers. *Modifiers* are lexical markers that increase (e.g., *very*, *the most*) or decrease

---

[2] https://github.com/sfu-discourse-lab/SO-CAL

(e.g., *slightly, somewhat*) the intensity of the next sentiment word. *Negations* (e.g., *not, nothing*) either invert the polarity of the next sentiment word, or shift its intensity towards the opposite polarity (for example, in SO-CAL, a shift is used, and in VADER – an inversion with a certain coefficient). *Irrealis markers* indicate that the sentiment score for a given sentence should not be taken into account. These markers are modal verbs (e.g., *could, should*), conditional words (e.g., *if*), some verbs (e.g., *expect, doubt*), a question mark, and quoted words.

Hutto and Gilbert proposed VADER[3] (Valence Aware Dictionary for sEntiment Reasoning) - a lexicon, method and tool for sentiment analysis of English texts [10]. The sentiment lexicon was built from the well-known dictionaries LIWC, ANEW, and General Inquirer and then crowdsourced in sentiment intensity. Also, emoticons, acronyms and slang were included into the lexicon. The VADER takes into account exclamation marks, capitalization, modifiers, negations and contrasts.

Pattern[4] is a web mining library that supports sentiment analysis in English and French [7]. For this, a lexicon of sentiment adjectives is used, often found in product reviews.

TextBlob[5] is a text processing library that includes two components for sentiment analysis - based on a naive Bayesian classifier and an implementation from the *Pattern* library.

Thelwall et al. developed SentiStrength[6], a tool for sentiment analysis of short social media texts based on the method bearing the same name [32]. The tool gives two scores for the input text: a negative score from -1 to -5 and a positive score from +1 to +5. The decision is based on a list of sentiment words with weights corresponding to the sentiment intensity. The method also takes into account modifiers, negations, question words, slang, idioms and emoticons.

Schmidt et al. developed SentText[7], a web-based sentiment analysis tool in the digital humanities [25]. The original version of SentText was developed for the German language using the SentiWS and BAWL-R dictionaries. This tool takes negations into account. SentText has the ability to visualize the results, including highlighting the sentiment words, information about the polarity of individual words and the text as a whole, as well as comparing texts by sentiment.

Of these tools, as far as we know, only SentiStrength has two adaptations for the Russian language[8], but we could not find a detailed description of their implementations.

In our work, we have adapted SO-CAL and SentiStrength for the Russian language. SO-CAL is the most advanced open source sentiment analysis tools. SentiStrength, despite being proprietary software, makes it easy to adapt to a new language. For this purpose, it is necessary to provide it with a sentiment

---

[3] https://github.com/cjhutto/vaderSentiment.
[4] https://github.com/clips/pattern.
[5] https://textblob.readthedocs.io.
[6] http://sentistrength.wlv.ac.uk.
[7] https://thomasschmidtur.pythonanywhere.com.
[8] Given on the website: http://sentistrength.wlv.ac.uk.

lexicon in the target language and other linguistic resources: lists of modifiers, negations, interrogative words, slang and idioms.

3   Materials and Methods

3.1   Lexicon-Based Methods Adaptation

The adaptation of SO-CAL and SentiStrength to the Russian language includes the following steps:

- morphological analysis of input texts based on RNNMorph[9];
- preparation of a Russian-language sentiment lexicon. The existing lexicons were used to form it and are described in Subsect. 3.2. A peculiarity of SO-CAL is to take into account only nouns, adjectives, verbs, adverbs;
- preparation of Russian-language lists of modifiers (e.g., *очень, едва, значительно*) and negations (e.g., *не, без, невозможно*). They were obtained by translating the corresponding SO-CAL and SentiStrength lists, as well as by adding Russian-language synonyms;
- preparation for SO-CAL of Russian-language lists of irrealis markers (e.g., *ожидать, можно, кто-нибудь*);
- modification of the SO-CAL source code for processing texts with the results of Russian morphological analysis;
- organization of programming interface with the desktop version of SentiStrength to submit input texts and process its results.

3.2   Sentiment Lexicons

The key resource for the considered sentiment analysis methods is the sentiment lexicon. The performance of sentiment analysis depends on the completeness and accuracy of such a lexicon. We have used 9 publicly available Russian sentiment lexicons (two of them - EmoLex and Chen-Skiena's - are Russian versions of multi-lingual lexicons) [13].

Each lexicon has been processed as follows:

- neutral words have been removed (if such were present in the lexicon);
- words that are both positive and negative in the lexicon have been removed (including the analysis of words with the spelling "e" and "ё");
- words containing Latin letters have been removed;
- all words have been converted to a lower case;
- words have been normalized using RNNMorph;
- only one occurrence of each element has been left (an element can be a separate word or phrase).

The characteristics of the lexicons are shown in Table 1.

---

[9] https://github.com/IlyaGusev/rnnmorph.

Table 1. The characteristics of sentiment lexicons.

Lexicon	Total	Positive elements		Negative elements	
		#	%	#	%
RuSentiLex [20]	12,560	3,258	25.9%	9,302	74.1%
Word Map [15]	11,237	4,491	40.0%	6,746	60.0%
SentiRusColl [14]	6,538	3.981	60.9%	2,557	39.1%
EmoLex [22]	4,600	1,982	43.1%	2,618	56.9%
LinisCrowd [11]	3,986	1,126	28.2%	2,860	71.8%
Blinov's lexicon [3]	3,524	1,611	45.7%	1,913	54.3%
Kotelnikov's lexicon [12]	3,206	1,028	32.1%	2,178	67.9%
Chen-Skiena's lexicon [4]	2,604	1,139	43.7%	1,465	56.3%
Tutubalina's lexicon [33]	2,442	1,032	42.3%	1,410	57.7%

We also used the "voting" procedure of these lexicons to build a set of 8 combined lexicons *Lex1..Lex8*: only those words that are included in at least *N* sentiment lexicons are included in the *LexN* lexicon. Thus, *Lex1* includes all sentiment words that occur in at least one lexicon. *Lex9* turned out to be empty - not a single item is included in all sentiment lexicons at the same time. The characteristics of the combined lexicons are shown in Table 2.

3.3 Text Corpora

For evaluation, we have used 16 public text corpora labelled by sentiment [26] (see Table 3), including:

- corpora of reviews about books, movies and cameras, as well as news articles from the ROMIP 2011 [5] and ROMIP 2012 [6] seminars;
- corpora of reviews about cars and restaurants, as well as tweets about banks and telecom companies of the SentiRuEval 2015 [21], SentiRuEval 2016 [19] and SemEval 2016 [23] seminars;
- the RuSentiment corpus containing posts on VKontakte [24];
- LinisCrowd corpus, including posts and comments from LiveJournal [11]. We have used texts labelled by one annotator as training data, and texts labelled by several annotators as test data.

4 Results

4.1 Experimental Setup

We have tested two lexicon-based methods of sentiment analysis adapted for the Russian language - SO-CAL and SentiStrength, as well as a deep neural network model RuBERT [16].

In general, lexicon-based methods can be used for sentiment analysis without training. However, since training corpora are available in our experiments, we have used them to tune the hyperparameters of the lexicon-based methods: we

Table 2. The characteristics of the combined sentiment lexicons.

Lexicon	Total	Positive elements		Negative elements	
		#	%	#	%
Lex1	33,080	13,443	40.6%	19,637	59.4%
Lex2	9,377	3,147	33.6%	6,230	66.4%
Lex3	4,325	1,521	35.2%	2,804	64.8%
Lex4	2,313	823	35.6%	1,490	64.4%
Lex5	1,266	475	37.5%	791	62.5%
Lex6	607	258	42.5%	349	57.5%
Lex7	240	114	47.5%	126	52.5%
Lex8	52	31	59.6%	21	40.4%

have chosen the optimal sentiment lexicons for both methods (out of 17 lexicons - see Subsect. 3.2) and have determined the thresholds for positive and negative sentiment classes.

The thresholds are defined as follows. SO-CAL returns the single sentiment score s for the text to be converted to a class label (positive, negative, or neutral). We have fit two thresholds on the training data - t_{pos} and t_{neg}. The decision about the sentiment of the text c is made on the basis of the following expression:

$$c = \begin{cases} neutral, \ if \ s < t_{pos} \ and \ s > t_{neg}, \\ positive, \ if \ s \geq t_{pos}, \\ negative, \ if \ s \leq t_{neg}. \end{cases}$$

SentiStrength returns two sentiment scores for the text - positive s_{pos} and negative s_{neg}. We select two coefficients k_{neut} and k such that:

$$c = \begin{cases} neutral, \ if \ s_{pos} \leq k_{neut} \ and \ s_{neg} \leq k_{neut}, \\ positive, \ if \ s_{pos} > ks_{neg}, \\ negative, \ otherwise. \end{cases}$$

The results of selecting lexicons and threshold values are given in Subsect. 4.2.

The pretrained RuBERT model was fine-tuned separately on each training corpus with the following hyperparameters: learning rate $2 \cdot 10^{-5}$, number of epochs 5, batch size 12. The results are given on average for five runs to reduce the influence of random weight initialization. The training has been carried out using the Google Colab Pro service on NVIDIA Tesla P100 and V100 video cards.

For all the corpora a three-class problem of sentiment analysis was solved - the classification of texts into positive, negative and neutral. We used the macro F1-score as the main performance metric.

Table 3. Corpora characteristics.

Corpus	Type	Split	Total	Positive	Negative	Neutral
LinisCrowd	Posts	Train	28,853	7.7%	42.5%	49.8%
		Test	14,260	9.5%	47.3%	43.2%
Romip 2011	Book reviews	Train	22,098	79.7%	9.3%	11.0%
		Test	228	64.0%	6.2%	29.8%
	Movie reviews	Train	14,808	70.6%	12.7%	16.7%
		Test	263	70.3%	10.7%	19.0%
	Camera reviews	Train	9,460	80.5%	10.6%	8.9%
		Test	207	61.8%	17.9%	20.3%
Romip 2012	Book reviews	Test	129	77.5%	7.0%	15.5%
	Movie reviews	Test	408	65.2%	15.4%	19.4%
	Camera reviews	Test	411	85.4%	1.7%	12.9%
	News	Train	4,260	26.2%	43.7%	30.1%
		Test	4,573	31.7%	41.3%	27.0%
SentiRuEval 2015	Car reviews	Train	203	56.6%	14.8%	28.6%
		Test	200	49.0%	13.0%	38.0%
	Restaurant reviews	Train	200	68.0%	14.0%	18.0%
		Test	203	71.9%	12.8%	15.3%
	Bank tweets	Train	4,883	7.2%	21.7%	71.1%
		Test	4,534	7.6%	14.4%	78.0%
	Telecom tweets	Train	4,839	18.8%	32.7%	48.5%
		Test	3,774	9.1%	22.4%	68.5%
SentiRuEval 2016	Bank tweets	Test	3,302	9.1%	23.1%	67.8%
	Telecom tweets	Test	2,198	8.3%	45.8%	45.9%
SemEval 2016	Restaurant reviews	Test	103	67.0%	14.6%	18.4%
RuSentiment	Posts	Train	24,124	38.0%	15.2%	46.8%
		Test	2,621	36.0%	9.8%	54.2%

4.2 Results of Experiments

We have run two series of experiments. In the first series, the training data were used to select the optimal hyperparameters for lexicon-based methods: lexicon and threshold values (see Subsect. 4.1). In the second series the lexicon-based methods with selected hyperparameters were compared on test data with the fine-tuned RuBERT model.

The results of the first series of experiments on selecting the optimal sentiment lexicon are shown in Fig. 1. Kotelnikov's lexicon has turned out to be the best lexicon for SO-CAL, *Lex1* and *Lex2* - for SentiStrength (*Lex2* was used as the optimal lexicon, being the smaller one).

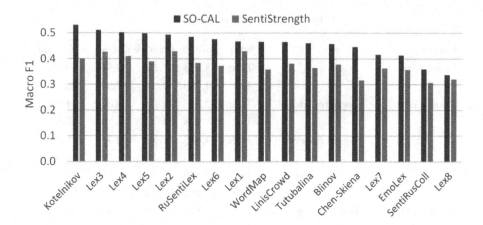

Fig. 1. The results of experiments on the selection of the optimal lexicon.

The mean values and standard deviations of the fitted thresholds for SO-CAL with Kotelnikov's lexicon turned out to be: $t_1 = -1.1 \pm 0.95$, $t_2 = 0.4 \pm 0.82$. For SentiStrength with $Lex2$, coefficient $k_{neut} = 0.6 \pm 0.65, k = 1.1 \pm 0.36$.

The results of the second series of experiments - the comparison of lexicon-based methods with RuBERT - are shown in Fig. 2. RuBERT is superior to lexicon-based methods: on average over all the corpora for RuBERT F1-score=0.5833, for SO-CAL F1 score=0.5310, for SentiStrength F1-score=0.4290.

For 12 corpora out of 16 RuBERT outperforms both lexicon-based methods. The most significant difference was for RuSentiment (28% points), Romip 2012 News (21 p.p.), tweets of SentiRuEval 2015 Banks and SentiRuEval 2016 Telecoms (20 p.p.). However, for four corpora out of 16, SO-CAL comes out on top, and for three of them the difference is quite significant: SentiRuEval 2015 Cars (32 p.p.), SentiRuEval 2015 Restaurants (29 p.p.), SemEval 2016 (25 p.p.), ROMIP 2012 Books (5 p.p.). SentiStrength, as a rule, loses to both methods, with the exception of the corpus SentiRuEval 2015 Cars.

In general, RuBERT analyzes all the corpora with short texts much better - an average number of symbols in the text less than 100 (the difference is from 13 to 28 p.p.). For medium-sized texts (700–900 symbols), the lexicon-based methods are better. For longer texts, the situation is ambiguous.

5   Discussion

We have compared sets of predictions on all the test corpora for all the methods. As a result, five subsets have been obtained: 1) the predictions of all methods matched, 2) the SO-CAL and SentiStrength's predictions matched, which did not match with RuBERT; 3) the predictions from RuBERT and SO-CAL matched, which did not match with SentiStrength; 4) RuBERT and SentiStrength's predictions matched, which did not match with SO-CAL; 5) the predictions of all the methods did not match. We have calculated the macro F1-score for each of these sets (Table 4).

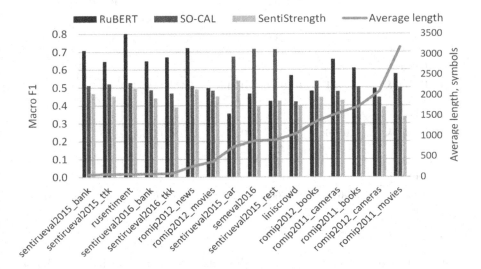

Fig. 2. Comparison results of lexicon-based methods and RuBERT on test corpora.

Table 4 shows that for the set of matching predictions (the first row - 38% of the test dataset), the performance turns out to be quite high (0.8100). On average, these are shorter texts (the average length is 481 characters). For the set of matching predictions from RuBERT and SO-CAL (the third row - 18% of the test dataset), the result is significantly higher than that of SentiStrength, which did not coincide with them - 0.7151 vs. 0.1955. For the set of identical predictions from RuBERT and SentiStrength (the fourth row - 17% of the test sample), the difference is smaller - 0.6576 vs. 0.2347 - SentiStrength is worse than SO-CAL. Finally, for the set of the unmatched predictions (the fifth row - 6%) RuBERT is far superior to both lexicon-based methods - 0.5617 vs. 0.1860 (SentiStrength) and 0.2074 (SO-CAL).

We analyzed in more detail the set of matching predictions of SO-CAL and SentiStrength that did not match RuBERT (the second row - 21% of the test dataset). The results of lexicon-based methods were significantly lower than RuBERT (0.3237 vs. 0.5625). In this case, lexicon-based methods recognize positive and negative texts poorly (0.2008 and 0.3395, respectively, versus 0.5228 and 0.6654 for RuBERT). For neutral texts, the difference is not so significant - 0.4308 for lexicon-based methods vs. 0.4993 for RuBERT. If we consider the results for individual corpora, then, in general, the picture remains the same as on Fig. 2 with a few exceptions. The lexicon-based methods show the best results for six corpora, and not for four - two new corpora (ROMIP 2012 Cameras and Movies) were added to the previous corpora (SentiRuEval 2015 Cars and Restaurants, SemEval 2016, and ROMIP 2012 Books). Also, the ROMIP 2012 Books

Table 4. Performance metrics (macro F1-score) on predictions sets.

Set	RuBERT	Senti-Strength	SO-CAL	Set size	Average text length, sym.
All matched	0.8100			14,310 (38%)	481
SentiStrength & SO-CAL matched	0.5625	0.3237		7,698 (21%)	519
RuBERT & SO-CAL matched	0.7151	0.1955	0.7151	6,755 (18%)	603
RuBERT & Senti-Strength matched	0.6576		0.2347	6,289 (17%)	686
All didn't match	0.5617	0.1860	0.2074	2,362 (6%)	602

corpus is recognized on this set of predictions much better with lexicon-based methods than with RuBERT: 0.5000 vs. 0.0833.

We also analyzed the reasons for the incorrect predictions of the lexicon-based methods. Errors most often arise due to an insufficient size of sentiment lexicon, the absence of sentiment words in the text, an incorrect recognition of negation and irrealis, an overbalance of words of the opposite sentiment, sarcasm, and erroneous identification of domain-oriented words.

As an illustration (see Fig. 3), we can give some examples that are incorrectly classified by lexicon-based methods (the first and the third examples) or RuBERT (the second and the fourth examples). In the first example lexicon-based methods didn't recognize that the phrase *settle trouble debts* has positive polarity. In the third example the word *wonder* led the lexicon-based methods to the wrong decision: they didn't take into account the phrase *ATMs do not work*. Unfortunately, RuBERT does not have the same good interpretability as lexicon-based methods, so we can't explain why RuBERT misclassified the second and fourth examples. But as we can see from the first and third examples, RuBERT can correctly classify texts even when they have words of opposite sentiment.

6   Conclusion

We have compared the lexicon-based methods SO-CAL and SentiStrength with a deep neural network model RuBERT on 16 Russian-language sentiment corpora for a three-class problem of sentiment analysis. On average, RuBERT shows a higher classification performance than lexicon-based methods, exceeding SO-CAL by an average of 5 p.p. SentiStrength lags behind SO-CAL by 10 p.p.

etalon: positive RuBERT: positive SentiStrength: negative SO-CAL: negative	До конца лета клиенты <bank_name> могут урегулировать проблемную задолженность на упрощенных условиях.	Until the end of summer, <bank_name> clients can settle troubled debts on simplified terms.
etalon: positive RuBERT: negative SentiStrength: positive SO-CAL: positive	В <bank_name>-онлайн появилась крутая функция анализа расходов по категориям. Респект!	<bank_name>-online has a cool function for analyzing expenses by category. Respect!
etalon: negative RuBERT: negative SentiStrength: positive SO-CAL: positive	Интересно почему это в Подмосковье через один не работают банкоматы <bank_name>. У нас в округе НИ ОДИН не выдает денег. В отделениях очереди.	I wonder why the <bank_name>'s ATMs do not work in the Moscow region through one. In our district NO ONE gives out money. There are queues in the offices.
etalon: negative RuBERT: positive SentiStrength: negative SO-CAL: negative	<bank_name> берет комиссию за оплату штрафов! Мелочь, а не приятно, особенно когда можно оплатить без комиссий.	<bank_name> takes commission for paying fines! It's a trifle, but not pleasant, especially when you can pay without commissions.

Fig. 3. Examples of classifiers' errors: the first and third examples are incorrectly classified by lexicon-based methods; the second and fourth examples are misclassified by RuBERT. The first column shows real class and the answers of the methods. The second column is in Russian, the third one – the translation into English. Sentiment words in examples are coloured: positive ones are green, negative ones are red.

However, for four corpora out of 16 (usually with medium-length texts) SO-CAL shows better results than RuBERT. This keeps us optimistic about the lexicon-based approach in general.

In the future, we intend to modify SO-CAL in order to more accurately take into account the peculiarities of the Russian language, such as negation and irrealis. Also, a promising area of research is the development of hybrid models that combine the ability to take into account the context of deep neural networks and linguistic knowledge contained in the sentiment lexicons.

Acknowledgement. This research is financially supported by The Russian Science Foundation, Agreement №17-71-30029 with co-financing of Bank Saint Petersburg.

References

1. Belinkov, Y., Gehrmann, S., Pavlick, E.: Interpretability and analysis in neural NLP. In: Proceedings of the 58th Annual Meeting of the Association for Computational Linguistics, pp. 1–5 (2020)
2. Birjali, M., Kasri, M., Beni-Hssane, A.: A comprehensive survey on sentiment analysis: approaches, challenges and trends. Knowl.-Based Syst. **226**, 107134 (2021)
3. Blinov, P.D., Klekovkina, M.V., Kotelnikov, E.V., Pestov, O.A.: Research of lexical approach and machine learning methods for sentiment analysis. In: Computational

Linguistics and Intellectual Technologies: Proceedings of the International Conference "Dialogue", vol. 12, no. 19, pp. 51–61 (2013)

4. Chen, Y., Skiena, S.: Building sentiment lexicons for all major languages. In: Proceedings of the 52nd Annual Meeting of the Association for Computational Linguistics, pp. 383–389 (2014)

5. Chetviorkin, I., Braslavskiy, P., Loukachevitch, N.: Sentiment Analysis Track at ROMIP 2011. In: Computational Linguistics and Intellectual Technologies: Proceedings of the International Conference "Dialog", vol. 2, pp. 1–14 (2012)

6. Chetviorkin, I.I., Loukachevitch, N.V.: Sentiment analysis track at ROMIP 2012. In: Computational Linguistics and Intellectual Technologies: Proceedings of the International Conference "Dialog", vol. 2, pp. 40–50 (2013)

7. De Smedt, T., Daelemans, W.: Pattern for Python. J. Mach. Learn. Res. **13**, 2063–2067 (2012)

8. Devlin, J., Chang, M.W., Lee, K., Toutanova, K.: BERT: pre-training of deep bidirectional transformers for language understanding. In: Proceedings of 7th Annual Conference of the North American Chapter of the Association for Computational Linguistics: Human Language Technologies (NAACL-HLT 2019), pp. 4171–4186 (2019)

9. Golubev, A., Loukachevitch, N.: Transfer Learning for Improving results on Russian Sentiment Datasets. In: Computational Linguistics and Intellectual Technologies: Proceedings of the International Conference "Dialog", pp. 268–277 (2021)

10. Hutto, C.J., Gilbert, E.: VADER: a parsimonious rule-based model for sentiment analysis of social media text. In: Proceedings of the International AAAI Conference on Web and Social Media, pp. 216–225 (2014)

11. Koltsova, O.Y., Alexeeva, S.V., Kolcov, S.N.: An opinion word lexicon and a training dataset for russian sentiment analysis of social media. In: Computational Linguistics and Intellectual Technologies: Proceedings of the International Conference "Dialog", pp. 277–287 (2016)

12. Kotelnikov, E., Bushmeleva, N., Razova, E., Peskisheva, T., Pletneva, M.: Manually created sentiment lexicons: research and development. In: Computational Linguistics and Intellectual Technologies: Proceedings of the International Conference "Dialog", vol. 15(22), pp. 300–314 (2016)

13. Kotelnikov, E., Peskisheva, T., Kotelnikova, A., Razova, E.: A comparative study of publicly available russian sentiment lexicons. In: Ustalov, D., Filchenkov, A., Pivovarova, L., Žižka, J. (eds.) AINL 2018. CCIS, vol. 930, pp. 139–151. Springer, Cham (2018). https://doi.org/10.1007/978-3-030-01204-5_14

14. Kotelnikova, A., Kotelnikov, E.: SentiRusColl: Russian collocation lexicon for sentiment analysis. In: Ustalov, D., Filchenkov, A., Pivovarova, L. (eds.) AINL 2019. CCIS, vol. 1119, pp. 18–32. Springer, Cham (2019). https://doi.org/10.1007/978-3-030-34518-1_2

15. Kulagin, D.: Russian word sentiment polarity dictionary: a publicly available dataset. In: Artificial Intelligence and Natural Language. AINL 2019 (2019)

16. Kuratov, Y., Arkhipov, M.: Adaptation of deep bidirectional multilingual transformers for Russian language. In: Computational Linguistics and Intellectual Technologies: Proceedings of the International Conference "Dialog", pp. 333–340 (2019)

17. Kuznetsova, E.S., Chetviorkin, I.I., Loukachevitch, N.V.: Testing rules for sentiment analysis system. In: Computational Linguistics and Intellectual Technologies: Proceedings of the International Conference "Dialog", vol. 2, pp. 71–80 (2013)

18. Li, H.: Deep learning for natural language processing: advantages and challenges. Natl. Sci. Rev. **5**(1), 24–26 (2018)

19. Loukachevitch, N.V., Rubtsova, Y.V.: SentiRuEval-2016: overcoming time gap and data sparsity in tweet sentiment analysis. In: Computational Linguistics and Intellectual Technologies: Proceedings of the International Conference "Dialog", pp. 416–426 (2016)

20. Loukachevitch, N., Levchik, A.: Creating a general Russian sentiment lexicon. In: Proceedings of Language Resources and Evaluation Conference (LREC), pp. 1171–1176 (2016)

21. Loukashevitch, N.V., Blinov, P.D., Kotelnikov, E.V., Rubtsova, Y.V., Ivanov, V.V., Tutubalina, E.V.: SentiRuEval: testing object-oriented sentiment analysis systems in Russian. In: Computational Linguistics and Intellectual Technologies: Proceedings of the International Conference "Dialog", vol. 2, pp. 2–13 (2015)

22. Mohammad, S.M., Turney, P.D.: Crowdsourcing a word-emotion association lexicon. Comput. Intell. **29**(3), 436–465 (2013)

23. Pontiki, M., et al.: SemEval-2016 task 5: aspect based sentiment analysis. In: Proceedings of the 10th International Workshop on Semantic Evaluation (SemEval), pp. 19–30 (2016)

24. Rogers, A., Romanov, A., Rumshisky, A., Volkova, S., Gronas, M., Gribov, A.: RuSentiment: an enriched sentiment analysis dataset for social media in Russian. In: Proceedings of the 27th International Conference on Computational Linguistics, pp. 755–763 (2018)

25. Schmidt, T., Dangel, J., Wolff, C.: SentText: a tool for lexicon-based sentiment analysis in digital humanities. In: Proceedings of the 16th International Symposium of Information Science (ISI), pp. 156–172 (2021)

26. Smetanin, S.: The applications of sentiment analysis for Russian language texts: Current challenges and future perspectives. IEEE Access **8**, 110693–110719 (2020)

27. Smetanin, S., Komarov, M.: Deep transfer learning baselines for sentiment analysis in Russian. Inf. Process. Manage. **58**, 102484 (2021)

28. Socher, R., et al.: Recursive deep models for semantic compositionality over a sentiment treebank. In: Proceedings of the 2013 Conference on Empirical Methods in Natural Language Processing (EMNLP), pp. 1631–1642 (2013)

29. Sun, Z., Fan, C., Han, Q., Sun, X., Meng, Y., et al.: Self-explaining structures improve NLP models (2020). https://arxiv.org/abs/2012.01786

30. Taboada, M.: Sentiment Analysis: An Overview from Linguistics. Ann. Rev. Linguist. **2**, 325–347 (2016)

31. Taboada, M., Brooke, J., Tofiloski, M., Voll, K., Stede, M.: Lexicon-based methods for sentiment analysis. Comput. Linguist. **37**(2), 267–307 (2011)

32. Thelwall, M., Buckley, K., Paltoglou, G., Cai, D., Kappas, A.: Sentiment strength detection in short informal text. J. Am. Soc. Inform. Sci. Technol. **61**(12), 2544–2558 (2010)

33. Tutubalina, E.V.: Extraction and summarization methods for critical user reviews of a product. Ph.D. thesis, Kazan Federal University, Kazan, Russia (2016)

34. Vaswani, A., et al.: Attention is All you Need. In: Proceedings of the 31st Conference on Neural Information Processing Systems (NeurIPS), vol. 30, pp. 5998–6008 (2017)

35. Yang, Z., Dai, Z., Yang, Y., Carbonell, J., Salakhutdinov, R.R., Le, Q.V.: XLNet: generalized autoregressive pretraining for language understanding. In: Proceedings of the 33rd Conference on Neural Information Processing Systems (NeurIPS), vol. 32 (2019)

36. Zhang, X., Zhao, J., LeCun, Y.: Character-level convolutional networks for text classification. In: Proceedings of the 29th Conference on Neural Information Processing Systems (NeurIPS), vol. 28 (2015)

Multilingual Embeddings for Clustering Cultural Events

Maria Kunilovskaya[1,2](✉) ⓘ and Elizaveta Kuzmenko[3] ⓘ

[1] University of Tyumen, Tyumen, Russia
mkunilovskaya@gmail.com
[2] University of Wolverhampton, Wolverhampton, UK
[3] University of Trento, Trento, Italy

Abstract. In the present paper we describe our approach to semi-automatic text annotation based on clustering. Given a large collection of announcements of cultural events from several websites, we group them based on their content and infer respective semantic categories that can be used for annotation (e.g. lecture, sports, food, music). We experiment with various models for vectorising the texts, including pretrained multilingual Sentence Transformers and multilingual ELMo models. The produced text embeddings are then clustered using K-means. We evaluate our clustering results using a stratified sample of texts with pre-existing categories (collected from websites listing the events) as well as intrinsic evaluation measures. The rationale behind this work is to produce a single categorisation covering texts from various sources and in two languages - English and Russian. The labelled collection of texts is intended for use in a Digital Humanities project aimed at describing cultural life in a selected location, for example, comparing types of events in Russian and British cities.

Keywords: Clustering · mSBERT · mELMo · Multilingual embeddings · Digital humanities

1 Introduction

The present paper describes our clustering experiments with the view to annotate a large collection of short texts announcing cultural events taking place across the world. We aim to produce a single categorisation for texts coming from various sources in two main languages - English and Russian. The labelled collection of texts is intended for use in a Digital Humanities project focused on cultural analytics. It seeks to explore cultural trends in select regions or cities, to detect trendsetters and cultural hotspots, to study patterns of cultural production following Wallerstein's theory of core and periphery [21].

This work has been partly supported by the Russian Foundation for Basic Research within Project *Cultural Trends in the Tyumen Region in the National and Global Contexts* No. 20-411-720010 p_a_Tyumen region.

E. Burnaev et al. (Eds.): AIST 2021, LNCS 13217, pp. 84–96, 2022.
https://doi.org/10.1007/978-3-031-16500-9_8

Although the collection of texts harvested from six websites includes some semantic categorisation based on the names of websites sections where each announcement appeared, it has two major drawbacks: it covers 80% of the items and it is not consistent across the web-resources and languages, with the number of section on the websites ranging from 0 to 107. Besides, mapping the categories from Russian websites to international websites is less than straightforward. Clustering-based re-annotation of this collection posed several problems, including making a principled choice of vectorisation and clustering options that would be practical, given the size of the data. This paper compares the performance of two multilingual vectorisation frameworks - multilingual BERT model (further mSBERT) and multilingual ELMo (further mELMo) in the context of Mini-Batch K-Means clustering of a bilingual text collection. In particular, we report the results of extrinsic (comparing with ground truth) and intrinsic (algorithm-specific) evaluation, as well as observations from manual analysis of the partitions generated by each language model.

The paper is structured as follows: Sect. 2 describes similar research aimed at clustering large collections of short texts and arising problems. Section 3 gives details on our data. Section 4 has the experimental setup and reports the results. Further on, the results of the manual analysis of two alternative data partitions are presented in Sect. 5, while Sect. 6 summarises our experience and findings.

2   Related Work

The task of clustering short texts to induce underlying categories is not infrequently called upon in Digital Humanities studies. The data processed in such projects includes news feed [2,9], tweets [12,17], social media posts [3]. The major challenge with such data is the sparsity of words in a single text. Therefore, several techniques were developed to overcome this problem. They include enriching the texts with external data like Wordnet [18], Wikipedia articles [2], word embeddings [5,22]. In our research we create numeric representations for texts using contextual word embeddings from Sentence-BERT using a model trained on sentence similarity task [15], available from https://huggingface.co/sentence-transformers/stsb-xlm-r-multilingual and ELMo [11] using a multilingual model described in [14].

Different approaches towards clustering short texts are described in [13]. The most widely used algorithms are K-means clustering, graph-based Affinity Propagation clustering, or Singular Value Decomposition and similar approaches like LSI (Latent Semantic Indexing) or PCA (Principal Components Analysis). While clustering is known to be a computationally- and memory-intensive task, and our data counts over two million texts, we were limited in the choice of clustering algorithm and experimented with various settings of Mini-Batch K-Means only.

This algorithm requires a pre-select number of clusters. To take this decision we relied on a range of methods described in [7]. This includes 'elbow'-based metrics such as distortion, inertia and extremum-based metrics such as silhouette score, Davies-Bouldin score and gap statistics.

The problem of evaluating the clustering results is covered in [1,4]. The major distinction is made between intrinsic and extrinsic measures. Intrinsic metrics evaluate various parameters of data partition, including inner cluster coherence (e.g. inertia, the k-means objective function), separation and compactness of the clusters (e.g. silhouette score, Calinski-Harabasz Index), comparison with random partitions (gap statistic).

Intrinsic measures do not allow to conclude how fair a partition is with regard to the expected semantic criteria. An algorithm can separate data based on uninteresting formal properties of texts. To compensate for this deficiency, intrinsic metrics are usually combined with extrinsic evaluation measures. This approach includes using labeled data, and a partition generated in an unsupervised manner is compared with some human-evaluated ground truth. In many practical settings, the ground truth for a clustering task might not be available. Our data is only partially annotated, and the existing labels are not very reliable. Nonetheless, we setup an extrinsic evaluation study, bearing in mind its limitations.

3   Data Description

The process of gathering and processing the data for this research is fully described in [8]. This section summarises the main points to provide the necessary information about the data used in this study.

To collect the announcements of cultural events, six[1] web resources aggregating cultural news were scraped. The sources included **social networks** (Meetup), **commercial and non-commercial platforms** (Timepad, e-flux, TED-local, Theory & Practice), **government-curated open databases** (Russian Ministry of culture). Most of the scraped collections contained texts in one language (Russian or English), and some sources (mainly Timepad) provided data in both languages. After preprocessing and lemmatisation, the dataset consists of 2.3 million datapoints from 16,960 locations across the globe (though among the locations only 383 are accompanied with more than 1000 records of events) and contains over 350 million tokens (after preprocessing). The collected events cover 10 years of records and are supplied with rich metadata, including date, characteristics of locations and textual attributes. All the data were cleaned from noise, which includes deleting URLs, emails, HTML code, longer than 15-character tokens, non-alphanumeric tokens, etc., and lemmatised with *UDPipe* [20]. Additionally, we performed automatic language detection using a *fasttext* model [6].

The size of the dataset after all preprocessing is overviewed in Table 1. As can be seen from this table, the major problems with the data include the imbalance of datapoints coming from different resources and the skewness of data towards

---

[1] Based on the previous results, the texts from *Behance* were excluded: They did not contain the descriptions of cultural events, but were mostly captions for graphics.

the English-speaking world. The statistics describing the distribution of lengths of the texts (in tokens/characters) is given in Table 2.

Table 1. Dataset size by language and data source

Data source	English	Russian
MEETUP	1.9 mln	-
Timepad	12k	228k
Mincult	-	84k
Theory&Practice	49	64k
TED	33k	50
e-flux	10k	-
Overall	1.9 mln	376k

Table 2. Dataset text sizes in number of words and number of characters.

Language	Min, max, mean num of words	Min, max, mean num of chars
English	1, 6600, 142	10, 40852, 799
Russian	1, 6720, 44	10, 50706, 326

Some data were provided with category labels induced from original websites. A stratified 10% sample of labeled data was used as gold standard for extrinsic evaluation. The stratification reflected the distribution of data across 20 most well-represented categories and the 1:5 ratio between languages. Therefore, test items come from 3 web-resources for each language: *e-flux, TED, MEETUP* for English and *Mincult, Theory&Practice, Timepad* for Russian. The categories represented in the test set include *socialising, festivals, travel, concerts, kids, lectures, self development, discussions, learning, sports, etc.*. The categories were manually mapped between resources and languages, which can be viewed as an additional limitation of the evaluation study design. Overall, the test set contains 191,205 entries for English and 38,000 entries for Russian.

4   Experimental Setup and Results

The purpose of this paper is not only solving a practical real-world task, but also exploring the behaviour of two language representation models, which are referred to as multilingual. To this end, we vectorised the texts using the two respective frameworks - Sentence BERT[2] and ELMo[3] - and ran a Mini-Batch K-Means algorithm [19] using scikit-learn implementation on each of the resulting matrices. The vectors were normalised to have unit L-2 norm and shuffled before all experiments.

Determining the best Mini-Batch K-Means setting involved experimental evaluation of such clusterer parameters as batch size, reassignment ratio and other settings controlling early stopping (we experimented with k=20, 25, 38). Due to consideration of space, we do not report these results, especially as they had very little impact on the metrics scores.

The final settings for our clustering algorithm are the follows:

- `init='k-means++'`: initial centers of the clusters are spread out instead of random choice;
- `batch_size=1024`: we have not seen any changes in the algorithm performance associated with the increased `batch_size` (up to 16204)
- `reassignment_ratio=0.1`: we used a higher than default value to avoid early convergence on local optima;
- `max_iter=500`
- `max_no_improvement=10`;
- `tol=0.0`: switching off early stopping based on this parameter

With regard to the key clustering parameter, i.e. the value of k, we decided to stick with $k = 20$. The results from intrinsic metrics pointed to various k-values ranging from 14 to 38 and were inconclusive (we visually analyzed the plots for the metrics such as inertia, silhouette, etc.). Generally, $k = 20$ seems to be a convenient number, a human-manageable level of granularity, suitable for manual qualitative analysis.

The intrinsic metrics for the chosen $k = 20$ calculated over the entire dataset being clustered with these settings are reported in Table 3.

Table 3. Intrinsic clustering metrics for $k = 20$

Framework	Inertia	Silhouette	Davies-Bouldin	Calinski-Harabasz	Gap stat
mSBERT	0.548	0.031	**570.086**	**1.001**	4.756
mELMo	0.174	**0.076**	679.873	0.813	**6.190**

Inertia scores are averaged per item and are given here for reference only, they depend on the values of vector components and can only be compared for different clusterings of items in the same vector space; they are not directly comparable here. The Silhouette scores, which evaluate within-cluster consistency,

[2] https://huggingface.co/sentence-transformers/stsb-xlm-r-multilingual, this model was chosen because it was fine-tuned on the semantic textual similarity task, and we want similar texts in two languages be placed in the same cluster.

[3] https://github.com/ltgoslo/simple_elmo (model #219 in the NLPL repository http://vectors.nlpl.eu/repository/).

are close to 0 indicating overlapping clusters for both frameworks (but better for mELMo). It can be explained by the propensity of KMeans for equal-sized clusters, which are not seen in this dataset. The difference in relative Davies-Bouldin Index indicates that clusters on mSBERT vectors are farther apart and more compact. According to Calinski-Harabasz Index they are also better balanced in terms of cluster size.

Additionally, we computed lexical coherence of the resulting clusters using lemmas of content words only and the c_v variant of the coherence measure, which was shown to perform better than other options [16]. This coherence score evaluates whether words (content lemmas) in a cluster tend to co-occur. Values from 0.55 to 0.7 are optimal. The clustering on mELMo vectors returned higher coherence 0.638, compared to mSBERT (0.529).

As was discussed in Sect. 2, evaluation of clustering results can be done with intrinsic or extrinsic evaluation metrics. Intrinsic metrics reported earlier mostly helped us to set down with the number of clusters and the algorithm parameters that lead to higher quality of the results. However, in order to get understanding on how mSBERT-based and mELMo-based partitions correspond to human interpretation, we turned to extrinsic evaluation (even though the ground truth available in our setting is less than reliable). We have calculated Adjusted Rand Index (ARI) and Adjusted Mutual Info Score (AMI). The results for the data partitions on mSBERT and mELMo can be found in Table 4.

Table 4. ARI and AMI scores calculated against a specifically constructed stratified 10% test set

	ARI	AMI
mSBERT	**0.19**	**0.28**
mELMo	0.05	0.19

As we can see from these results, our multilingual BERT model tends to yield higher ARI and AMI scores and thus supposedly provides better quality of the clustering. Given the fact that our classes are imbalanced (*business* is over 20% of the Russian part of the dataset, whereas *theatre* is about 3%), we AMI is better suited for our case than ARI[4].

The results of this extrinsic evaluation should be taken with a grain of salt. Some of the reasons behind low similarity between names of website sections (used as classes) and clusterings are:

- website sections from different resources have a different degree of granularity;
- sections from various resources do not share the classification approach: some reflect the type of activity (*talks, socialising, travel_tours*), while others feature the topic/sphere of social practice (*sports, business, theatre*), which leads to severe overlaps between gold categories;

---

[4] https://stats.stackexchange.com/questions/260487/adjusted-rand-index-vs-adjusted-mutual-information.

- to produce a set of categories/classes, shared between the two languages, we had to translate Russian names of website sections.

Figure 1 reflects the differences in the categories for English and Russian.

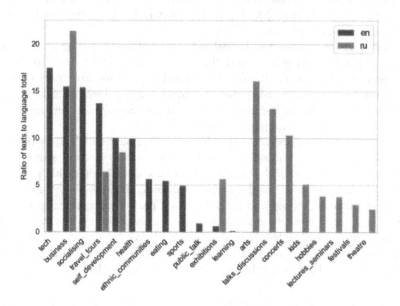

Fig. 1. Distribution of texts across classes in the 'golden' annotation test set

As we can see, English and Russian texts in the gold standard sets form two different worlds with low overlap between announcements categories. It is unlikely that this picture faithfully reflects the reality and there are no events in Russian in the *Tech* category and no English events about *Arts*. We rather assume that the gold classes reflect the limitations listed above.

In addition to evaluating each of the clusterings against the gold standard classes, we also compared mSBERT and mELMo results between themselves using the same approach. Contrary to expectations, the correlations between mSBERT and mELMo clusterings was very low: ARI score achieved 0.07, and AMI was 0.18. This result shows that the two clusterings were very different from each other (which is counter-intuitive - the two context embeddings frameworks were expected to return more comparable results for the same text collection). The difference between the two clustering results will be discussed in Sect. 5, where we test the topical clusters on their human interpretability.

As a sanity check we also evaluated embeddings in the supervised setting. To this end, we trained a neural classifier (`Sequential` model with one fully-connected hidden layer[5]) on each type of vectors, using website section as labels. The standard performance metrics for this classifier are reported in Table 5.

---

[5] Algorithm settings: stratified 5-fold cv, earlystop=True; balanced=True.

Table 5. Classification results

	Accuracy	Macro-F1
mSBERT	73.99%	0.73
mELMo	71.33%	0.71

As we can see from these results, the classifier performs quite well on the data, indicating that the vector representations do pick the differences between the classes when presented with labels. Therefore, low ARI and AMI scores for the clustering experiment do not point towards a poor quality of the vector representations, but rather prove that the clusterers followed the patterns that were not exactly reflected in the golden labels.

Overall, from our evaluation setup we can draw several conclusions with regard to using different language models for clustering. First, intrinsic evaluation metrics returned ambiguous results with regard to the quality of clusters using mSBERT or mELMo representations: Silhouette score, gap statistic and lexical coherence seemed to be better for mELMo-based clusters, while Davies-Bouldin and Calinski-Harabasz coefficients suggested better clusters from mSBERT representations. Extrinsic evaluation metrics indicated in favour of mSBERT-based partition. However, it remains unclear how much we can trust the gold classes and the extrinsic evaluation metrics. Second, we have noticed that unlike mSBERT, mELMo results are less affected by any changes in the clustering settings, which makes them too imposing and inflexible in changing contexts. Finally, it is clear that the mSBERT-based and mELMo clusterings capture different patterns in the data, which results in very different results between them.

5   Human Interpretation

This section presents the results of the manual analysis that was undertaken to test the clusters in two partitions on their human interpretability. This analysis helps to understand which text properties, mediated by the embeddings, were picked by the clusterer as useful for producing 20 groups with the least distance between data points and their closest cluster centers (minimum inertia). The analysis was based on the following information about each cluster extracted for each language:

1. number of texts in the cluster;
2. the text that is calculated as the cluster centre;
3. 20 texts most similar to the center;
4. 20 random texts from the cluster;
5. top 100 count-based keywords;
6. top 100 keywords identified from tf-idf-weighted frequencies, which treated each cluster as a single document.

Both types of keywords were calculated based on lemmatised cluster texts, with function words and stopwords filtered out. Additionally, the count-based keywords lists for each cluster were filtered for 20 lemmas that most often appear in other 19 clusters. The usefulness of these sources of information for inferring the content of the clusters varied across clusters and partitions. Generally, centroids and most-similar texts were not particularly useful as they usually were semantically under-specified and noisy texts (e.g. *Please see the details - they are the same, This week we will be at Pei Wei, #1 4-6 -$720 /-$220 RSL Ultimate*). Tf-idf keywords were more informative than count-based keywords, especially in well-defined mSBERT clusters, but could include mostly strange strings (e.g. *jangchng, lunchr, chiltern, dadar, lugsole, keertan, gpx, hillock*). Russian keywords were always more interpretable due to the higher quality of texts – the Russian part of the data has a lower proportion of user-generated content.

Using the six types of information above about each cluster, we managed to name the clusters with the overall concept shared by the texts. In this process we indeed discovered significant differences between mSBERT and mELMo models.

mELMo-Based Partition. The distribution of texts in two languages across the clusters named as a result of the manual analysis is shown in Fig. 2.

In mEMLo embeddings language differences override any semantic cross-linguistic similarities. The mELMo-based partition included five all-Russian clusters with zero texts in English, 12 clusters with a disproportionately small number of Russian text (60-278 texts compared to 59.5-229.7k texts for English), and one cluster where the overwhelming majority of texts were in Russian ('one-liners with numbers'). There were only two clusters that had a relative balance of texts in two languages: one contained texts that announced various training programmes, including *start-ups, interior design, business logistics and taxes, yoga and meditation* (175k and 34.5k texts for English and Russian respectively) and the other that attracted the texts with tokens in capital letters. Interestingly, the Russian texts in the clusters dominated by English always contained mixed-code messages that were not particularly thematically similar to the English texts. Texts like *Английский разговорный клуб: The importance of being snobbish* or *Дискуссия на английском: Privacy protection in social networks* could be found in any cluster.

Some clusters were difficult to generalise about semantically – they seemed to contain texts that were similar in structure and style rather than topical content. For example, one of the clusters had texts written mostly in capital letters, and another had one-line descriptions of events always including some numbers, whereas long descriptions of semantically similar events ended up in a separate cluster and similar texts without numbers were found in yet another cluster. The 'formal' cluster included announcement published by organisations rather than individuals; they covered a broad number of types of events: opera, theatre, concerts, talks, exhibitions, tours). There was a predominantly English cluster, where texts had an imperative construction (e.g. *Discuss IOT Protypes, Join our speaker, Suzie Jimenez, Start your day early with some great social*

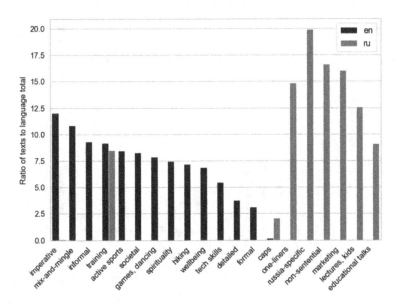

Fig. 2. mELMo-based partition: the distribution of texts across clusters by language

tennis, Learn how to program) and an all-Russian cluster, which included typical phrase-like headlines with no finite verbs.

Several clusters included texts inviting people to informal gatherings, usually around some activity mostly dancing and table games ('games, dancing') or a park clean-up, a theater group, candy making, scrapbooking ('mix-and-mingle'). A cluster dubbed 'informal' in Fig. 2 was only slightly different because the texts were mostly about meetups over food and drinks.

At the same time there were a few clusters that were easier to interpret. Apart from the 'training' cluster, clearly defined clusters were about hiking, health and well-being, social activism, including feminism.

mSBERT-Based Partition. The clustering on mSBERT embeddings is much more interpretable and agreeable with the idea of semantic categorisation, which is reflected in the names of the clusters (see Fig. 3).

Most importantly, mSBERT embeddings did not segregate the languages and returned clusters including semantically similar texts in English and Russian. As a result, the many mixed-code texts were treated with regard to their semantics: for example, most announcements of English clubs and courses were grouped into 'languages and speaking' together with training and events offering oratory opportunities. There were only two clusters that were almost exclusively English or Russian. The first one was dubbed 'socialising, hobbies and interests' and included the absolute minimum of Russian texts (*let us play board games today, Чем заняться вечером в среду?*), while the second one was about Russia-specific events (e.g. *How to deal with Russians? Fashion Bazaar на Красной Площади!*).

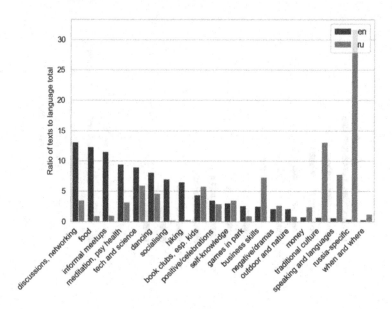

Fig. 3. mSBERT-based partition: the distribution of texts across clusters by language

This partition seemed to be less distracted by noisy texts and their formal properties. However, some of these tendencies were visible. Short texts with dates or place names were aggregated into a 'When and where' (e.g. *Where should we meet for June?*, *Details will be posted by Feb 1*, *Путешествие по маякам: 3 июня*, *Игры в Жан-Жаке 21 и 22 июля*). mSBERT model produced two sentiment-based clusters, one for positive sentiments and celebrations (*This place is so much fun!*) and another for negative sentiment (*Leave your ugly knitting at home*). These clusters were not singled out on the topic or type of the event, but rather on other text properties.

The bilingual nature of clusters does not hide culture-specific interpretations for the key topic of some clusters. For example, *food*-related advertisements in English are more often invitations to dine out including menu descriptions, while in Russian context they have to do with learning to cook and discussions of national cuisines. The advertisements about 'hiking' in English more often feature leisure strolls and walks to enjoy the neighbourhood, while texts in Russian mention rock-climbing, mountains, ridges, tents, glaciers.

Overall, although mELMo-based partition of the text collection can hardly be accepted as semantic grouping of texts, it can be used to filter out noise and to further refine the quality of this text collection. For example, we have noticed that several clusters contain meaningless strings of numbers with occasional words instead of true texts. mSBERT on the contrary offers a more semantic-oriented and accurate description of the underlying collection of texts. This result is much more in line with the task at hand.

6 Conclusion and Future Work

In this research project, we clustered a large collection of cultural events advertisements in the English and Russian languages using alternative pre-trained multilingual contextual embeddings models for text representation. The clustering results were evaluated on a range of intrinsic measures and (for a sample of texts) against pre-existing 'golden' categories.

An important finding of this work is that mELMO and mSBERT embeddings produced very dissimilar clustering results. The mSBERT-based partition was more semantically-oriented and offered a more accurate description of the underlying collection of ads according to the manual comparative analysis of the two clusterings. It returned higher correlation with the existing labels in the external evaluation study. The scores for intrinsic measures indicate that this partition was better globally: it represents the whole collection better. The mELMo-based clusters were largely guided by formal rather than semantic properties of texts, such as text register, presence of particular words or grammatical constructions (e.g. negations). This clustering returned better intrinsic scores for metrics that consider inner-cluster quality. This study demonstrates that artificial neural networks, when unsupervised, can pay attention to the properties of language that are not entirely illogical and uninterpretable, but are not necessarily most prominent or useful to a human perception. Namely, mELMo-based partition revealed patterns of formally identifiable noise in the documents.

The directions of future work lie in testing another algorithmic approaches to clustering as well as using other types of embeddings models, for example, static word embeddings from *word2vec* [10]. We also would like to improve the quality of the text collection by removing the types of noise identified in this study, and the reliability of the manual analysis by inviting more annotators to this task and calculating their agreement.

References

1. Amigó, E., Gonzalo, J., Artiles, J., Verdejo, F.: A comparison of extrinsic clustering evaluation metrics based on formal constraints. Inf. Retrieval **12**(4), 461–486 (2009)
2. Banerjee, S., Ramanathan, K., Gupta, A.: Clustering short texts using Wikipedia. In: Proceedings of the 30th Annual International ACM SIGIR Conference on Research and Development in Information Retrieval, pp. 787–788 (2007)
3. Comito, C., Forestiero, A., Pizzuti, C.: Word embedding based clustering to detect topics in social media. In: 2019 IEEE/WIC/ACM International Conference on Web Intelligence (WI), pp. 192–199. IEEE (2019)
4. Färber, I., et al.: On using class-labels in evaluation of clusterings. In: MultiClust: 1st International Workshop on Discovering, Summarizing and Using Multiple Clusterings Held in Conjunction with KDD, p. 1 (2010)
5. Hadifar, A., Sterckx, L., Demeester, T., Develder, C.: A self-training approach for short text clustering. In: Proceedings of the 4th Workshop on Representation Learning for NLP (RepL4NLP-2019), pp. 194–199 (2019)

6. Joulin, A., Grave, E., Bojanowski, P., Douze, M., Jégou, H., Mikolov, T.: Fast-text.zip: compressing text classification models. arXiv preprint arXiv:1612.03651 (2016)
7. Kodinariya, T.M., Makwana, P.R.: Review on determining number of cluster in k-means clustering. Int. J. **1**(6), 90–95 (2013)
8. Kunilovskaya, M., Plum, A.: Text preprocessing and its implications in a digital humanities project. In: Proceedings of the Student Research Workshop associated with the 13th International Conference on Recent Advances in Natural Language Processing (RANLP), pp. 85–93 (2021)
9. Marutho, D., Handaka, S.H., Wijaya, E., et al.: The determination of cluster number at k-mean using elbow method and purity evaluation on headline news. In: 2018 International Seminar on Application for Technology of Information and Communication, pp. 533–538. IEEE (2018)
10. Mikolov, T., Yih, W., Zweig, G.: Linguistic regularities in continuous space word representations. In: Proceedings of the 2013 Conference of the North American Chapter of the Association for Computational Linguistics: Human Language Technologies, pp. 746–751 (2013)
11. Peters, M.E., et al.: Deep contextualized word representations. arXiv preprint arXiv:1802.05365 (2018)
12. Poomagal, S., Visalakshi, P., Hamsapriya, T.: A novel method for clustering tweets in Twitter. Int. J. Web Based Communities **11**(2), 170–187 (2015)
13. Rangrej, A., Kulkarni, S., Tendulkar, A.V.: Comparative study of clustering techniques for short text documents. In: Proceedings of the 20th International Conference Companion on World Wide Web, pp. 111–112 (2011)
14. Ravishankar, V., Kutuzov, A., Øvrelid, L., Velldal, E.: Multilingual ELMo and the effects of corpus sampling. In: Proceedings of the 23rd Nordic Conference on Computational Linguistics (NoDaLiDa), pp. 378–384 (2021)
15. Reimers, N., Gurevych, I.: Sentence-BERT: sentence embeddings using Siamese BERT-networks. arXiv preprint arXiv:1908.10084 (2019)
16. Röder, M., Both, A., Hinneburg, A.: Exploring the space of topic coherence measures. In: Proceedings of the Eighth ACM International Conference on Web Search and Data Mining, pp. 399–408 (2015)
17. Rosa, K.D., Shah, R., Lin, B., Gershman, A., Frederking, R.: Topical clustering of tweets. Proceedings of the ACM SIGIR: SWSM, vol. 63 (2011)
18. Rudrapal, D., Das, A., Bhattacharya, B.: Measuring semantic similarity for Bengali tweets using wordnet. In: Proceedings of the International Conference Recent Advances in Natural Language Processing, pp. 537–544 (2015)
19. Sculley, D.: Web-scale k-means clustering. In: Proceedings of the 19th International Conference on World Wide Web, pp. 1177–1178 (2010)
20. Straka, M., Straková, J.: Tokenizing, POS tagging, lemmatizing and parsing UD 2.0 with UDPipe. In: Proceedings of the CoNLL 2017 Shared Task: Multilingual Parsing from Raw Text to Universal Dependencies, pp. 88–99 (2017)
21. Wallerstein, I.: World-Systems Analysis: An Introduction. Duke University Press, Durham and London (2004)
22. Wang, P., Xu, B., Xu, J., Tian, G., Liu, C.L., Hao, H.: Semantic expansion using word embedding clustering and convolutional neural network for improving short text classification. Neurocomputing **174**, 806–814 (2016)

Jokingbird: Funny Headline Generation for News

Nikita Login[1], Alexander Baranov[1], and Pavel Braslavski[1,2(✉)]

[1] HSE University, Moscow, Russia
`pbras@yandex.ru`
[2] Ural Federal University, Yekaterinburg, Russia

Abstract. In this study, we address the problem of generating funny headlines for news articles. Funny headlines are beneficial even for serious news stories – they attract and entertain the reader. Automatically generated funny headlines can serve as prompts for news editors. More generally, humor generation can be applied to other domains, e.g. conversational systems. Like previous approaches, our methods are based on lexical substitutions. We consider two techniques for generating substitute words: one based on BERT and another based on collocation strength and semantic distance. At the final stage, a humor classifier chooses the funniest variant from the generated pool. An in-house evaluation of 200 generated headlines showed that the BERT-based model produces the funniest and in most cases grammatically correct output.

Keywords: Computational humor · Headline generation · Humor generation

1 Introduction

In this study, we address the problem of automatic generation of humorous news headlines. Funny headlines can be advantageous even in case of serious articles, allowing them to stand out in the news stream and attract readers' attention. Automatic humor generation is a topical task: its relevance has increased recently due to the growing popularity of conversational systems. Virtual assistants with a sense of humor make communication more entertaining and engaging.

Almost all automatic humor generation methods rely on some predefined joke patterns. This is also mirrored in design of datasets for computational humor research. For example, humorous sentences in FunLines [14] and Humicroedit [13] datasets are generated by volunteers and crowd workers by changing only one token in the original headline. In this work, we are trying to automate the very same tactics. The proposed approach to headline modification nicknamed *Jokingbird* consists of three steps. First, a word for substitution is selected in the original news headline. Second, several candidate substitutions are generated. Third, the funniest variant is chosen based on a humor classifier.

E. Burnaev et al. (Eds.): AIST 2021, LNCS 13217, pp. 97–109, 2022.
https://doi.org/10.1007/978-3-031-16500-9_9

We conducted a manual evaluation of 200 modified headlines along two dimensions: *funniness* and *grammaticality* on a 4-point scale. Evaluation shows that Jokingbird is able to produce grammatically correct edits, however *funniness* of generated output is still quite low.

2   Related Work

Humor has been studied by many disciplines, including psychology, linguistics, and formal semantics. For example, Raskin stated in his classical work that a text carries a joke if it is "partially or fully compatible with two different scripts" [22]. A script in Raskin's theory is a meaning associated with a word or a linguistic unit. The opposition between scripts can be a universal dichotomy, such as real vs. unreal, good vs. bad, life vs. death, etc. Raskin also defines a 'script-switch trigger' – a linguistic unit that is 'responsible' for a humorous effect. Many computational humor studies follow this notion explicitly or implicitly. Attardo provides a comprehensive survey of linguistic theories of humor [3].

Computational humor has two major research directions – humor detection and humor generation. The former is usually formulated as a binary classification problem – assigning a text to either 'humorous' or 'serious' class – or as a task of ranking texts according to their funniness. In their pioneering work, Mihalcea and Strapparava [17] represent short texts using humor-specific features such as alliteration, antonymy and adult slang, and train Naïve Bayes and SVM classifiers. Yang et al. [31] employ various features, including those based on *word2vec* representations, and a random forest classifier. More recent studies use Transformer architecture and large pre-trained language models [2,28].

Most humor generation approaches start with a predefined joke pattern [1]. For instance, Stock and Strapparava developed a system HAHAcronym that produced a humorous acronym deciphering based on WordNet relations and domains [25]. Sjöbergh and Araki [23] proposed a humor generation approach based on polysemous words, a setup–punchline structure, and properties of English compound nouns. Valitutti et al. [26] generated jokes by substituting a word in real SMS messages by similarly sounding and taboo words. Blinov et al. [5] cast the problem of funny response generation in human-computer conversation as joke retrieval from a large collection of funny tweets.

Yu et al. [32] imposed minimal constraints on pun generation: they produced a pun-like sequence using a recurrent neural network (RNN) trained on a corpus with explicitly marked senses of polysemous words (e.g., *bass_1* for fish and *bass_2* for musical instrument). He et al. [11] proposed a pun generation method that starts with a pair of similarly sounding words. The method retrieves a sentence mentioning a member of such a pair, substitutes the word with its counterpart, and adjusts the context to amplify the surprise effect. The work by Horvitz et al. [12] is close to ours: it dealt with the same task of funny news headline generation. However, it departs from ours and other humor generation approaches, since is aimed at generating not pun-like jokes, but jokes involving world knowledge. The authors used data from satirical news site The Onion[1],

---

[1] https://www.theonion.com/.

related Wikipedia and CNN articles, and a BERT-based abstractive summarizer. Weller et al. [27] used a Transformer model trained on Humicroedit dataset (see Sect. 3) in a sequence-to-sequence manner. Our work is close to this study, but employs a modular architecture.

Winters et al. [29] proposed a unified design scheme for a computer humor system as a combination of a humor *Generator*, a set of relevant *Features*, a feature *Aggregator*, and a joke *Template* with *Keywords*. Jokingbird roughly follows the scheme: *Templates* are selected based on collocation strength, *Keyword* values are generated by either BERT [9] or *word2vec* [18]; a BERT-based humor classifier works as an *Aggreggator*. In our work, we aimed at modifying a collocation in the original headline to turn it into a 'humor trigger'. Collocations have been intensively studied in relation to humor [19,20,24].

3 Data

In this study, we use a large news corpus and two humor datasets.

News corpus is a result of merging three datasets: *Times front page news*[2], *All the News*[3], and *Harvard news articles*[4]. The combined dataset contains 361,261 articles in total spanning from 2013 to 2018. Only headlines of the articles are used in the current study.

RedditJokes is a dataset which, as its name suggests, contains jokes from the social media website *reddit*[5]. Chen and Soo [7] compiled the first version of the dataset of a Kaggle dataset of *reddit* jokes[6] (as positive class) and sentences from WMT16 English news corpus[7] (as negative class). Since the dataset wasn't made publicly available, Weller and Seppi [28] later re-created the negative class following the described approach. In our study we use the training set of this latter version that contains 347,486 items in both classes. The balanced dataset allowed Jokingbird to learn non-humorous collocations as well as humorous ones. The same dataset was used in the study by Annamoradnejad and Zoghi [2]; however, the authors additionally filtered the data and retained only 100K items from each class.

Humicroedit and FunLines are two datasets with similar design principles and the same structure: they consist of pairs of original/modified news headlines along with the funniness score of the latter. The modification was performed either by volunteers or crowd workers; their task was to make the headline funny

---

[2] https://components.one/datasets/above-the-fold/.
[3] https://www.kaggle.com/snapcrack/all-the-news.
[4] https://doi.org/10.7910/DVN/GMFCTR.
[5] https://www.reddit.com/.
[6] https://www.kaggle.com/abhinavmoudgil95/short-jokes.
[7] http://www.statmt.org/wmt16/translation-task.html.

by minimal edits (mostly by substituting a single word). We combined Humi-croedit [13] (15,095 items) and FunLines [14] (8,248 items)[8]. The funniness of edited sentences in Humicroedit/FunLines is rated on a scale from 0 to 3; items with score 2 and above are considered 'funny'. The average funniness score for the overall dataset is 1.05; the distribution of funniness is displayed in Fig. 1. As we wanted to develop models to predict funny modifications, we took only pairs with non-repeating original headlines having average humorousness score of no less than 2. Using this restriction, we retained 1,918 entries. The average funniness in the "cleaned" dataset is 2.15. We used Humicroedit/Funlines data for fine-tuning our BERT models.

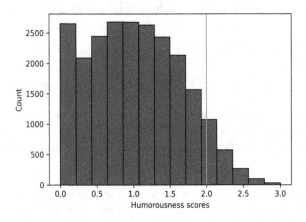

Fig. 1. Distribution of humorousness scores in the merged Humicroedit/FunLines dataset.

4   Methods

Our approach is based on the hypothesis that a word or a phrase semantically different from the original meaning of a sentence can potentially create a humor-ous effect. Selection of semantically distant items is in line with most humor generation studies, see for example the approach to pun generation by He et al. [11]. We also tried to follow this concept by using BERT-based models, albeit in a more relaxed manner – without explicit limitations on the (dis)similarity of the words.

4.1   Pipeline

The key idea behind Jokingbird is to replace a word in a news headline to cre-ate a humorous surprise effect. We broke our approach down into three major

---

[8] Both datasets are available from https://cs.rochester.edu/u/nhossain/funlines.html.

steps: 1) selection of the word in a headline to be replaced; 2) finding a word to be inserted instead, and 3) picking the 'funniest' edit from the pool of generated variants using a humor classifier. Our word replacement technique relies on collocation strength only, whereas word selection is performed either by a BERT-based model or using cosine distance between *word2vec* embeddings. As a humor classifier on top of our pipeline we employ a readily available ColBERT model [2]. Figure 2 illustrates the proposed pipeline.

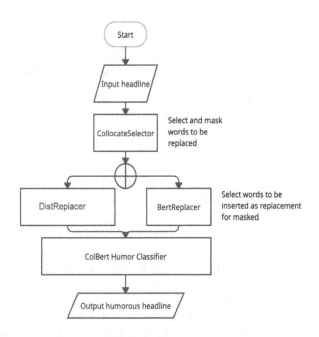

Fig. 2. The Jokingbird's pipeline.

4.2   Word Selection

To select a word to be replaced, Jokingbird searches for strong collocations in a given sentence. As a collocation strength measure we use Pointwise Mutual Information (PMI) [8]. We identified bi- and tri-gram collocations in News corpus headlines and RedditJokes. As News corpus items were of mixed case, we corrected their casing with the Truecase[9] package before collocation extraction. For News corpus we used SpaCy[10] for tokenization, POS-tagging, and lemmatization. Items from RedditJokes were tokenized and POS-tagged with NLTK [4]. Choosing between speed and performance, we decided on using SpaCy for the smaller dataset (News corpus) and NLTK for the larger one (RedditJokes).

To better illustrate the work of CollocateSelector, let's take an example:

---

[9] https://pypi.org/project/truecase/, based on [15].
[10] https://spacy.io/.

(1)   Will the Trump nomination change our polarized partisan patterns?

Let's suppose that we have set the collocation strength threshold to 3. Then the following n-grams exceed the threshold (PMI scores between the *left* and *right* part are in parentheses): *Trump nomination* **change** (6.24), *change our* **polarized** (10.73), *our polarized* **partisan** (10.47), *polarized* **partisan** (8.76), *polarized partisan* **patterns** (11.12), and *partisan* **patterns** (6.96). Thus, we get the sentences a)–d) as input for the word replacer:

a) Will the Trump nomination [MASK] our polarized partisan patterns?
b) Will the Trump nomination change our [MASK] partisan patterns?
c) Will the Trump nomination change our polarized [MASK] patterns?
d) Will the Trump nomination change our polarized partisan [MASK]?

In this step we also filter out collocations, where either the word in the right part or all the words in the left part are assigned a functional POS tag[11].

4.3   Word Substitution

We apply two techniques to replace words masked on the previous step: one based on collocation strength and semantic distance, and another based on the BERT model.

DistReplacer uses a *word2vec* embeddings trained on a lemmatized and POS-tagged dump of English Wikipedia from WebVectors repository [10]. DistReplacer retrieves words of the same POS that are semantically most distant from the masked word. Then, DistReplacer filters out words that don't exceed the collocation strength threshold set by previously retrieved candidates. It considers collocation strength between a candidate word and a unigram or bigram that precede the masked word. Next, all candidates are inflected to be in the same grammatical form as the masked word. This step is done using the LemmInflect package[12].

In the rest of the paper we will refer to two versions of DistReplacer depending on the source for collocation matrix learning (News Corpus vs. RedditJokes): DistReplacerNews and DistReplacerReddit.

BERT Models from Transformers library [30] serve as the basis for our second approach. We utilize two BERT models to produce a funny replacement of the masked word: BERTHumEdit is fine-tuned on unique sentences from Humicroedit/Funlines, while BERTJokes is fine-tuned on ShortJokes. BERTHumEdit was trained to predict words from humorous edits in place of [MASK]. BERTJokes was trained to predict random masked words in RedditJokes. During the BERTHumEdit training we only updated the weights of

---

[11] PART, CCONJ, SCONJ, ADP, AUX, DET, PRON, PUNCT, or NUM.
[12] https://github.com/bjascob/LemmInflect.

the last layer of the pre-trained model, while during the BERTJokes training all weights of the pre-trained model were updated. We sampled top-K predictions of the models and excluded one- and two-character tokens, pronouns, articles, prepositions, particles and numbers before passing the output to the next step. Upon examining preliminary results, we set K = 12 for BERTJokes and K = 3 for BERTHumEdit.

4.4 Threshold Selection

To select a threshold for collocation strength, we manually selected 25 sentences containing strong collocations from the News corpus. Then we ran BERTHumEdit on this sample with PMI threshold values from 0 to 6 and manually evaluated the output for humorousness. The best value was selected by the share of humorous sentences among 'successful' sentences[13]. Figure 3 illustrates results of this process.

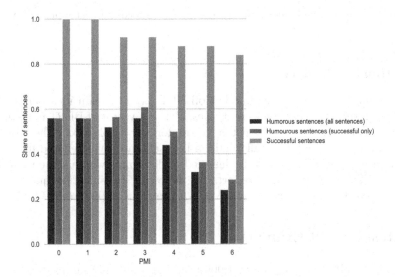

Fig. 3. The impact of different PMI thresholds on BERTHumEdit output.

To find an optimal semantic distance threshold, we manually selected 15 headlines and ran DistReplacerNews on them with semantic distance thresholds from 0.2 to 0.55 (using PMI threshold of 3 determined previously). As in the previous step, we selected the value that delivers the best ratio of humorous sentences among the successful ones, see Fig. 4.

---

[13] 'Successful' sentences are those, where at least one candidate word for replacement was found and at least one generated replacement passed the model's restrictions on the predicted word.

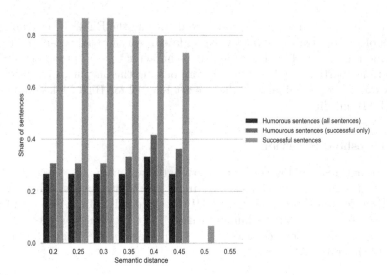

Fig. 4. The impact of different semantic distance thresholds on DistReplacerNews output.

4.5 Humor Classifier

The classifier atop of our pipeline is ColBERT [2]. The main idea behind Col-BERT is to split input text into sentences, obtain a vector of BERT embeddings for each sentence, and finally get a vector for the whole text. The resulting embeddings go through a parallel hidden layer and are concatenated into a single vector. The resulting vector is classified to determine whether the input text is a joke. The original paper reports the following classification quality on a subsample of RedditJokes: accuracy –0.982, precision –0.990, recall –0.974.

5   Results and Evaluation

We applied ColBERT classifier to 1,000 headlines randomly sampled from News Corpus and their modifications produced by all four variants of the Jokingbird pipeline (DistReplacerReddit, DistReplacerNews, BERTHumEdit, and BERT-Jokes)[14]. As Table 1 shows, all models' modifications led to increased ColBERT's funniness scores. BERTHumEdit has the highest share of successful sentences, while BERTJokes has the highest average funniness score. At the same time, ColBERT's scores on the Humicroedit/FunLines data suggest that these results must be taken with caution: original Humicroedit/FunLines headlines have a significantly higher average score compared to the headlines sampled from News corpus; the increase of the score in manually modified and presumably funny headlines is marginal.

---

[14] When sampling the sentences, we discarded headlines containing words indicating sensitive topics like violence, death, and religion.

Table 1. Average ColBERT classifier's score on original vs. modified headlines (higher values imply funnier texts); the second column cites the number of successfully modified sentences out of random thousand. ColBERT's scores on Humicroedit/FunLines dataset are added for reference.

Model	# Sentences	Original	Modified
DistReplacerReddit	148	0.016	0.033
DistReplacerNews	520	0.018	0.022
BERTHumEdit	724	0.018	0.093
BERTJokes	636	0.019	0.114
Humicroedit/FunLines	1,918	0.043	0.054

Table 2. Human evaluation of Jokingbird's 200 headlines on a 0..3 scale (funniness: 0–not funny at all, 3–hilarious; grammaticality: 0–very bad, 3–very good).

Model	Funniness	Grammaticality
DistReplacerReddit	0.48	1.58
DistReplacerNews	0.43	1.45
BERTHumEdit	0.95	1.80
BERTJokes	0.44	2.15
Weller et al. [27]	0.28	0.83

200 sentences (50 items sampled from each model's output) were evaluated manually by two authors of the paper. In addition, they evaluated the output of the model by Weller et al. [27] on the same 200 original news headlines. Each modified headline was evaluated by two persons on two aspects: funniness and grammaticality. Each aspect was evaluated on a 4-point scale (0..3). Resulting scores are averages over assessors and evaluated items, see Table 2. We can see from the Table that BERTHumEdit produced most humorous output, close to the 'somewhat funny' grade. BERTJokes did the best job on grammaticality. Generally, grammaticality score of the BERT-based models was higher than of the collocation/distance-based models. All our models outperformed the model by Weller et al. [27] both in terms of funniness and grammaticality. However, as Table 3 shows, the annotators' agreement is very low; agreement on funniness in somewhat higher in case of DistReplacerReddit and BERTHumEdit, but still quite poor[15].

Humor evaluation is known to be a hard task, since humor perception is highly subjective and is influenced by many factors such as cultural background and language proficiency [6]. In order to obtain more objective scores we conducted a pilot evaluation study on Toloka crowdsourcing platform[16]. BERTHum-

---

[15] We used Cohen's Kappa with linearly decreasing weights as implemented in Scikit-Learn [21] package.

[16] https://toloka.ai/.

Table 3. Weighted Cohen's kappa between two annotators on Jokingbird results.

Model	Funniness	Grammaticality
DistReplacerReddit	0.33	0.06
DistReplacerNews	0.07	0.05
BERTHumEdit	0.37	0.07
BERTJokes	0.06	0.01
Weller et al. [27]	0.22	0.11

Edit results on 1,000 headlines formed 718 tasks in total. For entry qualifications and continuous control we used additional 41 tasks: 21 headlines, where manual annotation on the previous step by two authors fully matched, and 20 original news headlines (for them we expected funniness score 0 and grammaticality 3). We selected top 50% Toloka performers for the task; each task was annotated by four workers. Unfortunately, the agreement among crowd workers was even lower and we decided not to conduct a large-scale evaluation. We will elaborate on design of crowd-sourced evaluation and hope to obtain more reliable results in the future.

Below are examples of BERTHumEdit, where two annotators' humorousness scores differ by no more than one point (averaged scores are in parentheses):

(2)   Embattled Pennsylvania Attorney General ~~resigns~~ **sacked** after perjury conviction (0)

(3)   The Epipen ~~boss~~ **salesman** tried to defend price HIKES to Congress. No one bought it (0)

(4)   Will the Trump nomination change our polarized ~~partisan~~ **sleeping** patterns? (1)

(5)   With town hall ~~script~~ **stool** flipped on GOP, will history repeat itself? (1)

(6)   Lindsey Graham's ~~political~~ **disco** career (2)

(7)   UK leaders must let the Brexit ~~vote~~ **sandwich** stand (2)

(8)   As he moves campaign to battlegrounds, which Donald ~~Trump~~ **duck** will show up? (2.5)

(9)   Wall Street dips before French ~~election~~ **toast**, but up for week (2.5)

As we can see from example (2), simple prediction of synonyms results in the least humorous output. Examples (3–6) have incompatible scripts and thus are humorous. The humorousness increases with higher incompatibility of the scripts (**political** vs **disco**) and commonness of the modified collocation (**sleeping patterns, Donald Duck, French toast**).

6 Conclusion

Based on the obtained results, we can conclude that while being able to produce mostly grammatically correct sentences, our models still fail to induce comic effect in most cases. Moreover, relying on strong collocations heavily reduces the number of sentences suitable for generating humorous edits, as not all of them contain such items. Using a state-of-the-art classifier for selecting the best edit also does not guarantee it will be able to detect humor in automatically generated examples.

Possible improvements to Jokingbird may include additional restriction on the predicted word to be orthographically or phonetically similar to the replaced item, as seen in the study by Valitutti et al. [26]. Another improvement may include joining a BERT model and humor classifier into a generative adversarial network (GAN), so the model will try to predict words the insertion of which will be rated higher by the classifier (a similar approach is implemented in the study by Luo et al. [16]). It may also be useful to utilize a supervised humor ranker instead of a binary classifier, as our model selects the best edit among several alternatives. Another possible enhancement might be inserting words instead of just replacing them.

Acknowledgments. The described experiments were partly conducted using HPC facilities of the HSE University. We thank Daria Overnikova for useful comments and suggestions on a draft of this paper. Pavel Braslavski thanks Exactpro company (https://exactpro.com/) for supporting the project.

References

1. Amin, M., Burghardt, M.: A survey on approaches to computational humor generation. In: Proceedings of the The 4th Joint SIGHUM Workshop on Computational Linguistics for Cultural Heritage, Social Sciences, Humanities and Literature, pp. 29–41 (2020)
2. Annamoradnejad, I., Zoghi, G.: ColBERT: using BERT sentence embedding for humor detection. arXiv preprint arXiv:2004.12765 (2020)
3. Attardo, S.: Linguistic Theories of Humor. Walter de Gruyter (1994)
4. Bird, S., Klein, E., Loper, E.: Natural Language Processing with Python: Analyzing Text with the Natural Language Toolkit. O'Reilly Media, Inc. (2009)
5. Blinov, V., Mishchenko, K., Bolotova, V., Braslavski, P.: A pinch of humor for short-text conversation: an information retrieval approach. In: International Conference of the Cross-Language Evaluation Forum for European Languages, pp. 3–15 (2017). https://doi.org/10.1007/978-3-319-65813-1_1
6. Braslavski, P., Blinov, V., Bolotova, V., Pertsova, K.: How to evaluate humorous response generation, seriously? In: Proceedings of the 2018 Conference on Human Information Interaction & Retrieval, pp. 225–228 (2018)
7. Chen, P.Y., Soo, V.W.: Humor recognition using deep learning. In: Proceedings of the 2018 Conference of the North American Chapter of the Association for Computational Linguistics: Human Language Technologies, Volume 2 (Short Papers), pp. 113–117 (2018)

8. Church, K.W., Hanks, P.: Word association norms, mutual information, and lexicography. Comput. Linguist. **16**(1), 22–29 (1990)
9. Devlin, J., Chang, M.W., Lee, K., Toutanova, K.: BERT: pre-training of deep bidirectional transformers for language understanding. In: Proceedings of the 2019 Conference of the North American Chapter of the Association for Computational Linguistics: Human Language Technologies, Volume 1 (Long and Short Papers), pp. 4171–4186 (2019)
10. Fares, M., Kutuzov, A., Oepen, S., Velldal, E.: Word vectors, reuse, and replicability: Towards a community repository of large-text resources. In: Proceedings of the 21st Nordic Conference on Computational Linguistics, pp. 271–276 (2017)
11. He, H., Peng, N., Liang, P.: Pun generation with surprise. In: Proceedings of the 2019 Conference of the North American Chapter of the Association for Computational Linguistics: Human Language Technologies, Volume 1 (Long and Short Papers), pp. 1734–1744 (2019)
12. Horvitz, Z., Do, N., Littman, M.L.: Context-driven satirical news generation. In: Proceedings of the Second Workshop on Figurative Language Processing, pp. 40–50 (2020)
13. Hossain, N., Krumm, J., Gamon, M.: President vows to cut <Taxes> hair: dataset and analysis of creative text editing for humorous headlines. In: Proceedings of the 2019 Conference of the North American Chapter of the Association for Computational Linguistics: Human Language Technologies, Volume 1 (Long and Short Papers), pp. 133–142 (2019)
14. Hossain, N., Krumm, J., Sajed, T., Kautz, H.: Stimulating creativity with funlines: a case study of humor generation in headlines. In: Proceedings of the 58th Annual Meeting of the Association for Computational Linguistics: System Demonstrations, pp. 256–262 (2020)
15. Lita, L.V., Ittycheriah, A., Roukos, S., Kambhatla, N.: tRuEcasIng. In: Proceedings of the 41st Annual Meeting of the Association for Computational Linguistics, pp. 152–159 (2003)
16. Luo, F., Li, S., Yang, P., Chang, B., Sui, Z., Sun, X., et al.: Pun-GAN: generative adversarial network for pun generation. arXiv preprint arXiv:1910.10950 (2019)
17. Mihalcea, R., Strapparava, C.: Making computers laugh: investigations in automatic humor recognition. In: Proceedings of the Conference on Human Language Technology and Empirical Methods in Natural Language Processing, pp. 531–538 (2005)
18. Mikolov, T., Chen, K., Corrado, G., Dean, J.: Efficient estimation of word representations in vector space. arXiv preprint arXiv:1301.3781 (2013)
19. Nagy, T.: Raising collocational awareness with humour. Acta Univ. Sapientiae Philologica **12**(2), 99–113 (2020)
20. Partington, A.: 'Kicking the Habit': the exploitation of collocation in literature and humour. Linguistic Approaches to Literature, English Language (1995)
21. Pedregosa, F., et al.: Scikit-learn: machine learning in Python. J. Mach. Learn. Res. **12**, 2825–2830 (2011)
22. Raskin, V.: Semantic Mechanisms of Humor (1985)
23. Sjöbergh, J., Araki, K.: A measure of funniness, applied to finding funny things in WordNet. In: Proceedings of the Conference of the Pacific Association for Computational Linguistics 2009, pp. 236–241 (2009)
24. Skalicky, S.: Lexical priming in humorous satirical newspaper headlines. Humor **31**(4), 583–602 (2018)
25. Stock, O., Strapparava, C.: HAHAcronym: a computational humor system. In: ACL (demo), pp. 113–116 (2005)

26. Valitutti, A., Toivonen, H., Doucet, A., Toivanen, J.M.: Let everything turn well in your wife: generation of adult humor using lexical constraints. In: Proceedings of the 51st Annual Meeting of the Association for Computational Linguistics (Volume 2: Short Papers), pp. 243–248 (2013)
27. Weller, O., Fulda, N., Seppi, K.: Can humor prediction datasets be used for humor generation? humorous headline generation via style transfer. In: Proceedings of the Second Workshop on Figurative Language Processing, pp. 186–191 (2020)
28. Weller, O., Seppi, K.: Humor detection: a transformer gets the last laugh. In: Proceedings of the 2019 Conference on Empirical Methods in Natural Language Processing, pp. 3621–3625 (2019)
29. Winters, T., Nys, V., De Schreye, D.: Towards a general framework for humor generation from rated examples. In: Proceedings of the 10th International Conference on Computational Creativity, pp. 274–281 (2019)
30. Wolf, T., et al.: Transformers: state-of-the-art natural language processing. In: Proceedings of the 2020 Conference on Empirical Methods in Natural Language Processing: System Demonstrations, pp. 38–45 (2020)
31. Yang, D., Lavie, A., Dyer, C., Hovy, E.: Humor recognition and humor anchor extraction. In: Proceedings of the 2015 Conference on Empirical Methods in Natural Language Processing, pp. 2367–2376 (2015)
32. Yu, Z., Tan, J., Wan, X.: A neural approach to pun generation. In: Proceedings of the 56th Annual Meeting of the Association for Computational Linguistics (Volume 1: Long Papers), pp. 1650–1660 (2018)

Learning to Rank with Capsule Neural Networks

Anna Nesterenko and Anastasia Ianina

Moscow Institute of Physics and Technology, Dolgoprudny, Russian Federation
{nesterenko.aa,yanina}@phystech.edu

Abstract. Exploratory search scenarios are becoming common in recommender systems. Exploratory search is aimed at self-education and knowledge acquisition, its query often cannot be formulated in a short phrase but can be represented as one or several topically coherent documents. The results of exploratory search can be very extensive, but the ranking algorithms mostly use simple statistics which are not sufficient to rank documents in the convenient for further investigation and refinement order. In this work, we study capsule neural networks in combination with various textual representations applied to document ranking. The proposed model is tested on ArXiv triplets and Microsoft News Datasets and compared to classical approaches. Experiments show that capsule networks can significantly improve the quality of rankings in terms of search Presicion and Recall.

Keywords: Recommendation systems · Exploratory search · Document-by-document search · Capsule neural networks · Ranking algorithms

1 Introduction

Exploratory search is a paradigm of information retrieval aimed at self-education, acquisition, systematization of knowledge and therefore is often associated with the search for scientific literature. An example would be the search for articles that are semantically similar to an article recently studied by the user. An important feature of such a scenario is that a search demand cannot be formulated as a short query and also a user may be uncertain of their information needs. We consider typical exploratory search tasks formulated by long text queries (e.g. several topically consistent documents). In this case, the user does not formulate queries iteratively, but instead immediately sets a wide search direction by submitting a whole document or a fragment of a document to the exploratory search system. Exploratory search scope is wider that the one for well-formulated, navigational and close information requests.

At the moment, there are many systems specifically designed for scientific literature search, e.g. Google Scholar. The ranking algorithms for this system

© The Author(s), under exclusive license to Springer Nature Switzerland AG 2022
E. Burnaev et al. (Eds.): AIST 2021, LNCS 13217, pp. 110–123, 2022.
https://doi.org/10.1007/978-3-031-16500-9_10

are unknown. However, [12] states that the most important features for ranking are the number of citations of a scientific article, the presence of words from the query in the article title and the author's or journal's popularity. The study [12] emphasizes that this system (like most others) is aimed at short and well-formulated queries and therefore does not work well with document-by-document search. The search results to such queries can be extensive, so ranking quality becomes even more important for discovery-oriented applications. Placing the documents in the right, easy-to-process order lowers the cognitive burden and helps to achieve search goals faster.

The main way to rank exploratory search results is to sort texts by semantic similarity. It can be measured using simple similarity measures or by machine learning models. For a long time search engines have used methods of the first type, for example, the cosine similarity. This paper discusses the application of capsule neural networks to the problem of text ranking. Such models start to play an increasing role in document ranking due to its ability to capture hidden semantics and relations between query and suggested documents. In situations with not enough data for training classical learn-to-rank algorithms, neural similarity measures become a good alternative to conventional cosine similarity or dot product.

The rest of the paper is organized as follows: the related work is discussed in the next section. Then we explain the proposed ranking algorithm. The main focus is put upon the designed neural-based similarity measure. We demonstrate its effectiveness within different document representations, including Doc2Vec [15], FastText [5] and BERT-based embeddings [7] in the Experiments section (Subsect. 3.4). We also provide a comparison to baseline methods (Subsect. 3.5). All the experiments are conducted on two datasets: ArXiv triplets [6] and Microsoft News Dataset [23]. The paper closes with a brief conclusion and future work discussion.

2   Related Work

Exploratory search systems aim at satisfying search requests formulated very broadly. Solving a knowledge acquisition task, it may be hard to skip irrelevant results by reading just the titles (like we sometimes do when googling for something in particular). To prevent a user from wasting time reading poorly related documents, the results of an exploratory search system should re-ranked to better match user's needs, being highly personalized ideally. Incorporating user-related information into ranking formula is a conventional well-known approach [2,25]. An interesting approach is presented in [21] where user feedback is incorporated explicitly to re-rank results according to the user's intent.

Another promising research direction is interactive exploratory search systems. They process user's feedback in real time, altering search results on the go [21]. Such search engines may be used not only for scientific or highly specialized information demands, but also in mass user products, like online shopping [14].

Another approach to solve exploratory search problem is to leverage topical features in order to find semantically similar documents [8,10,11,19]. In such cases the trained topic model is used to get topic features for every document in the collection and compare it to the topic features of the query. Topic model may be hierarchical [9] making the searching task cascading.

However, in all those systems the semantic relevance of a document to the query is usually measured by non-customized simple measures, like cosine similarity or dot product. Since neural networks are powerful tools for uncovering relations between texts, they can easily substitute conventional similarity measures within exploratory search scenarios, providing more accurate and personalized results. One of the steps in such a direction is Capsule network based model for Rating Prediction with user reviews (CARP, [16]) that extracts viewpoints and aspects from the reviews and infer their corresponding sentiments. This approach allows one to resonate about specific attributes that make a user like or dislike an item.

There is a set of works exploiting capsule networks for recommendations [16, 22]. Some of them use deep reinforcement learning based on capsule networks [24] formulating the task of sequential interactions between users and recommender systems as a Markov Decision Process. Other approaches use capsule networks to model relationships within knowledge graph entities [22]. We contribute to this direction by introducing a capsule-based document similarity measure designed specifically for ranking exploratory search results.

3   Ranking Model

Our method is based on several design choices from [22] which can be also modified for solving search personalization task. Following this modification, the network from [22] reflects the relationship between the query, the user, and the search results. In our problem setting similarity measure should be universal, meaning that user information is redundant. Thus, we train our network on pairs "query - result", representing both as long texts (documents from the dataset). Trained in such a way, the network may serve as a score function, providing relevance scores for unseen queries and documents.

Following [22] we represent every document as a k-dimensional embedding. Contrary to [22] we do not learn such embeddings from scratch, but instead initialize them with previously learned representations (Doc2vec, fasttext or BERT-based).

Our architecture consists of two capsule layers. The first one includes k capsules, the output of each capsule is then multiplied by a weight matrix and then summed up to produce an input to a second capsule. The capsule then performs squashing and produces the final vector e.

$$squash(s) = \frac{||s||^2}{1 + ||s||^2} \frac{s}{||s||}$$

Please refer to [22] for a more detailed architecture overview.

We learn the aforementioned net using Adam optimizer [13] by minimizing the loss function:

$$L = \sum_{(query, doc)} \log(1 + \exp(\text{rel}(query, doc) * ||e||)),$$

where rel(query, doc) = 1 if doc is related to the query, and 0 otherwise.

4 Experiments

4.1 Datasets

For the experiments we chose two datasets: ArXiv triplets and Microsoft News Datasets (MIND) [23].

ArXiv Triplets Datasets. This dataset includes 963564 papers from https://arxiv.org/, an open repository of scientific articles. The average document length is 4000 words. 20000 triplets are constructed from this documents in such a way that each triplet $[id_1, id_2, id_3]$ has papers $id1$ and $id2$ semantically and topically close to each other and papers id1 and id3 completely opposite.

Microsoft News Dataset (MIND). Microsoft News Dataset (MIND) [23] is a dataset collected from anonymous user logs from Microsoft News. For this study we chose its subset called MIND small that includes 50000 user logs with information about clicks, suggested news, the overall user experience and also detailed information about the news articles itself. The average news articles length is 500 words.

Each user log includes: $history = [id_{h_1}, id_{h_2}, \ldots, id_{h_k}]$ - ids of news articles that the user clicked recently, $suggestion = [id_{s_1}, id_{s_2}, \ldots, id_{s_l}]$ - ids of suggested to the user articles, $impression = [i_{s_1}, i_{s_2}, \ldots, i_{s_l}]$ - user's impression over the articles where $i_{s_j} = 1$, if the user clicked on id_{s_j} and 0 otherwise.

4.2 Data Preprocessing

All the texts were reduced to lower case, stripped of punctuation marks and converted to vectors in several ways:

- Doc2Vec and FastText
  Each of this models was pretrained on ArXiv papers dataset. The embedding size is 200.
- BERT-based representations
  The model was pretrained on English Wikipedia. The embedding size is 768.

Suggested news from each MIND log were sorted in such a way that clicked news appeared in the beginning of the list. Then we deleted the logs with: $|suggestion| < 10$, or $|\{i\,|\,i \in impression, i = 1\}| < 3$, or $|\{i\,|\,i \in impression, i = 0\}| < 3$. Thus, only logs containing a sufficient number of both relevant and irrelevant responses remained in the dataset.

The neural net takes as an input:

$$\begin{bmatrix} [q, a_1, r_1] \\ [q, a_2, r_2] \\ \dots \\ [q, a_n, r_n] \end{bmatrix}$$

where q is a query, $a = (a_1, a_2, \dots, a_n)$ - search result of length n, $r = (r_1, r_2, \dots, r_n)$ - search result relevance (1 if relevant, 0 otherwise).

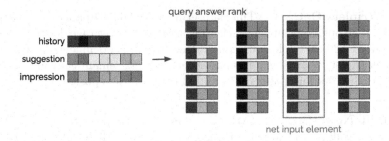

Fig. 1. MIND logs preprocessing

4.3   Evaluation

The model was validated based on ranking metrics that were developed separately for each dataset as well as traditional ranking quality metrics. These metrics take into account special aspects of the data and its structure.

Let E be the set of search elements, S - neural network, $S(e) = S((q, a, r)) = (a', r')$, where (a', r') - re-ranked search results with the corresponding values of relevance.

Validation metrics for arXiv triplets:

– *Recall@1* - the proportion of search answers, for which the relevant text is the first in the network-ranked list.

$$Recall@1(S, E) = \frac{\sum_{e \in E} I\{r'_0 = 1\}}{|E|}$$

Validation metrics for MIND:
Let $k(e)$ - the figure of relevant texts in search output of search element s.

- *Recall@k* - the proportion of answers for which all relevant texts are higher that irrelevant in network-ranked list.

$$Recall@K(S, E) = \frac{\sum\limits_{e \in E} I\{\forall k, l : r'_k = 1, r'_l = 0 \rightarrow k < l\}}{|E|}$$

Metric *Recall@k* generalizes *Recall@1* for any length of search output.
- *Average precision@k* - average value of proportion of relevant texts, ranked by the net for the top of relevant texts.

$$Average\, precision@k(S, E) = \frac{1}{|E|} \sum_{e \in E} \frac{|\{k|r'_k = 1, k < \sum\limits_{j=1}^{n} r'_j\}|}{\sum\limits_{j=1}^{n} r'_j}$$

- *NDCG@M* - normalized discounted cumulative gain at M (used for $M = 5$, 10).

$$NDCG@M = \frac{DCG@M}{IDCG@M}$$

$$IDCG@M = \sum_{m=1}^{M} \frac{1}{log_2(m + 1)} \text{ - because relevance } r \text{ takes only binary values}$$

$$DCG@M = \sum_{m=1}^{M} \frac{r'_m}{log_2(m + 1)}$$

4.4 Training Details

During training on both datasets we used the following parameters: Training parameters:

- Optimiser Adam [13]
- Learning rate $\in \{5e^{-6}, 1e^{-5}, 5e^{-5}, 1e^{-4}, 5e^{-4}\}$
- Number of filters $N \in \{50, 100, 200, 400, 500\}$
- Epoch number is 50 for arXiv triplets and 100 for MIND

The training was performed on one GPU. The speed of training depends on number of filters N and on embedding length. One epoch takes 3–10 min for different sets of parameters on a graphic processor and up to 40 min for CPU. The training was conducted in Google Colaboratory. This service has a limitation for GPU use. Therefore the number of epochs, the range of figures with configured parameters and dimension of embeddings were selected ultimately due to specified limitations.

4.5   Results: ArXiv Triplets Dataset

For arxiv triplets dataset we measured the quality with *Recall@1* metric.

Doc2Vec. Table 1 shows that the quality of the model is better for higher number of filters N and learning rate from considered range. During the training we achieved *Recall@1* $\sim$ 0.9. The best value is equal to 0.905. Learning curves for learning rate = 0.0001 are presented in the Fig. 2.

Table 1. Recall@1, arXiv triplets - Doc2Vec, N - number of filters

	N = 50	N = 100	N = 200	N = 400	N = 500
$5e^{-6}$	0.777	0.819	0.851	0.889	0.890
$1e^{-5}$	0.844	0.882	0.895	0.897	0.896
$5e^{-5}$	0.897	0.904	0.902	0.900	0.897
$1e^{-4}$	0.887	**0.905**	0.902	0.898	0.900
$5e^{-4}$	0.902	0.900	0.900	0.901	0.901

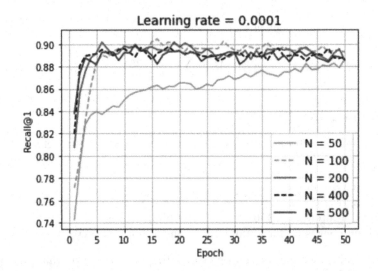

Fig. 2. Learning curve Recall@1 for ArXiv triplets, DocVec

FastText. For the embedding model FastText the best value of *Recall@1* is also observed for higher values of learning rate and N from the considered range (Table 2). The quality of the model is increasing along with the number of filters N because the capsules are able to capture more significant and common text features with a higher number of filters. We achieved *Recall@1* $\sim$ 0.92, the best value is 0.929. Learning curves for learning rate = 0.0005 are shown in Fig. 3.

Table 2. Recall@1, arXiv triplets - FastText

	N = 50	N = 100	N = 200	N = 400	N = 500
$5e^{-6}$	0.713	0.713	0.719	0.763	0.827
$1e^{-5}$	0.722	0.743	0.828	0.828	0.856
$5e^{-5}$	0.792	0.882	0.896	0.910	0.916
$1e^{-4}$	0.840	0.894	0.915	0.923	0.925
$5e^{-4}$	0.919	0.919	0.925	**0.929**	0.928

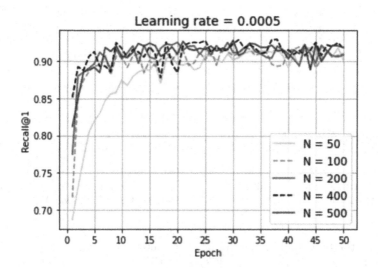

Fig. 3. Learning curve Recall@1 for ArXiv triplets, FastText

4.6   Results: MIND Dataset

Doc2Vec. For dataset MIND and embeddings Doc2Vec we succeed to reach the precision ~ 1. It is worth mentioning that unlike arXiv tripltes here the model trains faster with the fewer amount of filters N. It happens due to the shorter length of MIND texts, allowing capsules to detect and generalize their features faster.

Table 3. MIND - Doc2Vec

	Recall@K				Average precision@K			
	N = 50	N = 100	N = 200	N = 400	N = 50	N = 100	N = 200	N = 400
$1e^{-5}$	0.998	0.991	0.941	0.837	0.999	0.998	0.985	0.954
$5e^{-5}$	**1.0**	0.993	0.892	0.751	**1.0**	0.998	0.972	0.925
$1e^{-4}$	**1.0**	0.997	0.771	0.523	**1.0**	0.999	0.931	0.836
$5e^{-4}$	**1.0**	0.999	0.999	0.99	**1.0**	0.999	0.999	0.999

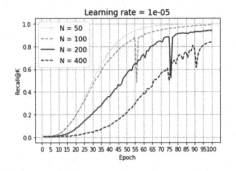

Fig. 4. Learning curve Recall@K for MIND, Doc2Vec

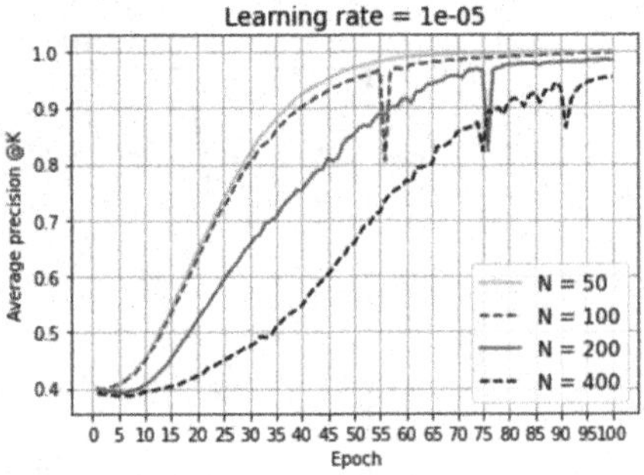

Fig. 5. Learning curve Average precision@K for MIND, Doc2Vec

One of the best combinations of net parameters is *learning rate* $= 1e^{-4}, N = 50$. For this combination the $NDCG@5, NDCG@10$ were calculated as well. Results are - $NDCG@5 = 0.83, NDCG@10 = 0.55$.

However, it is incorrect to compare this results with existing once from MIND official leaderboard, because in this article another problem is solved and the data from MIND dataset was transformed.

FastText. Model reaches *Recall@K* $= 1$ and *Average precision@K* $= 1$. The trend is identical to Doc2Vec model.

Table 4. MIND - FastText

	Recall@K				Average precision@K			
	N = 50	N = 100	N = 200	N = 400	N = 50	N = 100	N = 200	N = 400
$1e^{-5}$	**1.0**	0.996	0.995	0.992	**1.0**	0.999	0.999	0.997
$5e^{-5}$	**1.0**	**1.0**	0.996	0.994	**1.0**	**1.0**	0.999	0.998
$1e^{-4}$	**1.0**	**1.0**	0.999	0.993	**1.0**	**1.0**	0.999	0.998
$5e^{-4}$	**1.0**	**1.0**	**1.0**	**1.0**	**1.0**	**1.0**	**1.0**	**1.0**

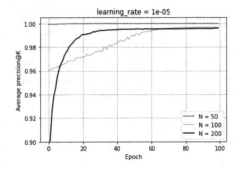 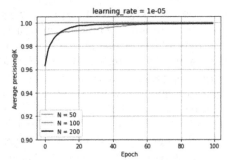

Fig. 6. Learning curve Recall@K for MIND, FastText

Fig. 7. Learning curve Average precision@K for MIND, FastText

BERT. The BERT embeddings are several times longer than Doc2Vec and FastText ones. The training of capsule neural net with such input parameters takes $4-5$ times more time. Therefore, not all combinations of learning rate and N mentioned above were considered. However, we succeeded to get the highest metric value for the selected parameters. For combination of *learning rate* $= 1e^{-4}, N = 50$: $NDCG@5 = 0.528, NDCG@10 = 0.80$.

Table 5. MIND - BERT

	Recall@K		Average precision@K	
	N = 50	N = 100	N = 50	N = 100
$5e^{-6}$	0.946	0.557	0.984	0.839
$1e^{-4}$	**0.999**	0.886	**0.999**	0.965

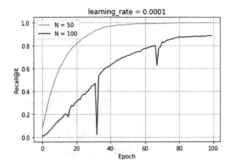 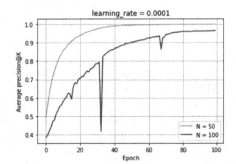

Fig. 8. Learning curve Recall@K for MIND, BERT

Fig. 9. Learning curve Average precision@K for MIND, BERT

The results described above show that capsule neural nets allow to reach high quality of ranking. The difference of the quality of ranking for two considered datasets is also worse noting. The search elements for MIND include a longer search output. Moreover, $Recall@K$ and $Averageprecision@K$ metrics are more strict compared to $Recall@1$. The key difference between datasets stems from text average length: for MIND it is 500, for arXiv is up 4000 words. Due to the limited computational resources, the lengths of embeddings for Doc2Vec and Fasttext were chosen to be 200 for both datasets. One may recall that vectors of such length are not sufficient to successfully encode the information about a long text. We leave this issue for future study.

4.7   Comparison to Alternative Approaches

As a main baseline we used cosine similarity and BM25 ranking function. Let E be the multiple search elements, $C(e) = C((q, a, r)) = (a', r')$ - search output re-ranked in descending order of baseline function to search demand. The quality metrics are calculated identically to capsule net: Recall@1 for ArXiv triplets and Recall@K, AP@K, NDCG@5, NDCG@10 - for MIND dataset. Here K is different for each search element ($K \geq 3$ in all the cases). Unlike the conventional Recall@k metric denotion, we use a slightly different formulation. In our case Recall@K is a proportion of cases for which all relevant texts are ranked higher than irrelevant ones, making K the number of relevant texts in the search output. We discuss metrics in more detail in Sect. 4.3. Final results are presented in Table 6 for ArXiv triplets dataset and in Table 7 for MIND dataset .

It should be pointed out that for arXiv triplets the length of search results is 2, where exactly one text is relevant and one is irrelevant. Therefore, from the metrics listed above only Recall@1 can be calculated for this dataset (Table 6).

Table 6. ArXiv triplets, Capsule Net vs. cosine similarity, Recall@1

	Doc2Vec	Fasttext
Cosine Similarity	0.786	0.855
Capsule Net	**0.901**	**0.928**

We further compare our capsule net based similarity measure with BM25. BM25 function does not use embeddings at all, relying on bag-of-words representation. For arxiv triplets dataset we achieved $Recall@1 = 0.730$, BM25 results for MIND dataset are presented in Table 8. All in all, our proposed similarity measure outperforms BM25 with a huge margin on both datasets. Moreover, BM25 function applied to our task gets a little bit worse results compared even to cosine similarity.

The tables above show that rank quality of cosine similarity and BM25 (Recall@1) is significantly better for arXiv triplets than for MIND. This can be explained by average length of texts from arXiv triplets being much longer

Table 7. MIND, Capsule Net based similarity measure vs. cosine similarity

	Recall@K	AP@K	NDCG@5	NDCG@10
Doc2Vec, cos.sim	0.018	0.449	0.446	0.633
FastText, cos.sim	0.020	0.447	0.440	0.630
BERT, cos.sim	0.021	0.422	0.422	0.593
Doc2Vec, capsule net	**1.0**	**1.0**	**0.552**	**0.831**
FastText, capsule net	**1.0**	**1.0**	**0.551**	**0.831**
BERT, capsule net	0.999	0.999	0.528	0.802

Table 8. MIND, BM25 vs. Capsule Net

	Recall@1	Recall@K	AP@K	NDCG@5	NDCG@10
BM25	0.428	0.015	0.416	0.416	0.620
Capsule Net	**0.928**	**1.0**	**1.0**	**0.552**	**0.831**

compared to MIND dataset. It indicates that cosine similarity is more effective for large texts (e.g. several thousands words).

Besides, we can see the plummeting of ranking quality in terms of $Recall@K$ with the increase of parameter k. When $k = 3$ the precision of ranking decreases to several percents. While cosine similarity is a good baseline, it cannot cope with ranking of long search output up to 1000 words.

The comparison results show that the neural network has a good quality regardless of number of search results that needed to be ranked. Our proposed ranking algorithm is significantly better than methods like cosine similarity or BM25 ranking function.

5 Conclusion

We propose an algorithm for ranking exploratory search results by semantic proximity using capsule neural networks. The presented similarity measure can be seen as a step towards exploratory search systems improvements. We show that a capsule network usage significantly improves the quality of ranking compared to classical approaches, like cosine similarity or BM25 ranking function.

As for the future work, one of the possible directions is studying how the self-attention mechanism in combination with capsule neural networks can solve exploratory search re-ranking problem. Also we plan to further optimize the architecture proposed in [22] for ranking text documents by semantic similarity. Our code is open-sourced and available at https://github.com/AnnNesterenko/capsule_net_ranking.

Acknowledgements. The work is funded by RFBR, project number 20–37-90025.

References

1. Янина, А., Воронцов, К.: Мультимодальные тематические модели для разведочного поиска в коллективном блоге. Машинное обучение и анализ данных **2**(2), 173–186 (2016)
2. Agichtein, E., Brill, E., Dumais, S.: Improving web search ranking by incorporating user behavior information. In: Proceedings of the 29th Annual International ACM SIGIR Conference on Research and Development in Information Retrieval, pp. 19–26 (2006)
3. Beel, J., Gipp, B.: Google scholar's ranking algorithm: an introductory overview. In: Proceedings of the 12th International Conference on Scientometrics and Informetrics (ISSI 2009), vol. 1, pp. 230–241. Rio de Janeiro (Brazil) (2009)
4. Benajiba, Y., Sun, J., Zhang, Y., Jiang, L., Weng, Z., Biran, O.: Siamese networks for semantic pattern similarity. In: 2019 IEEE 13th International Conference on Semantic Computing (ICSC), pp. 191–194. IEEE (2019)
5. Bojanowski, P., Grave, E., Joulin, A., Mikolov, T.: Enriching word vectors with subword information. Trans. Assoc. Comput. Linguist. **5**, 135–146 (2017)
6. Dai, A.M., Olah, C., Le, Q.V.: Document embedding with paragraph vectors. arXiv preprint arXiv:1507.07998 (2015)
7. Devlin, J., Chang, M.W., Lee, K., Toutanova, K.: Bert: Pre-training of deep bidirectional transformers for language understanding. arXiv preprint arXiv:1810.04805 (2018)
8. Ianina, A., Vorontsov, K.: Regularized multimodal hierarchical topic model for document-by-document exploratory search. In: 2019 25th Conference of Open Innovations Association (FRUCT), pp. 131–138. IEEE (2019)
9. Ianina, A., Vorontsov, K.: Hierarchical interpretable topical embeddings for exploratory search and real-time document tracking. Int. J. Embed. Real-Time Commun. Syst. (IJERTCS) **11**(4), 134–152 (2020)
10. Ianina, A., Vorontsov, K.: Multimodal topic modeling for exploratory search in collective blog. J. Mach. Learn. Data Anal. **2**(2), 173–186 (2016)
11. Jiang, T.: Exploratory search: a critical analysis of the theoretical foundations, system features, and research trends. In: Library and Information Sciences, pp. 79–103. Springer, Berlin, Heidelberg (2014). https://doi.org/10.1007/978-3-642-54812-3_7/cover/
12. Beel, J., Gipp, B.: Google scholar's ranking algorithm: an introductory overview, pp. 230–241 (2009)
13. Kingma, D.P., Ba, J.: Adam: a method for stochastic optimization. arXiv preprint arXiv:1412.6980 (2014)
14. Koike, E., Itoh, T.: An interactive exploratory search system for on-line apparel shopping. In: Proceedings of the 8th International Symposium on Visual Information Communication and Interaction, pp. 103–108 (2015)
15. Le, Q., Mikolov, T.: Distributed representations of sentences and documents. In: International Conference on Machine Learning, pp. 1188–1196. PMLR (2014)
16. Li, C., Quan, C., Peng, L., Qi, Y., Deng, Y., Wu, L.: A capsule network for recommendation and explaining what you like and dislike. In: Proceedings of the 42nd International ACM SIGIR Conference on Research and Development in Information Retrieval, pp. 275–284 (2019)
17. Mazzia, V., Salvetti, F., Chiaberge, M.: Efficient-capsnet: capsule network with self-attention routing. arXiv preprint arXiv:2101.12491 (2021)

18. Punjabi, A., Schmid, J., Katsaggelos, A.K.: Examining the benefits of capsule neural networks. arXiv preprint arXiv:2001.10964 (2020)
19. Rahman, M.: Search engines going beyond keyword search: a survey. Int. J. Comput. Appl. **75**(17), 1–8 (2013)
20. Sabour, S., Frosst, N., Hinton, G.E.: Dynamic routing between capsules. arXiv preprint arXiv:1710.09829 (2017)
21. Singh, V., Singh, A.: Learn-as-you-go: feedback-driven result ranking and query refinement for interactive data exploration. Procedia Comput. Sci. **125**, 550–559 (2018)
22. Vu, T., Nguyen, T.D., Nguyen, D.Q., Phung, D., et al.: A capsule network-based embedding model for knowledge graph completion and search personalization. In: Proceedings of the 2019 Conference of the North American Chapter of the Association for Computational Linguistics: Human Language Technologies, Volume 1 (Long and Short Papers), pp. 2180–2189 (2019)
23. Wu, F., et al.: Mind: a large-scale dataset for news recommendation. In: Proceedings of the 58th Annual Meeting of the Association for Computational Linguistics, pp. 3597–3606 (2020)
24. Zhao, C., Hu, L.: CapDRL: a deep capsule reinforcement learning for movie recommendation. In: Nayak, A.C., Sharma, A. (eds.) PRICAI 2019. LNCS (LNAI), vol. 11672, pp. 734–739. Springer, Cham (2019). https://doi.org/10.1007/978-3-030-29894-4_59
25. Zhuang, Z., Cucerzan, S.: Re-ranking search results using query logs. In: Proceedings of the 15th ACM International Conference on Information and Knowledge Management, pp. 860–861 (2006)

Building a Bilingual QA-system
with ruGPT-3

Tatiana Shavrina[1,2,3(✉)] , Dina Pisarevskaya[3] , and Valentin Malykh[4,5]

[1] SberDevices, Sberbank, Moscow, Russia
rybolos@gmail.com
[2] National Research University Higher School of Economics, Moscow, Russia
[3] AI Research Institute, Moscow, Russia
[4] Huawei Noah's Ark lab, Moscow, Russia
valentin.malykh@phystech.edu
[5] Kazan Federal University, Kazan, Russia

Abstract. In this work, we present an approach of cross-lingual transfer learning for English and Russian languages within the QA task. Our approach implies using a generative transformer model that has seen Wikipedia texts in both languages during the pretraining phase and then fine-tuning it with a special token of language, forcing the model to generate texts in a particular language. We are focusing on SQuAD data (English) and updated SberQuAD data (Russian) plus their translations for training and testing, and use ruGPT-3 XL model, which is forced to answer questions in English based on Russian paragraphs and reverse: can answer in Russian when provided information in English. Monolingual QA-abilities of the model are also preserved.

Our results show that the fine-tuned model demonstrates bilingual ability and can generate answers that are close to correct in fuzzy metrics: model generates answers in Russian when based on English texts: 75% named entities ratio, 28% Levenshtein Distance string matching, 28% ROUGE-L; model generates answers in English when based on Russian data: 51% named entities ratio, 27% Levenshtein Distance string matching, 27% ROUGE-L; monolingual Russian quality: 83% named entities ratio, 59% Levenshtein Distance string matching, 57% ROUGE-L; monolingual English quality: 52% named entities ratio, 24% Levenshtein Distance string matching, 25% ROUGE-L.

Keywords: SQuAD · SberQuaD · Cross-lingual QA · Question answering · ruGPT-3

1 Introduction

Question-answering (QA) domain has demonstrated striking progress recently. The usage of large pretrained language models (LMs, e.g. foundation models [9]) has opened up new horizons for creating systems that do not explicitly contain the indexing parts but operate with the knowledge learnt during the pretraining

E. Burnaev et al. (Eds.): AIST 2021, LNCS 13217, pp. 124–136, 2022.
https://doi.org/10.1007/978-3-031-16500-9_11

phase on the large corpora. The foundation models commonly being multilingual (like mBERT [14], mT5 [37], etc.) and competing in their multilingual abilities in various benchmarks (XNLI [13], XGLUE [22], XTREME [33]), can significantly expand the boundaries of what is possible in the main scope of QA - in the search for relevant information. Using information that is easily found in a more resourceful language can significantly increase the availability of the correct answer in another language.

In this work, we demonstrate the results of the bilingual QA-modeling for two distantly relative languages – Russian and English[1].

Using English data as the additional source, we extend the ability of the bilingual generative pretrained transformer model (ruGPT-3) to perform the SQuAD task both in answering in Russian or English and basing both on English or Russian paragraphs. The resulting model shows the ability to generate answers correctly for both languages, yet demonstrates few features inherent in generative readers - answers the question not with an exact citation from the paragraph, but in its "own" words:

- **paragraph:** For adequate monetary policy, the Bank of Canada uses a monetary policy index that combines interest rates and exchange rates. A 3-point exchange rate change is equivalent to a 1-point change in interest rates. In response to economic development, it makes decisions based on these indices. For example, when the Bank has to directly influence monetary policy, the exchange rate may be affected by a change in the daily target financing rate. Thus, an increase in the interest rate often means an increase in capital investment to Canada from abroad as well as a significant dollar position. The Bank of Canada is also trying to maintain monetary conditions consistent with long-term monetary policy objectives. From the time when monetary policy is determined to have a real impact on the economy, there is a slight delay of 18 to 24 months. This delay is due to a long process of consistent interaction, and it is necessary to limit monetary conditions and influence inflation.
- **lang:** rus
- **question:** Какая задержка наблюдается от момента определения денежно-кредитной политики до времени её реального воздействия на экономику?
- **golden answer:** от 18 до 24 месяцев
- **model answer:** 18–24 месяца

The resulting system evaluation is presented in Sect. 5, while training details and data are described in Sects. 4 and 3 correspondingly.

Our data and code are available at https://github.com/RussianNLP/RusEnQA.

---

[1] By common computational phylogenic approach, the Slavic language family has detached from other European families, including Germanic one, at least 4 thousand years ago [29].

2 Previous Work

Usage of large pretrained language models for QA tasks, mostly for English language, was in the research focus in the recent years. Petroni et al. [28] focused on the task of open-domain QA and examined that such language models store an amount of world knowledge: without any fine-tuning, BERT does remarkably well against a supervised baseline. Roberts et al. [32] investigated how to use large pretrained generative models, namely T5 models, with fine-tuning but without using additional knowledge, for open domain QA. Izacard & Grave [17] studied how such models can benefit from retrieving text passages, potentially containing evidence. Kassner & Schütze [20] combined BERT with a traditional information retrieval step and a kNN search over a large datastore of an embedded text collection, it outperformed BERT on cloze-style QA by large margins without any further training, and let recover more factually correct answers. Kassner et al. [19] examined which knowledge is contained in pretrained multilingual language models without any kind of supervised fine-tuning. Wang et al. [35] investigated the performance of BART as knowledge base on closed-book QA dataset that was constructed using SQuAD. Heinzerling & Inui [16] sorted out entity-based limitations of language models as knowledge bases, and Cao et al. [12] explored underlying predicting mechanisms of masked language models as factual knowledge bases. Liu et al. [23] fine-tuned English GPT-2 small model for question-answer generation.

Current studies on question-answering tasks are focused on implementing cross-lingual transfer learning techniques. Ahn et al. [1] translated an answer to the target language for cross-lingual QA; Bos & Nissim [10] used similar approach. Mitamura et al. [26] used translation of keywords extracted from question to get an answer in the target language. Lee & Lee [21] applied transfer learning for cross-lingual QA using training on English data and evaluation on Chinese data. As to generating cross-lingual questions and answers from raw texts, Shakeri et al. [34] proposed a method to generate multilingual question and answer pairs by a generative model (a fine-tuned multilingual T5 model). It included automatically translated samples from English to the target domain. Artetxe et al. [2] experimented on the Cross-lingual Question Answering Dataset (XQuAD) benchmark and studied cross-lingual transferability of monolingual representations of a transformer-based masked language model. XQuAD consists of a subset of 240 paragraphs and 1190 question-answer pairs from SQuAD v1.1 (by Rajpurkar et al. [31]) together with their translations into ten languages. Longpre et al. [24] presented Multilingual Knowledge Questions and Answers (MKQA), an open-domain question answering evaluation set for 26 languages, and, moreover, suggested Multilingual BERT, XLM and XLM-RoBERTa baselines on it, in zero-shot and translation settings. M'hamdi et al. [25] examined a cross-lingual optimization-based meta-learning approach (meta-training from the source language to the target language(s) and meta-adaptation on the same target language(s) for more language-specific adaptation), to learn to adapt to new languages for question answering. Zhou et al. [38] studied unsupervised bilingual lexicon induction for zero-shot cross-lingual transfer for multilingual

QA, in order to map training questions in the source language into those in the target language as augmented training data, which is important for zero-resource languages.

To the best of our knowledge, cross-lingual transfer learning for English and Russian within the question-answering task, using ruGPT-3, was never investigated before.

3   Data

We are consolidating our efforts on an SQuAD task. Therefore, we use SQuAD 2.0 datasets (English) [30] and an updated version of SberQuAD (for Russian) [15] as a starting point[2]. In addition, we also use the machine translation described in Sect. 3.1, as well as the data formatting described further in Sect. 3.2.

Thus, the short formulation of our task boils down to the following: model how to process paragraphs in any (Russian or English) language, answering questions in a language that is required by a special token before the question.

3.1   Data Translation

We have used both datasets, SQuAD and SberQuAD, in a direct fashion for model fine-tuning, still we have conducted additional experiments to improve cross-lingual ability of the system. Thus, we cross-translated SQuAD from English into Russian and SberQuAD into English in the opposite direction, including train and dev parts. We have used publicly available models from University of Helsinki[3] available in HuggingFace Transformers library [36]. These models were trained using MarianNMT [18].

En-Ru model was evaluated on News Translation Task corpora from WMT conference of 2012, 2013, 2015–2019 years, and Tatoeba corpus. The results on these corpora are presented in Table 1[4].

3.2   Data Formatting

Fine-tuning transformer models often requires special text formatting. We perform language model fine-tuning, turning each SQuAD example into the JSON-like line starting with a special beginning-of-sequence token ($\langle s \rangle$), a paragraph in English, special language token of the question&answer (rus/eng), then question and answer, followed with the end-of-sequence special token ($\langle /s \rangle$).

See the resulting formatting scheme below:

---

[2] The SberQuAD dataset was updated in 2021, the major changer included better data preprocessing, QA-validation and extension of the dataset. The version used is available at https://huggingface.co/datasets/sberquad.

[3] https://huggingface.co/Helsinki-NLP.

[4] The results are adopted from: https://huggingface.co/Helsinki-NLP/opus-mt-en-ru and https://huggingface.co/Helsinki-NLP/opus-mt-ru-en.

Table 1. BLEU scores for Helsinki-NLP models for En-Ru and Ru-En pairs on different corpora.

Corpus\Direction	En-Ru	Ru-En
NewsTest2012 [11]	31.1	34.8
NewsTest2013 [6]	23.5	27.9
NewsTest2015 [8]	27.5	30.4
NewsTest2016 [7]	26.4	30.1
NewsTest2017 [27]	29.1	33.4
NewsTest2018 [5]	25.4	29.6
NewsTest2019 [4]	27.1	31.4
Tatoeba [3]	48.4	61.1

⟨s⟩ + "paragraph: " + eng_paragraph+"n nlang: "rus"+ ": " +rus_question+ " answer: "+ rus_answer + ⟨/s⟩

For the baseline model pretraining, the standard SQuAD format was used[5] with the mutual substitution of English and Russian QA-pairs.

4   Modeling Bilingual QA

4.1   Model Training

For the training procedure, we followed the classical train/dev split of the SberQuAD and SQuAD datasets. All the samples of Ru-Ru, En-Ru, Ru-En, En-En QA were shuffled, resulting in 157388 training formatted lines. The training procedure included 8k paragraphs from both languages, with different number of questions per paragraph. The open data for evaluating the models (dev set) included 3858 QA pairs for each language pair: Ru-Ru, En-Ru, Ru-En, En-En.

The base generative model chosen was ruGPT-3 XL[6], which includes 1.3 billion parameters. Unlike smaller models, the model was pretrained on bilingual corpus of Russian and English. At the same time, the English language was represented only by Wikipedia (in the state current at this paper preparation time) and by occasional citations naturally occur in Russian-language texts.

The fine-tuning of the model is described by the following parameters:

- max-files-per-process 20000
- num-layers (number of layers) 24
- seq-length 2048
- max-position-embeddings (maximum sequence length for the model) 2048
- train-iters 20000
- lr (learning rate) 0.000015
- weight-decay 1e-2
- sparse-mode alternating

---

[5] https://simpletransformers.ai/docs/qa-data-formats/.
[6] https://github.com/sberbank-ai/ru-gpts.

4.2   Baseline

We have chosen ruBERT model[7] for the baseline solution in order to provide a cross-lingual comparison.

The ruBERT model is pre-trained using Russian data, with the usage of multilingually-pretrained BERT for initialization. This creates the most similar conditions to compare the model's ability to make cross-lingual generalizations with fine-tuning. The fine-tuning procedure included simpletransformers pipeline[8] with default parameters. A single dataset was passed for model tuning: SQuAD and SberQuAD mutually translated and multiplied with cross-lingual pairs "English paragraph + translated QA in Russian", "Russian paragraph + translated QA in English". After the training, the ruBERT model has demonstrated less cross-lingual ability, extracting the answers (frequently the correct ones) in the language of the paragraph, and not in the language of the question:

- **paragraph:** For adequate monetary policy, the Bank of Canada uses a monetary policy index that combines interest rates and exchange rates. A 3-point exchange rate change is equivalent to a 1-point change in interest rates. In response to economic development, it makes decisions based on these indices. For example, when the Bank has to directly influence monetary policy, the exchange rate may be affected by a change in the daily target financing rate. Thus, an increase in the interest rate often means an increase in capital investment to Canada from abroad as well as a significant dollar position. The Bank of Canada is also trying to maintain monetary conditions consistent with long-term monetary policy objectives. From the time when monetary policy is determined to have a real impact on the economy, there is a slight delay of 18 to 24 months. This delay is due to a long process of consistent interaction, and it is necessary to limit monetary conditions and influence inflation.
- **question:** Что использует Банк Канады для адекватной денежно-кредитной политики?
- **golden answer:** индекс денежно-кредитных условий
- **model answer:** a monetary policy index

The evaluation of the baseline model on the developer set is presented in the next section.

5   Metrics and Evaluation

We applied a set of various evaluation metrics presented in Table 2. In addition to classical strict metrics, such as Exact Match, ROUGE-L, METEOR, we also propose multiple fuzzy metrics that might present different aspects of a generated answer relevancy.

---

[7] base model, cased https://huggingface.co/DeepPavlov/rubert-base-cased.
[8] https://simpletransformers.ai/docs/qa-model/.

In En-Ru setup, question, gold and generated answers are in Russian, and text is in English. In Ru-En setup, question and both answers are in English, and text is in Russian. Ru-Ru and En-En setups are represented by only one language (Russian and English respectively).

As to classical strict metrics, Ru-Ru setup yields the highest exact match values between gold answer and generated answer. Both cross-lingual setups provide better results than the simple En-En setup. ROUGE-L and METEOR metrics results also demonstrate the same patterns. Is worth mentioning that, among string matching metrics, exact match has the lowest scores: generated answers, especially for En-En and cross-lingual setups, may be not presented as an exact citation from the paragraph. Therefore, fuzzy metrics are also implemented.

Among fuzzy string matching metrics, Ru-Ru setup also demonstrates the highest results for matching gold answer and generated answer using Levenshtein Distance[9] (Fuzz ratio in Table 2). Cross-lingual setups have better scores than En-En setup. The same peculiarity also concerns the fuzzy lemmas intersection metric with threshold over 0.7.

We also measured spatial cosine distance between gold answers and generated answers using BERT large model (uncased) for Sentence Embeddings in Russian language[10] for Ru-Ru and En-Ru setups and paraphrase-MiniLM-L12-v2 model[11] for En-En and Ru-En setups. If we compare two latter results, En-En results are lower than cross-lingual Ru-En results, as in other aforementioned metrics. We tried a variety of thresholds, and the option remains with more strict and less strict thresholds.

Finally, we implemented fuzzy metrics based on named entities intersection in gold answers and generated answers. Named entities were extracted using Natasha python package[12]. We estimated named entities ratio for the whole dev set. In addition, we also measured more strictly precision only for generated examples with named entities: an example was considered as true positive if minimum one named entity from the generated answer was presented in the gold answer, and the whole number of examples with named entities in generated answers was considered as true positives + false positives. As in other metrics, Ru-Ru setup yields the best results. But, in contrary to other metrics, En-En setup results for named entities precision are better than cross-lingual setups results.

We compared results for ruGPT-3 XL with baseline results for ruBERT base cased model for the same setups (see Table 3). Although Ru-Ru results are higher, cross-lingual results are reasonably low for this monolingual model. In cross-lingual setups, most of the model answers were presented in the text language (not in the gold answer language).

---

[9] https://github.com/seatgeek/fuzzywuzzy.
[10] https://huggingface.co/sberbank-ai/sbert_large_nlu_ru.
[11] https://huggingface.co/sentence-transformers/paraphrase-MiniLM-L12-v2.
[12] https://github.com/natasha/natasha.

Table 2. Evaluation metrics for Ru-Ru, En-Ru, En-En, Ru-En setups for ruGPT-3 XL.

Metrics for ruGPT-3 XL	Ru-Ru	En-Ru	En-En	Ru-En
Lemmas intersection ratio over 0.7	0.39	**0.12**	0.09	**0.1**
Named entities ratio	0.83	0.75	0.52	0.51
Named entities precision	0.52	0.2	0.32	0.27
Fuzz ratio	0.59	**0.28**	0.24	**0.27**
Exact match	0.32	**0.1**	0.08	**0.1**
METEOR mean	0.48	0.21	0.22	0.22
METEOR ratio over 0.7	0.36	0.11	0.1	0.11
ROUGE-L mean	0.57	**0.28**	0.25	**0.27**
ROUGE-L ratio over 0.7	0.46	0.16	0.12	0.14
SBERT cosine distance 0.4 or less	0.56	0.59	0.31	0.35
SBERT cosine distance 0.1 or less	0.41	0.19	0.13	0.15

Table 3. Evaluation metrics for Ru-Ru, En-Ru, En-En, Ru-En setups for RuBERT.

Metrics for ruBERT base cased	Ru-Ru	En-Ru	En-En	Ru-En
Lemmas intersection ratio over 0.7	0.61	0.02	0.14	0.02
Named entities ratio	0.90	0.74	0.55	0.38
Named entities precision	0.71	0.03	0.3	0.12
Fuzz ratio	0.73	0.03	0.31	0.05
Exact match	0.56	0.02	0.12	0.02
METEOR mean	0.66	0.03	0.24	0.04
METEOR ratio over 0.7	0.58	0.01	0.13	0.01
ROUGE-L mean	0.73	0.06	0.31	0.06
ROUGE-L ratio over 0.7	0.66	0.02	0.18	0.02
SBERT cosine distance 0.4 or less	0.84	0.2	0.42	0.1
SBERT cosine distance 0.1 or less	0.61	0.02	0.22	0.03

6   Discussion and Impact

The following features of the obtained simulation results can be attributed to the subject of discussion:

– The results of these experiments leave two possibilities for the interpretation: has the model learned to translate texts from one language into another, or has it learned to compare two similar sources, two Wikipedias in different languages? To test this hypothesis, we have asked non-wiki questions based on news paragraphs[13]:

---

[13] Source: wikinews shorturl.at/yCSW3.

paragraph: Как передает RegioNews, об этом сообщил первый замести-. тель директора Одесского припортового завода Николай Щуриков. «Завод уже в процессе остановки. На сегодня причины две — переполненный состав карбамида и заоблачная цена сырья (природного газа)», — сообщил он. По его словам, «окно возможностей» для стабилизации работы завода закрылось. Как отмечает Щуриков, пока остается лишь шанс провести приватизацию завода. Напомним, Одесский припортовый завод выпускал свою продукцию с 1978 года. Специализация предприятия — производство аммиака, карбамида и другой химической продукции. Предприятие возобновило свою работу два года назад после длительного простоя.

question: What are two main reasons of the plant shutdown?

ruGPT-3 answer: odessa carbamid and over-billing for raw materials

ruBERT answer: переполненный состав карбамида и заоблачная цена сырья

gold answer: overfilled composition of urea and sky-high price of raw materials (natural gas)

Both models demonstrate QA-abilities, ruBERT making some sort of translation and extracting the correct answer from the text in the source language, while ruGPT-3 also makes the translation and then generates the correct answer in the language of the question. This point should be investigated further using diverse data.

– A separate topic for discussion remains the issue of metrics: of course, an exact match with the answer extracted from the paragraph remains a good metric. However, such a metric turns out to be too strict for generative models, which often give answers that are correct in meaning, but inaccurate in terms of exact match. Therefore, fuzzy metrics can be also implemented, in order to estimate answers that might be considered as correct, but are not exact match.

– Cross-lingual setups presented slightly better results than monolingual English setup, although the model was pre-trained on English texts too. It shows that such setups with English can help improving quality of systems that use Russian language. In further studies, other languages can be used for better search and information retrieval. Other existing models for monolingual experiments on the task might be also explored, to compare the results.

7   Conclusion

We present a cross-lingual approach to question-answering for Russian and English. Using the bilingually pre-trained transformer model, we obtain new results for Russian-language QA as the new potential sources of information emerge. We state the following results for the models obtained:

– the SberQuAD based on Russian data, questions in English - 51% named entities ratio, 27% Levenshtein Distance string matching, 27% ROUGE-L; the SberQuAD based on Russian data, questions in Russian - 83% named entities ratio, 59% Levenshtein Distance string matching, 57% ROUGE-L;

the SQuAD based on English data, questions in Russian - 75% named entities ratio, 28% Levenshtein Distance string matching, 28% ROUGE-L;
the SQuAD based on English data, questions in English - 52% named entities ratio, 24% Levenshtein Distance string matching, 25% ROUGE-L.

– usage of English-exclusive resources for multilingual information extraction purposes can now become possible without direct translation module. This fact is especially important for processing resources presented exclusively in one language, monolingual resources with unique genres and styles, as well as facts that are present in the texts in one language exclusively and for some reason absent in others.

We hope that our work can be useful for the community and we welcome researchers to reproduce our approach for other tasks and implementations outside of the QA field.

Acknowledgements. We would like to express our sincere gratitude to Sarah Caitlin Bennett for very valuable comments and advice on the text structure.

References

1. Ahn, K., Alex, B., Bos, J., Dalmas, T., Leidner, J.L., Smillie, M.B.: Crosslingual question answering using off-the-shelf machine translation. In: Peters, C., Clough, P., Gonzalo, J., Jones, G.J.F., Kluck, M., Magnini, B. (eds.) CLEF 2004. LNCS, vol. 3491, pp. 446–457. Springer, Heidelberg (2005). https://doi.org/10.1007/11519645_44
2. Artetxe, M., Ruder, S., Yogatama, D.: On the cross-lingual transferability of monolingual representations. In: Proceedings of the 58th Annual Meeting of the Association for Computational Linguistics, pp. 4623–4637. Association for Computational Linguistics, Online (2020)
3. Artetxe, M., Schwenk, H.: Massively multilingual sentence embeddings for zeroshot cross-lingual transfer and beyond. Trans. Assoc. Comput. Linguist. **7**, 597–610 (2019). https://doi.org/10.1162/tacl_a_00288, https://aclanthology.org/Q19-1038
4. Barrault, L., et al.: Findings of the 2019 conference on machine translation (wmt19). In: Proceedings of the Fourth Conference on Machine Translation (Volume 2: Shared Task Papers, Day 1), pp. 1–61 (2019)
5. Bojar, O., et al.: Findings of the 2018 conference on machine translation (wmt18)
6. Bojar, O., et al.: Findings of the 2013 workshop on statistical machine translation. In: Proceedings of the Eighth Workshop on Statistical Machine Translation, pp. 1–44. Association for Computational Linguistics, Sofia, Bulgaria (2013). http://www.aclweb.org/anthology/W13-2201
7. Bojar, O., et al.: Findings of the 2016 conference on machine translation. In: Proceedings of the First Conference on Machine Translation, pp. 131–198. Association for Computational Linguistics, Berlin, Germany (2016). http://www.aclweb.org/anthology/W/W16/W16-2301
8. Bojar, O., et al.: Findings of the 2015 workshop on statistical machine translation. In: Proceedings of the Tenth Workshop on Statistical Machine Translation, pp. 1–46. Association for Computational Linguistics, Lisbon, Portugal (2015). http://aclweb.org/anthology/W15-3001

9. Bommasani, R., et al.: On the opportunities and risks of foundation models (2021)
10. Bos, J., Nissim, M.: Cross-lingual question answering by answer translation. In: CLEF (Working Notes). Citeseer (2006)
11. Callison-Burch, C., Koehn, P., Monz, C., Post, M., Soricut, R., Specia, L.: Findings of the 2012 workshop on statistical machine translation. In: Proceedings of the Seventh Workshop on Statistical Machine Translation, pp. 10–51. Association for Computational Linguistics, Montréal, Canada (2012). http://www.aclweb.org/anthology/W12-3102
12. Cao, B., et al.: Knowledgeable or educated guess? revisiting language models as knowledge bases. In: Proceedings of the 59th Annual Meeting of the Association for Computational Linguistics and the 11th International Joint Conference on Natural Language Processing (Volume 1: Long Papers), pp. 1860–1874. Association for Computational Linguistics (2021). https://doi.org/10.18653/v1/2021.acl-long.146, https://aclanthology.org/2021.acl-long.146
13. Conneau, A., et al.: Xnli: Evaluating cross-lingual sentence representations. arXiv preprint arXiv:1809.05053 (2018)
14. Devlin, J., Chang, M.W., Lee, K., Toutanova, K.: Bert: Pre-training of deep bidirectional transformers for language understanding. In: Proceedings of the 2019 Conference of the North American Chapter of the Association for Computational Linguistics: Human Language Technologies, Volume 1 (Long and Short Papers), pp. 4171–4186 (2019)
15. Efimov, P., Chertok, A., Boytsov, L., Braslavski, P.: Sberquad-russian reading comprehension dataset: description and analysis. In: International Conference of the Cross-Language Evaluation Forum for European Languages, pp. 3–15. Springer (2020). https://doi.org/10.1007/978-3-030-58219-7_1
16. Heinzerling, B., Inui, K.: Language models as knowledge bases: on entity representations, storage capacity, and paraphrased queries. In: Proceedings of the 16th Conference of the European Chapter of the Association for Computational Linguistics: Main Volume, pp. 1772–1791. Association for Computational Linguistics (2021). https://aclanthology.org/2021.eacl-main.153
17. Izacard, G., Grave, E.: Leveraging passage retrieval with generative models for open domain question answering. In: Proceedings of the 16th Conference of the European Chapter of the Association for Computational Linguistics: Main Volume, pp. 874–880. Association for Computational Linguistics (2021), https://aclanthology.org/2021.eacl-main.74
18. Junczys-Dowmunt, M., et al.: Marian: fast neural machine translation in C++. In: Proceedings of ACL 2018, System Demonstrations, pp. 116–121. Association for Computational Linguistics, Melbourne, Australia (2018). http://www.aclweb.org/anthology/P18-4020
19. Kassner, N., Dufter, P., Schütze, H.: Multilingual LAMA: Investigating knowledge in multilingual pretrained language models. In: Proceedings of the 16th Conference of the European Chapter of the Association for Computational Linguistics: Main Volume, pp. 3250–3258. Association for Computational Linguistics (2021). https://aclanthology.org/2021.eacl-main.284
20. Kassner, N., Schütze, H.: BERT-kNN: adding a kNN search component to pretrained language models for better QA. In: Findings of the Association for Computational Linguistics: EMNLP 2020, pp. 3424–3430. Association for Computational Linguistics (2020). https://doi.org/10.18653/v1/2020.findings-emnlp.307, https://aclanthology.org/2020.findings-emnlp.307
21. Lee, C.H., Lee, H.Y.: Cross-lingual transfer learning for question answering. arXiv preprint arXiv:1907.06042 (2019)

22. Liang, Y., et al.: Xglue: a new benchmark dataset for cross-lingual pre-training, understanding and generation. arXiv:abs/2004.01401 (2020)
23. Liu, B., Wei, H., Niu, D., Chen, H., He, Y.: Asking questions the human way: scalable question-answer generation from text corpus. In: Proceedings of The Web Conference 2020, pp. 2032–2043. WWW 2020. Association for Computing Machinery, New York, NY, USA (2020). https://doi.org/10.1145/3366423.3380270
24. Longpre, S., Lu, Y., Daiber, J.: MKQA: A linguistically diverse benchmark for multilingual open domain question answering. In: TACL (2020)
25. M'hamdi, M., Kim, D.S., Dernoncourt, F., Bui, T., Ren, X., May, J.: X-METRA-ADA: cross-lingual meta-transfer learning adaptation to natural language understanding and question answering. In: Proceedings of the 2021 Conference of the North American Chapter of the Association for Computational Linguistics: Human Language Technologies, pp. 3617–3632. Association for Computational Linguistics (2021)
26. Mitamura, T., Wang, M., Shima, H., Lin, F.: Keyword translation accuracy and cross-lingual question answering in chinese and Japanese. In: Proceedings of the Workshop on Multilingual Question Answering-MLQA 2006 (2006)
27. Ondrej, B., et al.: Findings of the 2017 conference on machine translation (wmt17). In: Second Conference onMachine Translation, pp. 169–214. The Association for Computational Linguistics (2017)
28. Petroni, F., et al.: Language models as knowledge bases? In: Proceedings of the 2019 Conference on Empirical Methods in Natural Language Processing and the 9th International Joint Conference on Natural Language Processing (EMNLP-IJCNLP), pp. 2463–2473. Association for Computational Linguistics, Hong Kong, China (2019). https://doi.org/10.18653/v1/D19-1250, https://aclanthology.org/D19-1250
29. Petroni, F., Serva, M.: Language distance and tree reconstruction. J. Stat. Mech. Theory Exp. **2008**(08), P08012 (2008)
30. Rajpurkar, P., Jia, R., Liang, P.: Know what you don't know: unanswerable questions for squad. arXiv preprint arXiv:1806.03822 (2018)
31. Rajpurkar, P., Zhang, J., Lopyrev, K., Liang, P.: Squad: 100,000+ questions for machine comprehension of text (2016)
32. Roberts, A., Raffel, C., Shazeer, N.: How much knowledge can you pack into the parameters of a language model? In: Proceedings of the 2020 Conference on Empirical Methods in Natural Language Processing (EMNLP), pp. 5418–5426. Association for Computational Linguistics (2020). https://doi.org/10.18653/v1/2020.emnlp-main.437, https://aclanthology.org/2020.emnlp-main.437
33. Ruder, S., et al.: Xtreme-r: towards more challenging and nuanced multilingual evaluation. arXiv preprint arXiv:2104.07412 (2021)
34. Shakeri, S., Constant, N., Kale, M.S., Xue, L.: Multilingual synthetic question and answer generation for cross-lingual reading comprehension. arXiv preprint arXiv:2010.12008v1 (2020)
35. Wang, C., Liu, P., Zhang, Y.: Can generative pre-trained language models serve as knowledge bases for closed-book QA? In: Proceedings of the 59th Annual Meeting of the Association for Computational Linguistics and the 11th International Joint Conference on Natural Language Processing (Volume 1: Long Papers), pp. 3241–3251. Association for Computational Linguistics (2021). https://doi.org/10.18653/v1/2021.acl-long.251, https://aclanthology.org/2021.acl-long.251

36. Wolf, T., et al.: Transformers: State-of-the-art natural language processing. In: Proceedings of the 2020 Conference on Empirical Methods in Natural Language Processing: System Demonstrations, pp. 38–45. Association for Computational Linguistics (2020). https://doi.org/10.18653/v1/2020.emnlp-demos.6, https://aclanthology.org/2020.emnlp-demos.6
37. Xue, L., et al.: mt5: a massively multilingual pre-trained text-to-text transformer. arXiv preprint arXiv:2010.11934 (2020)
38. Zhou, Y., Geng, X., Shen, T., Zhang, W., Jiang, D.: Improving zero-shot cross-lingual transfer for multilingual question answering over knowledge graph. In: Proceedings of the 2021 Conference of the North American Chapter of the Association for Computational Linguistics: Human Language Technologies, pp. 5822–5834. Association for Computational Linguistics (2021)

Sculpting Enhanced Dependencies for Belarusian

Yana Shishkina[1]([✉]) and Olga Lyashevskaya[1,2]

[1] The National Research University Higher School of Economics,
Myasnitskaya Ulitsa 20, Moscow 101000, Russia
yaashishkina@edu.hse.ru

[2] Vinogradov Russian Language Institute RAS, Moscow,
Russia Volkhonka Street 18/2, Moscow 119019, Russia

Abstract. Enhanced Universal Dependencies (EUD) are enhanced graphs expressed on top of basic dependency trees. EUD support representation of deeper syntactic relations in constructions such as coordination, gapping, relative clauses, argument sharing through control and raising. The paper presents experiments on the EUD parsing of the low-resource Belarusian language, for which no corpus with enhanced annotations was available.

Models trained on the Universal Dependencies treebanks of two closely related Slavic languages, Russian and Ukrainian, were used to parse sentences translated from Belarusian. After that, enhanced dependencies were projected to the original sentences, which gave us ELAS (Enhanced Labeled Attachment Score) 78.1% for both Russian and Ukrainian in evaluation. We also trained the model of one of the IWPT 2020 Shared Task participants on obtained annotations in Belarusian and achieved ELAS 83.4%. Analysis shows that the most common mistakes of cross-lingual parsing root in different theoretical perspectives and practice approaches to the annotation of particular types of clauses in the three Slavic treebanks. Both Russian and Ukrainian EUD transfer models tend to make mistakes when dealing with the predicate argument relations, which are hard to identify without understanding the semantics of the sentence. The alignment method decreases the quality of the annotation by confusing tokens that occur in a sentence more than once.

Keywords: Dependency parsing · Enhanced dependencies · Universal dependencies · Annotation projection · Belarusian

1 Introduction

Enhanced dependencies were introduced to Universal Dependencies (UD) in [1], as deeper syntactic relations which can't be represented by a syntactic tree structure and require a graph. Four main goals of the enhanced universal dependencies (EUD) include adding ellided nodes, propagating relations to conjoined tokens, adding the subject to controlled verbs and specify the information about case, prepositions or conjunctions where needed.

© The Author(s), under exclusive license to Springer Nature Switzerland AG 2022
E. Burnaev et al. (Eds.): AIST 2021, LNCS 13217, pp. 137–147, 2022.
https://doi.org/10.1007/978-3-031-16500-9_12

Not all of the UD treebanks have all or even some types of enhanced dependencies annotated. Two East Slavic treebanks, Russian and Ukrainian, have such annotations; they were improved recently during the IWPT 2020 [2] and 2021 [3] Shared Task. Our goal was to create enhanced annotations for Belarusian, a language closely related to Russian and Ukrainian. In this paper we compare several approaches, including making use of data of a related language, and creating rules that take into account basic Belarusian syntactic dependencies and training a model on top of obtained Belarusian annotations.

2   Related Work

Two methods of enhanced dependency annotation, rule-based and data-driven, were compared in [4]. The research showed that both approaches are applicable for annotation, but the scores of the second approach may be increasing with the introduction of multilingual transformer models for retrieving word embeddings. The majority of IWPT participants such as [5–7] and others, chose to use data-driven methods. Hybrid methods that use rules are still able to show high quality as was proven in [8]. However, all kinds of methods encounter some difficulties while parsing constructions such as coordination, control & raising, and relative clauses [3].

A methods of cross-lingual dependency parsing within closely related language groups were surveyed in [10], for West and South Slavic languages, and in [9] for Scandinavian languages. In [9], authors suggested two different ways of using related language data, delexicalization and annotation projection, with the second approach showing better results. Our work uses an annotation projection very similar to one described in [9], but it is aimed at enhanced graphs instead of base dependency trees.

3   Method

3.1   Rules

We prepared the gold standard by annotating the Belarusian treebank with relatively simple rules, which were similar to those described in [8].

- Tokens with advcl and acl base dependencies, which denote clausal modifiers of nouns and predicates respectively, get additional information about conjunction, which is expressed by mark dependent. If conjunction has fixed dependent, i.e. is a complex (multi-word) conjunction, it is taken as a whole.
- Prepositions are added to oblique nominals and nominal modifiers (obl and nmod) by the same scheme, but instead of mark lemma is taken from case dependent.
- Obl and nmod tokens also get additional information about their case, which is usually extracted from token grammar. If there is nummmod:gov dependent, i.e. the token has a quantifier, case is extracted from quantifier.

- All conjoined tokens marked with `conj` dependency also get the same dependency as the first token in the sequence.
- If a verb has a controlled verb (`xcomp` dependent) and a subject, then the relation between the subject and the controlled verb is also added.
- `Ref` dependency is added to the tokens in a relative clause that have *Pron-Type=Rel* grammatical feature or that belong to the specific set of Belarusian words and the specific part of speech.

We decided to leave the annotation of ellipsis for future research because it is an extensive and self-sufficient work to design rules for the elided node retrieval in a new corpus.

3.2 Cross-Lingual Transfer

Translation. We compared three machine translation services that support Belarusian, Google translate [11], Yandex translate, and Apertium [12]. We manually checked three variants of translations of 50 sentences with length above average and found that Yandex translate suits our task best. Apertium does not translate out-of-vocabulary words, and our task does not allow leaving some tokens untranslated. Google translate tends to drastically change the structure and word order of the sentence, which can decrease the quality of the alignment.

Annotation of the Translated Sentences. The sentences were annotated with basic dependencies using UDPipe [13]. To add enhanced annotations, we chose Alibaba-NLP [14], a model which showed ELAS (Enhanced Labeled Attachment Score) 92.3% for Russian and ELAS 88.0% for Ukrainan in the IWPT Shared Task 2020 coarse post-evaluation.

Alignment. There are some tools for language alignment such as *simalign* [15]. However, experiments on a sample of sentences showed that these tools have a high probability of not giving a word any pair, which is inconvenient, since we know that the three chosen languages are closely related and there are rarely words in one language that will be represented by none in the other.

We aligned tokens based on the cosine similarity of their mBERT [16] embeddings. Next we had to align enhanced dependencies to the original sentences. If there were more than one source language token aligned to the target, we chose the dependency of the head in the established group. If a syntactic group was not established we chose the token with maximum cosine similarity. Conversely when there were less tokens to align, we could assign dependency to the head of the Belarusian group but had to copy the base dependency to other tokens.

Additional Rules. It is clear that conjuncts and prepositions are different in all three languages, so the transfer doesn't help with enhancement of oblique nominals *obl*, nominal modifiers *nmod* and clausal modifiers *advcl, acl*, which require lemma of dependent conjunct or preposition added to the dependency.

To add this enhancement we used the same rules as we used to create the gold corpus. We applied a set of rules to fix the obvious errors:

- punctuation is always labeled *punct*;
- a sentence must have a *root*, i.e. the head of the sentence;
- a token cannot have itself as a head.

For Ukrainian as a source language, language specific relation subtypes such as *xcomp:sp* denoting secondary predication or *nsubj:rel* which is used for the subjects of relative clauses were converted to basic labels.

3.3   Training Belarusian Model

We used Alibaba-NLP system, one of IWPT 2020 participants, to train Belarusian model on data annotated using above rules. The Alibaba-NLP team used neural networks for predicting if the relation is present between a pair of tokens, which were represented by multilingual contextual embeddings XLMR. Neural networks were also used to predict the type of the relation. As the result a connected graph with the most probable relations is chosen.

IWPT 2020 participants aimed to create a universal tool for the majority of languages, so no changes were needed to train the model for annotating enhanced dependencies in the Belarusian treebank. Annotated with rules Belarusian data was used for training the model.

4   Data

We used the following three UD treebanks.

UD_Russian-SynTagRus v.2.7 (1106 k tokens, 62 k sentences), automatically converted to UD [17];

UD_Ukrainian-IU v.2.7 (122 k tokens, 7 k sentences) with native UD annotation [18];

UD_Belarusian-HSE v.2.8 (305 k tokens, 25 k sentences) with native UD annotation.

Enhanced dependencies in the development part (1300 sentences) of the Belarusian treebank were checked manually and used for the analysis.

5   Results

Two metrics are used to evaluate enhanced dependencies. The ELAS (Enhanced Labeled Attachment Score) is F1-metric over arcs and all labels. The EULAS is F1-metric over arcs and universal for all languages labels, so EULAS does not consider errors in language specific labels.

The ELAS and EULAS scores are shown in the Table 1. The results of the transfer from both languages are very similar. The model trained on the rule-based annotated Belarusian data shows better results than both transfer methods. In Sect. 6, we will consider the sources of errors that cause the transfer approach to have lower quality than the the within-language model.

Table 1. Results

Method	ELAS	EULAS
Russian-Belarusian transfer	78.13	79.80
Ukraininan-Belarusian transfer	78.11	79.74
Belarusian: rules + AlibabaNLP	83.43	84.86

We also estimate the precision of enhanced annotations sensu stricto not taking into account those that are copied from basic dependencies. The ratio of dependencies adding new information to the base structure is approximately 30% in all of the three outputs. We calculated the number of EUDs obtained by transfer that matched EUDs in the gold annotations of the treebank and divided it by the number of all enhanced dependencies that do not copy basic ones. The results for Russian-Belarusian and Ukrainian-Belarusian transfer are nearly the same: 0.544 and 0.541 respectively. The metric for the Belarusian model is 0.6. These results confirm that the model trained on Belarusian data performs better than transfer and suggest that it is more difficult to choose the correct label of EUD than identify whether additional dependency is needed at all.

6   Analysis

We divide the mistakes into three groups by their cause. First, we will analyze the errors which were caused by the unique features of a specific treebank. Then we will describe the mistakes which were mostly the expenses of the model or the transfer method.

6.1   Common Mistakes for Russian-Belarusian and Ukrainian-Belarusian Transfer

Rules that we used for annotating Belarusian data add a relation between all of the conjoined tokens and the head of the first token in sequence. Analyzing the differences between rule annotated and transfer annotated data we found out that with transfer annotation conjoined tokens also get the relations with the children of the first token in sequence, not only its head. It seems to us that this additional rule is useful for enhanced annotation and avoiding it in our rules was a mistake.

Transfer approach also revealed that in the Russian and Ukrainian treebanks, modifiers such as nmod get the same dependency as its head, i.e. establish the relation between its heads head, as shown on Fig. 1. The benefits of this rule are not completely clear to us, so adding it to our set of rules is open for discussion.

Our rules help to create a relation between a controlled verb (xcomp) and its semantic subject, which is the same as the syntactic subject of its head verb. It is quite hard to detect whether token is a semantic subject or an object of the controlled verb, so we chose to always use the most probable alternative. Clearly

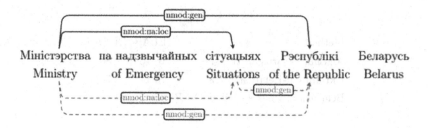

Fig. 1. The relation between `nmod` and its head head

it causes some amount of mistakes, but using the model to add the relations between controlled verb and both its objects and subjects does not help to avoid this problem completely. Both Ukrainian and Russian model tend to confuse object and subject roles when applying this rule. An example is shown on the Fig. 2.

Fig. 2. The relation between controlled verb `xcomp` and its semantic subject

There are also some less important mistakes, since both of the alternatives can be chosen without violating the enhanced UD rules.

The inventory of multi-word prepositions and other bound expressions differs among the three treebanks. As a result, the relation `fixed`, which is used to link words in such syntactic idioms, is not always present in the representation of one language whereas it is there in another in the translation of the same phrase.

The heads of clauses which come after a dash or semicolon, and clauses in parentheses are usually marked as `parataxis` in the Belarusian treebank, but in Russian and Ukrainian they are usually appositional modifiers `appos`. These dependency types show very similar relations according to the UD guidelines, so the confusion of these types will not make a great difference to the understanding of the syntactic structure.

URLs and some emphatic text decoration HTML-tags are deliberately preserved in the texts of the Belarusian treebank, being attached with a special tag for unspecified dependencies, `dep`. After applying the UED transfer the tags get the relation `punct`. Even though HTML-tags often have similar functions as punctuation, it is better to separate those two kinds of tokens.

6.2   The Mistakes of Russian-Belarusian Transfer

Not only modifiers and conjoined tokens get the relations of their head in the Russian treebank. Tokens with `appos` are also connected to its heads head when the enhanced annotation is applied as on the Fig. 3.

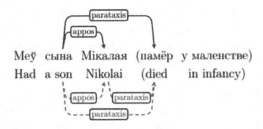

Fig. 3. The relation between `nmod` and its head head

Dependency `acl` is never used to mark the head of the relative clause in the Russian treebank, but it can be seen in some cases in the Ukrainian and Belarusian treebanks. This fact can influence the ability of the transfer method from Russian annotation to add `ref` dependency in all the expected cases in Belarusian. In our opinion, it would be better to unify all of the relative clauses under one dependency `acl:relcl` as it is done in the Russian treebank.

Numerals in parentheses have a `nummod` relation with its head, while in the Belarusian treebank they have the same relation as other clauses in parentheses - `parataxis`. Russian variant is not always accurate since the date or the year in parentheses is not a numeric modifier of its head.

6.3   The Mistakes of Ukrainian-Belarusian Transfer

There are participles that have a `amod` dependencies in Ukrainian but their translations in Belarusian would have an `acl` dependencies. This is related to the fact that in the Ukrainian treebank participles are considered as adjectives and whereas they are verbs in Belarusian and Russian data. So it is clear that adjectives would be adjectival modifiers and verbs would be heads of the clauses.

In the context of passive voice an auxiliary verb in the Belarusian treebank is marked as `aux:pass` since its function is to demonstrate some grammatical features of the main verb. After Ukrainian-Belarusian transfer it is common to find a `cop` dependency in the same context. This kind of dependency is used as a linking verb with non-verb predicates. The cause of this difference is similar to the previous one and has to do with the part of speech of the participles in different treebanks.

In Ukrainian the word *ščo* '*that*' is not considered a referent word in relative clauses, so it does not get `ref` dependency, it is always seen as `mark`. This does not apply to Russian or Belarusian.

One more small but interesting difference is that in the Ukrainian treebank, conjunctions can be the parts of conjoined sequence, too.

6.4   The Mistakes of Alignment Method

Some mistakes occurred because we chose to use the cosine similarity and some of the words vector representations in two languages were similar despite them not being the translations of each other. Most common of them is that punctuation has the wrong head after the transfer.

The same punctuation sign can be used more than once in one sentence and their vectors would be almost the same, so it is clear why their dependencies are confused while aligning the sentence and its translation. Similar mistake happens with other tokens, which tend to be used frequently in one sentence, such as tokens of relations `case`, `mark`, which denote prepositions and conjunctions.

Sometimes it does not have to be exactly the same word form to confuse the alignment script. Dependencies can be assigned wrong for the different forms of the same lemma. We could have avoided this if we used grammatical information that already existed, because we had no goal to convert raw text to enhanced dependency trees.

Our analysis shows that alignment is a part of transfer that is most responsible for producing mistakes. The method of alignment consists of several systems and each of them has its flaws.

Although it is difficult to report an exact ratio of the alignment mistakes due to the fact that the same kind of mistake can be caused by multiple factors, we can roughly estimate their contribution. In general, the wrong choice of the dependency head accounts for ca. 60% of all mistakes. About one half of such mistakes can be explained by the alignment issues. A group of errors in which the predicted label lacks one or more dependency consists almost entirely of alignment mistakes. Summarizing we can say that nearly 40% of all errors are caused by inappropriate alignment.

6.5   The Mistakes of the Model

Part of the differences was not caused by language features or alignment errors, they occurred because the annotation model is not able to predict everything perfectly.

The models for both Russian and Ukrainian tend to choose subject wrong in nominal clauses. It is not always clear what is a predicate and what is a subject in such clauses even for a person, so this kind of contexts is a problematic place for the automatic annotator. One of such cases is illustrated on the Fig. 4.

The models often fail to choose the right subject or object for controlled verbs `xcomp`, because the context can be ambiguous and that can not be fixed without understanding of the semanitcs. Same also applies for the contexts when there must be found a referent for a relative clause - there is not always only one candidate for that role, as on the Fig. 5.

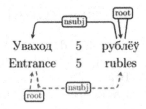

Fig. 4. The confusion between `root` and `nsubj`

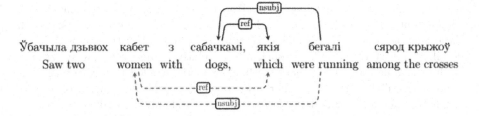

Fig. 5. The incorrectly chosen `ref`

6.6   The Mistakes of the Model Trained on Belarusian Data

We analyzed the differences between the rule annotated data and the data that was annotated by the model trained directly on Belarusian data.

The most common difference was in the head of the punctuation. There are some alternatives: punctuation can depend on the head of the clause or its last token; it can be dependent of the previous clause or the next one, etc. The model chose different alternatives from those in the original Belarusian treebank, but we do not consider this a problem for the understanding of the syntax of the sentence. It is also quite interesting, that punctuation can have more than one head in the model annotation, although it does not occur in any of the Slavic treebanks.

As we mentioned earlier, our rules do not add the relation between all of conjoined tokens and dependents of the first one, but the model does. Moreover, the relations between modifiers such as `nmod` and `amod` and children of their head are added by the model.

We discovered that model struggles to detect bound expressions and does not add the relation `fixed` in all the places needed.

The rules are more successful in adding cases, prepositions and conjunctions, since it is impossible to extract wrong information if it is given. The model predictions are not so accurate and the information can be missed. However, the model is able to predict the case of the words, that do not have this grammatical information, such as abbreviations.

7   Discussion and Concluding Remarks

We have run three experiments: transferred enhanced dependencies from Russian and Ukrainian into Belarusian and annotated enhanced dependencies by the model trained directly on the Belarusian data and compared their results.

The transfer method underperforms the trained model by 5%. The performance of the transfer heavily depends on the accuracy of alignment. Other methods of alignment can be applied in the future.

The limitation of the transfer method lies in the similarity of the language structures and annotations in general. Transfer is best applied to languages with similar syntactic structure, such as three Slavic languages in our case. Although these languages are closely related, annotation in their UD treebanks differ in detail. Inconsistent use of tags such as the relative clause relation `acl:relcl` in the Ukrainian and Belarusian treebanks can be avoided in native basic UD annotation. There are some differences in views on the annotation of certain linguistic phenomena in Slavic languages, which should be considered by the corpus developers and researchers in the future. Such cases include coverage of multi-word expressions, the choice of a part of speech for participles, and more.

Nevertheless, with small adjustments, cross-lingual parsing can be seen a reasonable way of creating the native EUD annotation for Belarusian, for which deeply annotated corpora have not been available so far. This method can also be leveraged in the annotation of the basic UD structures.

Training the model on Belarusian data and using it for annotation showed to be a more efficient method than transfer. The structure and origins of mistakes are not always clear. The annotation with the transfer can be adjusted more easily.

The specific of enhanced dependencies is their diversity, so defining the set of enhancement rules was not an easy task and it is still open for discussion. In our work we studied some of the ambiguous cases of annotation of the Russian, Ukrainian and Belarusian treebanks. Most of the differences were not exactly mistakes but varied views on the language features. We also described some weaknesses of the alignment method and the annotator model. Taking these mistakes and differences into account can help to improve future works on the cross-lingual syntactic annotation.

The annotation resulting from our experiments was made publicly available in the UD_Belarusian-HSE treebank v.2.8.

Acknowledgments. This research was supported in part through computational resources of HPC facilities at HSE University [19].

References

1. Schuster, S., Manning, C.D.: Enhanced English universal dependencies: an improved representation for natural language understanding tasks. In: Proceedings of the Tenth International Conference on Language Resources and Evaluation (2016)

2. Bouma, G., Seddah, D., Zeman, D.D.: Overview of the IWPT 2020 shared task on parsing into enhanced Universal Dependencies. In: Proceedings of the 16th International Conference on Parsing Technologies and the IWPT 2020 Shared Task on Parsing into Enhanced Universal Dependencies (2020)
3. Bouma, G., Seddah, D., Zeman, D.: From raw text to enhanced universal dependencies: the parsing shared task at IWPT 2021. In: Proceedings of the 17th International Conference on Parsing Technologies (2021)
4. Nivre, J., et al.: Enhancing universal dependency treebanks: a case study. In: Proceedings of the Second Workshop on Universal Dependencies (2018)
5. He, H., Choi, J.D.: Adaptation of multilingual transformer encoder for robust enhanced universal dependency parsing. In: Proceedings of the 16th International Conference on Parsing Technologies and the IWPT 2020 Shared Task on Parsing into Enhanced Universal Dependencies (2020)
6. Kanerva, J., Ginter F., Pyysalo S.: Turku enhanced parser pipeline: from raw text to enhanced graphs in the IWPT 2020 shared task. In: Proceedings of the 16th International Conference on Parsing Technologies and the IWPT 2020 Shared Task on Parsing into Enhanced Universal Dependencies (2020)
7. Dehouck, M., Anderson, M., Gómez-Rodríguez, C.: Efficient EUD parsing (2020)
8. Heinecke, J.: Hybrid enhanced Universal Dependencies parsing. In: Proceedings of the 16th International Conference on Parsing Technologies and the IWPT 2020 Shared Task on Parsing into Enhanced Universal Dependencies (2020)
9. Tyers, F., Sheyanova, M., Martynova, A., Stepachev, P., Vinogorodskiy, K.: Multi-source synthetic treebank creation for improved cross-lingual dependency parsing. In: Proceedings of the Second Workshop on Universal Dependencies (2018)
10. Agić, Ž., Tiedemann, J., Merkler, D., Krek, S., Dobrovoljc, K., Moze, S.: Cross-lingual dependency parsing of related languages with rich morphosyntactic tagsets. Association for Computational Linguistics (2014)
11. Wu, Y., et al.: Google's neural machine translation system: bridging the gap between human and machine translation (2016)
12. Forcada, M.L., Tyers, F.: Apertium: a free/open source platform for machine translation and basic language technology. In: Proceedings of the 19th Annual Conference of the European Association for Machine Translation: Projects/Products (2016)
13. Straka, M., Strakova, J.: Tokenizing, pos tagging, lemmatizing and parsing UD 2.0 with UDPipe. In: Proceedings of the CoNLL 2017 Shared Task: Multilingual Parsing from Raw Text to Universal Dependencies (2017)
14. Wang, X., Jiang, Y., Tu, K.: Enhanced universal dependency parsing with second-order inference and mixture of training data (2020)
15. Sabet, M.J., Dufter, P., Yvon, F., Schütze, H.: Simalign: high quality word alignments without parallel training data using static and contextualized embeddings (2021)
16. Devlin, J., Chang, M.W., Lee, K., Toutanova, K.: Bert: Pre-training of deep bidirectional transformers for language understanding (2018)
17. Droganova, K., Lyashevskaya, O., Zeman, D.: Data conversion and consistency of monolingual corpora: Russian UD treebanks. In: Proceedings of the 17th International Workshop on Treebanks and Linguistic Theories (2018)
18. Kotsyba, N., Moskalevskyi, B.: Using transitivity information for morphological and syntactic disambiguation of pronouns in Ukrainian. Visnyk Natsionalnoho universytetu "Lvivska politekhnika". Informatsiyni systemy tamerezhi (2019)
19. Kostenetskiy, P.S., Chulkevich, R.A., Kozyrev, V.I.: HPC resources of the higher school of economics. J. Phys. Conf. Ser. **1740**(1), 012050 (2021). https://doi.org/10.1088/1742-6596/1740/1/012050

Improving Morpheme Segmentation Using BERT Embeddings

Alexey Sorokin[1,2]([✉]) [iD]

[1] Yandex Research, Moscow, Russia
alexey.sorokin@list.ru
[2] Moscow State University, Moscow, Russia

Abstract. We offer a method of incorporating BERT embeddings into neural morpheme segmentation. We show that our method significantly improves over baseline on 6 typologically diverse languages (English, Finnish, Turkish, Estonian, Georgian and Zulu). Moreover, it establishes a new SOTA on 4 languages where language-specific models are available. We demonstrate that the key component of the performance is not only the BPE vocabulary of BERT, but also the embeddings themselves. Additionally, we show that a simpler pretraining task optimizing subword word2vec-like objective also reaches state-of-the-art performance on 4 of 6 languages considered.

1 Introduction

Morpheme segmentation is a task of dividing a word into meaningful segments called morphemes. For example, an English word *pretraining* is segmented as *pre/train/ing*. Typically, in languages like English, Spanish or Russian there is a closed set of affixes (*pre-* and *-ing* in the example above) and large open class of roots (*train*). Being important by itself for theoretical linguistic studies, morpheme segmentation can be helpful for machine translation [7,18] and speech recognition [24]. Recent studies [10] also show that morpheme-based subword segmentation is beneficial for downstream tasks in comparison to conventional byte-pair encoding (BPE) [23].

As any segmentation task, morpheme segmentation can be reduced to sequence labeling by means of BIO (begin-inside-out) or BMES (begin-middle-end-single) schemes. Thus, in principle, any architecture for sequence labeling, such as conditional random fields, convolutional or recurrent neural networks can be used. However, all the supervised approaches suffer from the lack of annotated data: the typical size of training dataset even for widely spoken languages such as English, Finnish or Turkish is about 1000 words[1]. It is partially explained by annotation difficulty: correct location of morpheme boundaries is often problematic and requires linguistic proficiency. Additionally, since the early years [1,8] of NLP morpheme segmentation was traditionally viewed as an unsupervised [12] or minimally supervised task [29]. The extensive usage of unlabeled data

---

[1] See http://morpho.aalto.fi/events/morphochallenge2010/datasets.shtml as example.

E. Burnaev et al. (Eds.): AIST 2021, LNCS 13217, pp. 148–161, 2022.
https://doi.org/10.1007/978-3-031-16500-9_13

is also the main paradigm of modern NLP, however, it is usually realized in a completely different way. Recent large models, such as BERT [3], are pretrained to restore the source text from its incomplete or corrupted version, while unsupervised morpheme segmentation approaches usually optimize the parameters of predefined probabilistic distribution over observed data.

In this paper we treat morpheme segmentation as a supervised labeling task. The most successful previous approaches to this problem include CRFs [21] and CNNs [26]. These methods were also applied in a semi-supervised way by using successor statistics [22] and character language modeling [25]. Both these methods does not take into account word semantics and the structure of related words, which contrasts the way how humans detect morpheme boundaries. The simplest way to invoke semantics is to use a pretrained model, such as BERT [3]. Since the entire model is very large and quickly overfits to small datasets, we only use the freezed subtoken embeddings as an additional input to our convolutional tagging model. It means that we use BERT only as subword encoder and do not utilize its transformer layers. To the best of our knowledge, we do not know any examples of using large pretrained models for morpheme segmentation problem.

We test our approach on 6 typologically diverse languages and observe significant improvements both when using the multilingual and language-specific BERTs. We also perform several ablation studies. The main conclusion of these additional experiments is that the superiority of our approach cannot be attributed solely to the BPE segmentation algorithm of BERT, but the weights of BERT embeddings themselves convey useful information for our problem. We also show, that for several languages a lightweight word2vec-like pretraining achieves comparable scores, also reaching state-of-the-art quality.

2   Task Description

Our training data contain words and their morpheme segmentations, for example

pretrains   pre/train/s.

Segmentation is reduced to sequence labeling by BIO or BMES schemes (Fig. 1). Note that any BIO-string beginning with B corresponds to a sequence of morphemes, while a BMES-sequence requires to satisfy the regular expression $(BM^*E|S)^+$ to encode a segmentation. The latter scheme imposes a restriction on possible model outputs and therefore may potentially rule out incorrect segmentations. Therefore we use the second variant throughout our work.

p r e t r a i n s
B I I B I I I B
B M E B M M M E S

Fig. 1. The BIO (above) and BMES (below) encoding for segmentation pre/train/s

3   Model Description

3.1   Model Architecture

As our basic architecture we use multilayer convolutional neural network[2]. This network takes as input a sequence of letter indexes and then converts it to a sequence of letter embeddings. These embeddings are passed through several convolutional layers. Each such layer includes the convolution operation followed by ReLU activation and optional batch normalization and dropout layers. Thus, the convolutional networks outputs a sequence of context-dependent encodings for each position of the word. Then a linear layer (or a multilayer perceptron) with softmax on its top is used to output a sequence of label distributions. The structure of this model is given in Table 1.

Table 1. The structure of convolutional network. Here D is the letter vocabulary and K is the number of output labels.

Layer	Output shape	Description
Input	$L_{word} * |D|$	Input letters
Embedding	$L_{word} * d_{emb}$	Letter embeddings
Convolutional layer 1	$L_{word} * d_{conv,1}$	Convolution
...	...	...
Convolutional layer k	$L_{word} * d_{conv,k}$	Convolution
Projection layer	$L_{word} * K$	Projection to class logits
Softmax	$L_{word} * K$	Final label probabilities

When using BERT, we additionally encode each letter with BERT-based representation of its context. This representation is obtained as following: we use subtoken embeddings of all character ngrams of length up to $k = 5$ beginning in current position either ending immediately before it. The ngrams which are not present in BERT vocabulary obtain a vector of all zeros. We extend this set with the longest ngram starting in current position or ending in the previous one provided their length exceeds k. If there are no such ngrams, we use the same zero vectors. The described $2k + 2$ embeddings are then concatenated and projected to lower dimension.

Example 1. Consider the Finnish word *adoptioilta*. Finnish BERT representation of its initial position before projection is

$$[\overline{0}_{768}, \overline{0}_{768}, \overline{0}_{768}, \overline{0}_{768}, \overline{0}_{768}, v_a, v_{ad}, v_{ado}, v_{adop}, \overline{0}_{768}, \overline{0}_{768}, v_{adoptio}],$$

where 768 is BERT embedding dimension, $\overline{0}_{768}$ denotes a vector of 768 zeros and [] stands for concatenation. First five vectors are all-zeros since there is no

---

[2] The hyperparameters are in the Appendix.

left context for the first symbol of the word. Since BERT vocabulary does not contain subtoken *adopt*, the tenth position is also zeroed. 11-th vector contains zeros since it stands for the longest ngram ending before current symbol and *adoptio* is the longest ngram which starts in the position under consideration.

Analogously, the vector for the second i in *adoptioilta* is

$$[v_{\#\#o}, v_{\#\#io}, v_{\#\#tio}, \overline{0}_{768}, \overline{0}_{768}, v_{\#\#i}, v_{\#\#il}, \overline{0}_{768}, v_{\#\#ilta}, \overline{0}_{768}, v_{adoptio}, \overline{0}_{768}].$$

Note the $\#\#$ symbol, which denotes the non-initial position in BERT WordPiece tokenization.

To complete this subsection, we would like to emphasize that there is nothing specific to BERT in our feature extraction algorithm, so it could be replaced by any other subword embedder. We explore several variants further in the paper.

3.2 Datasets

We evaluate our approach on 6 languages from different language families used in previous studies on morpheme segmentation: English, Finnish, Turkish (the data from [17]), Estonian, Georgian and Zulu (MorphaGram data from [4]). Since MorphaGram labeled data is not partitioned into train and test sets, we construct the partition by ourselves. We do not select a development set due to small amounts of data available. Dataset parameters are given in Table 2.

Table 2. The size of datasets used for training and evaluation.

	Finnish	Turkish	English	Estonian	Georgian	Zulu
Unlabeled	2928030	617298	878036	49621	50000	50000
Train	1000	1000	1000	1000	750	750
Test	835	763	694	492	250	250

3.3 Implementation and Training

The models are implemented using PyTorch library[3]. Since there is no development set, we train our models for 25 epochs with batch size 32. Cross entropy loss and Adam optimizer are used, all other training parameters are set to default. BERT models are from HuggingFace Transformers[4] repository, except for FinEst BERT [28]. Exact specification is given in Table 7.

As mentioned in Sect. 2, we use BMES encoding as model target. As the network may produce syntactically incorrect output, we extract the valid segmentation encoding with highest probability using Viterbi algorithm.

---

[3] It is freely available https://github.com/AlexeySorokin/MorphemeBert.
[4] https://huggingface.co/models.

4   Experiments and Results

4.1   The Effect of BERT

Our primary goal is to verify that embeddings from BERT are helpful for morpheme segmentation. So we compare the baseline convolutional architecture and the model with subtoken embeddings provided. As strong external baselines we select two models: the CRF model with Harris features from [22], which generally shows the highest performance among non-neural models, and the one-side convolutional network from [25][5]. We also compare with the state-of-the-art semi-supervised extension of the latter model, which was additionally trained on next and previous character prediction. As unsupervised and semi-supervised approaches to morpheme segmentation are widespread, we also compare against the semi-supervised Morfessor[6]. To isolate the role of particular BERT embeddings we also test our architectire with SBEmb subword embeddings from [9]. We evaluate two metrics: the percentage of words with correct segmentation and the F1-measure over internal morpheme boundaries. The results are given in Table 3. All our experiments are averaged over 5 independent runs.

Table 3. Word segmentation accuracy (left) and boundary F1 (right) for baselines and BERT-enhanced segmentation model on different languages. We also report the standard deviation in parentheses below the metric where it is applicable and available. The upper block contains external models, CNN + Character LM is current state-of-the-art.

Model	FIN	TUR	ENG	EST	GEO	ZUL	Average
CRF [22]	39,5 78,6	50,7 82,5	61,1 80,4	60,4 73,5	43,6 77,3	52,0 87,0	51,2 79,9
Morfessor [29]	14,1 53,8	15,6 47,6	42,9 62,6	48,3 59,8	28,8 59,3	4,4 43,2	25,7 54,4
CRF+Harris features [22]	48,9 81,4	51,0 83,2	60,2 78,3	64,8 74,4	39,3 74,8	53,6 88,2	53,0 80,1
Oneside CNN [25]	49,3 83,5	54,3 83,3	61,1 79,5	65,9 75,7	55,8 83,4	59,4 90,2	57,6 82,6
	(1,06) (0,28)	(1,3) (0,57)	(1,1) (0,8)	(0,97) (0,81)	(1,67) (0,57)	(1,59) (0,5)	(1,28) (0,59)
CNN + Character LM [25]	55,1 85,9	58,6 85,4	64,2 82,7	69,1 78,4	**58,5 84,9**	**62,0 91,5**	61,2 84,8
	(1,14) (0,43)	(0,84) (0,19)	(0,97) (0,5)	(0,75) (0,61)	(1,78) (0,8)	(1,5) (0,31)	(1,16) (0,47)
CNN (baseline, [26])	48,4 82,6	55,9 83,9	60,7 80,3	64,8 75,0	55,4 82,5	57,8 89,9	57,2 82,4
	(1,33) (0,51)	(1,02) (0,58)	(1,1) (0,3)	(0,47) (0,46)	(1,97) (0,97)	(1,30) (0,43)	(1,20) (0,54)
+multilingual BERT	54,2 85,6	56,0 84,2	*66,7 83,3*	*70,6 79,5*	56,1 83,4	60,6 90,2	60,7 84,4
	(1,20) (0,47)	(0,69) (0,23)	(0,46) (0,24)	(0,88) (0,85)	(0,64) (0,41)	(1,15) (0,39)	(0,83) (0,43)
+language BERT	**60,3 87,9**	**60,1 85,8**	**67,0 83,6**	**72,5 81,5**	NA NA	NA NA	**62,8 85,4**
	(0,92) (0,24)	(0,68) (0,18)	(0,3) (0,21)	(0,5) (0,26)			(0,7) (0,28)
+BPEmb	45,4 81,4	52,5 83,6	61,5 80,9	65,2 76,0	55,8 83,7	55,3 88,9	55,9 82,4

The key result of our evaluation is that multilingual BERT embeddings actually enhance model performance and language-specific BERT, where available, produces further substantial improvement. The gain for English are smaller due to its abundance in the multilingual BERT training data. Additionally, on all 4

---

[5] https://github.com/AlexeySorokin/NeuralMorphemeSegmentation The models are retrained for 50 epochs with the parameters provided in the repository.

[6] https://morfessor.readthedocs.io/en/latest/.

languages, where the language-specific BERT can be applied, we beat the semi-supervised SOTA. Hence, the usefulness of BERT subtoken embeddings for morpheme segmentation is justified. On the contrary, SBEmb subword embeddings yield no improvement over baseline.

5 Ablation Studies

5.1 Embeddings or Tokenization?

A careful reader may notice that our model benefits from BERT embeddings even in the case of Zulu, which was not present in multilingual BERT training data. Therefore, there is no evidence on the actual source of improvement. There are three main alternatives:

- The model simply benefits from having more parameters.
- BERT dictionary contains frequent letter combinations that often are morphemes.
- BERT embedding of a subtoken carries some morpheme-related information.

To check which explanation is correct we make ngram embeddings random, sampling them from normal distribution tuned on pretrained BERT embeddings. Other ngrams are still mapped to zeros.

Table 4. Evaluation for different variants of BERT embeddings. 'Multi' refers to multilingual BERT and 'Language' to language-specific BERT models. All results are averaged across 5 runs.

Subtokens	Embed.	FIN	TUR	ENG	EST	GEO	ZUL	Average
No	No	48,4 82,6	55,9 83,9	60,7 80,3	64,8 75,0	55,4 82,5	57,8 89,9	57,2 82,4
Multi	Random	51,0 84,2	53,6 83,3	61,0 80,8	67,9 77,1	54,4 82,9	57,9 89,9	57,6 83,0
Multi	Multi	54,2 85,6	56,0 84,2	66,7 83,3	70,6 79,5	56,1 83,4	60,6 90,2	60,7 84,4
Language	Random	54,0 85,2	55,6 83,9	59,5 80,3	68,7 78,0	NA NA	NA NA	58,3 83,4
Language	Language	60,3 87,9	60,1 85,8	67,0 83,6	72,5 81,5	NA NA	NA NA	62,8 85,4

As shown in Table 4, random embeddings perform much worse than pretrained ones, demonstrating that some knowledge about morphemes is encoded in BERT embeddings. However, even multilingual BERT tokenization with random embeddings improves the results for Finnish and Estonian, which implies that the presence in multilingual vocabulary is an important feature for morpheme segmentation at least for these languages. Language-specific BERT vocabulary has slightly stronger effect, as expected.

5.2   BERT or Other Embeddings?

BERT model is pretrained on the task of restoring the missing subtokens and next sentence prediction. So it rarely operates on subword level except for the case when one of the word subtokens is masked and others are not. However, the unlabeled data allows either to train subword embeddings from scratch, finetune the embeddings extracted from the pretrained BERT or finetune the whole BERT model. We consider the following variants:

1. Pass the corpus of words divided into subtokens and train standard word2vec [19] model on it[7]. As in the case of BERT, the word-initial and non-initial versions of the same token (e.g., *it* and *##it* in *commitment*) are treated as different elements of vocabulary.
2. Train word2vec with embedding layer of BERT used as initialization. We do not report the results as they are inferior compared to the previous variant.
3. Finetune BERT on morphology-related task such as morphological tagging to make its embeddings more morphology-oriented. That approach also failed and we do not provide the results.
4. Finetune BERT on the sequences of word subtokens by masking some of them. We decide to mask exactly one subtoken of each word. During finetuning all the layers except the embedding one are frozen.
5. Train word2vec as in the first variant, but also train BPE subword tokenization on the same corpus.

However, the BERT subword tokenizer is suboptimal for this task, as it always tries to produce the shortest segmentation. In particular, a word present in BERT vocabulary is never segmented. To deal with this issue we apply BPE-dropout[8] [20] that flips a coin to decide, whether to merge two shorter subtokens into a larger one. We set the dropout (rejection probability) to 0.3.

Table 5. Evaluation of models finetuned on subword prediction. 'Tokenization' means whether the original BERT is used or we retrain BPE vocabulary from scratch on the unlabeled data. 'Embeddings' refer to subword embeddings training algorithm. 'Finetune' column denotes whether the subword embeddings are used 'as is' (NO) or they are trained on subword prediction task, either during embeddings training in the case of word2vec algorithm or during additional pretraining as in the case of BERT.

Token	Embed.	Fine-tune	FIN	TUR	ENG	EST	GEO	ZUL	Average
BERT	BERT	NO	**60,3 87,9**	**60,1 85,8**	67,0 83,6	**72,5** 81,5	56,1 83,4	60,6 90,2	**62,8 85,4**
BERT	w2v	Subword	59,9 87,7	56,4 84,2	64,6 82,9	72,0 **81,7**	56,2 83,6	59,5 90,1	61,4 85,0
BERT	BERT	Subword	57,3 86,7	58,8 85,5	66,9 84,1	71,4 81,1	54,2 82,7	57,8 90,2	61,1 85,1
BPE	w2v	Subword	57,5 87,0	55,9 83,9	**67,7 84,5**	72,2 80,5	**56,6 83,5**	65,0 91,8	62,5 85,2

Evaluation results are in Table 5. In contrast to our intuition, finetuning BERT on subword prediction either gives no improvement or deteriorates perfor-

---

[7] Training parameters are in the Appendix.
[8] https://github.com/VProv/BPE-Dropout.

mance. Training subword word2vec model is also behind pretrained BERT, however, it produces comparable results despite having less computational requirements[9]. Additionally, when we replace the pretrained BERT vocab with the byte-pair encoding trained from scratch, we achieve significant improvement on English and Zulu, while for other languages the result changes only slightly. Comparing word2vec performance to previous SOTA of [25] from Table 3, we observe that it is beneficial on Finnish, English, Estonian and Zulu.

6 Discussion

The purpose of our ablation studies was to isolate the influence of two factors: subword vocabulary and subword embeddings. Intuitively, the vocabulary is more suitable for morpheme segmentation task if the probability of vocabulary subword to be a morpheme is higher than for an arbitrary subword of the same length. Obviously, the byte-pair encoding contains lots of non-morphs: for example, to obtain a 4-letter morpheme like -*ness* or -*able* by a sequence of merges we need to include into the dictionary some of its substrings, which are not morphemes.

We measure the *morphemeness* of vocabulary by the following approximate probe: we extract all substrings from the words in training and test data and treat the parts of morpheme segmentation as positive instances and other occurrences as negative. Then we simply evaluate the performance of the classifier that returns 1 for the substrings in the vocabulary and 0 otherwise. The results are presented in Table 6. We restrict the evaluation to morphemes of length between 3 and 7 because for shorter segments most of the matches are spurious.

Table 6. Quality metrics for naive vocabulary-based classifiers, classifying as morpheme occurrences only the elements of the corresponding vocabulary.

Language	Length	None			Multilingual			Language			Word2Vec		
		P	R	F1	P	R	F1	P	R	F1	P	R	F1
Finnish	3...7	4.3	100.0	8.2	5.7	34.0	9.8	9.3	61.8	16.2	9.9	49.2	**16.5**
Turkish	3...7	6.0	100.0	11.3	11.2	45.3	18.0	16.4	90.9	**27.8**	16.6	53.3	25.3
English	3...7	5.7	100.0	10.8	10.5	62.1	17.9	13.6	66.2	**26.6**	11.8	53.1	19.3
Estonian	3...7	5.7	100.0	10.8	5.7	30.1	9.6	9.3	72.4	**16.4**	8.2	43.4	13.7
Georgian	3...7	6.1	100.0	11.5	9.1	9.3	9.2	NA	NA	NA	12.2	49.5	**12.5**
Zulu	3...7	4.5	100.0	6.9	36.7	11.7	14.1	NA	NA	NA	12.0	48.3	**19.2**

Coverage evaluation does not provide a complete explanation of our results, however, several conclusions can be made. First, consider the two languages without language-specific BERT, Georgian and Zulu. In the case of Zulu word2vec

---

[9] Obviously, when the BERT weights are available, training is rather cheap. However, the pretraining cost of BERT is several orders of magnitude higher than the one of word2vec-like embeddings.

vocabulary has much higher precision and recall compared to the vocabulary of multilingual BERT. That explains the gains in performance observed in Table 5. On the contrary, language-specific vocabulary for Georgian does not improve much over the multilingual model, thus the gains of word2vec are marginal. It implies that BPE tokenization is suboptimal for Georgian as it does not respect morpheme boundaries. Probably, the relative ineffectiveness of word2vec embeddings for Turkish can be explained by the largest difference in recall between the language-specific BERT and word2vec vocabularies.

However, as our results for random embeddings show, that is not only the vocabulary that is significant: the vectors themselves play even a more important role. The most straightforward way to measure their quality is to use probing method, training a logistic classifier to predict whether a given subword is a morpheme or not. However, this approach is problematic due to the fact that subtoken may occur both as a morpheme or not. One may set a threshold on the percentage of morpheme occurrences of a given subtoken to reduce the task to a binary classification, however, the exact value of a threshold seems to be language- and frequency-dependent. In our preliminary probing experiments we found no clear pattern, therefore we leave this investigation for future work.

Comparing the two proposed algorithms for feature extraction, BERT and word2vec, we note that for truly low-resource languages the second one is beneficial: while training of BERT requires at least millions of sentences and subword embeddings are only its by-product, training the subword embeddings using word2vec objective requires a wordlist of modest size (50000 words in our experiments) and takes several minutes without using GPU. Since word2vec embeddings are of lower dimension (300 instead of 768), the neural model using them is also less prone to overfitting on small datasets.

At the end of this section we point that, according to Table 6, for most languages our subword vocabulary is far from perfect coverage. It arises the question about the applicability of subword tokenization algorithms to morphology-related tasks, since the existing methods were designed basing on purely statistical principles and thus could be suboptimal.

7   Related Work

Unsupervised morpheme segmentation has a long history going back to work of [1,8]. Newer studies include, among others, Morfessor [2,29], Adaptor grammars [4,12] and their semi-supervised extensions. Supervised approaches are less popular due to the lack of annotated data, the best non-neural architecture is probably CRF used in [21,22]. Neural models include, among others, seq2seq [13], LSTMs [5] and variants of CNNs [25,26]. The latter work also utilizes unsupervised pretraining on character language modeling, while [13] uses autoencoding objective for analogous purpose.

Since modern language models operate on the level of subtokens, the related problem of subword-based tokenization has very high importance. The most common approaches are different variants of byte-pair encoding [23] and SentencePiece [16]. The stochastic variants of these methods include BPE-Dropout [20] and subword regularization [15]. As demonstrated in [6], the SentencePiece algorithm is actually a special case of Morfessor algorithm which uses token-based training instead of the type-based and does not impose any prior distribution over subword lexicon.

In contrast to typical "Western" (say, Indo-European or Finno-Ugric) languages, for hieroglyphic languages, such as Chinese, the task of word tokenization is not straightforward. Therefore a similar problem of word segmentation arises. The early neural approaches include LSTM networks [30], however, modern methods are mostly based on pretrained transformers, such as BERT [14] or Roberta [11]. However, the abundance of Chinese word segmentation data (the largest datasets contain several millions characters, which is two-three orders of magnitude larger than the datasets used in these paper) allows the researchers to finetune the whole Transformer model on it and apply complex neural architectures, such as wordhood networks [27], on its top. It completely contrasts the typical setup of morpheme segmentation.

8   Conclusion

We propose two feature-based algorithms for morpheme segmentation, one based on BERT subtoken embeddings and another on word2vec-style subword prediction. They establish a new state-of-the-art for this task, reducing the error by up to 10%. When a language-specific BERT is available, either the BERT embeddings show better performance (for Finnish and Turkish) or the two algorithms perform comparably (on Estonian and English). On the contrary, the word2vec pretraining improves results for Zulu, which was absent in multilingual BERT training data.

Given the failure of our initial attempts to further improve BERT performance by additional pretraining, the question about optimal pretraining algorithm for morpheme-related task is still open. In ideal, such a method should learn morphology in an unsupervised way by means of network weights, not the cooccurrence counts or other raw statistics. We hope that our word2vec-style subword prediction task could be an initial step towards such a method.

Also important, but somewhat orthogonal question is whether downstream tasks that use the segmentation output can benefit from better morpheme boundaries detection. We leave this problem for future work.

Acknowledgements. The author thanks Alexander Panin for helpful discussions and ideas. He is also grateful to Natalia Loukachevitch for communication during preparing this paper.

Appendix

8.1   Preprocessing Details

We remove accents in Finnish training data as it was done during training of Finnish BERT. In all the wordlists we keep only the words containing alphabet characters and hyphens with at least 5 letters.

8.2   BERT Models

Table 7. BERT models used in the paper.

Model name	Languages
bert-base-multilingual-uncased	All
bert-base-finnish-uncased	Finnish
bert-base-english-uncased	English
bert-base-turkish-uncased	Turkish
FinEstBERT	Estonian

All the BERT models used in the paper are either language-specific or mulitilingual BERT-base models. In particular, the embeddings dimension is 768 (Fig. 8).

8.3   Convolutional Network Parameters

Table 8. Model parameters

Parameter	Value	Explored values
Letter embedding size	96	32, 64, 96
BERT projection size	256	128, 256, 384
ReLu after embeddings	Yes	No, Yes
Number of layers	3	2, 3
Layer structure	convolution+batch normalization+ReLU+dropout	No BatchNorm No dropout
Window size	5	3, 5, 7
Number of filters	256	128, 192, 256, 384
Dropout	0.2	0.0, 0.1, 0.2, 0.3

Network parameters are obtained by manual search using word accuracy on Finnish data and are transferred to other languages without change.

8.4   Ablation Studies Details

We train the word2vec model with default parameters from Gensim[10] for 10 epochs on small datasets (50000 words) and for 5 epochs on larger ones. In particular, the algorithm is CBOW, embeddings dimension is 300 and window size

---

[10] https://radimrehurek.com/gensim/models/word2vec.html.

2. We learn subword vocabularies by BPE [23] algorithm with SentencePiece[11] library. We use the command below to train the model:

```
spm_train --input ${INFILE} --model_prefix ${MODEL_PREFIX} \
--vocab_size=8000 --character_coverage=1.0 --model_type bpe
```

References

1. Andreev, N.D. (ed.): Statictical and combinatorial language modelling (Statistiko-kombinatornoe modelirovanie iazykov, in Russian). Nauka (1965)
2. Creutz, M., Lagus, K.: Unsupervised models for morpheme segmentation and morphology learning. ACM Trans. Speech Lang. Process. (TSLP) **4**(1), 3 (2007)
3. Devlin, J., Chang, M.W., Lee, K., Toutanova, K.: Bert: pre-training of deep bidirectional transformers for language understanding. In: Proceedings of the 2019 Conference of the North American Chapter of the Association for Computational Linguistics: Human Language Technologies, Volume 1 (Long and Short Papers), pp. 4171–4186 (2019)
4. Eskander, R., Callejas, F., Nichols, E., Klavans, J.L., Muresan, S.: Morphagram, evaluation and framework for unsupervised morphological segmentation. In: Proceedings of The 12th Language Resources and Evaluation Conference, pp. 7112–7122 (2020)
5. Grönroos, S.A., Virpioja, S., Kurimo, M.: North sámi morphological segmentation with low-resource semi-supervised sequence labeling. In: Proceedings of the Fifth International Workshop on Computational Linguistics for Uralic Languages, pp. 15–26 (2019)
6. Grönroos, S.A., Virpioja, S., Kurimo, M.: Morfessor em+ prune: improved subword segmentation with expectation maximization and pruning. arXiv preprint arXiv:2003.03131 (2020)
7. Grönroos, S.A., Virpioja, S., Kurimo, M.: Transfer learning and subword sampling for asymmetric-resource one-to-many neural translation. arXiv preprint arXiv:2004.04002 (2020)
8. Harris, Z.S.: Morpheme boundaries within words: report on a computer test. In: Papers in Structural and Transformational Linguistics. Formal Linguistics Series, pp. 68–77. Springer, Dordrecht (1970). https://doi.org/10.1007/978-94-017-6059-1_3
9. Heinzerling, B., Strube, M.: BPEmb: Tokenization-free pre-trained subword embeddings in 275 languages. In: chair, N.C.C., et al. (eds.) Proceedings of the Eleventh International Conference on Language Resources and Evaluation (LREC 2018), 7–12 May 2018. European Language Resources Association (ELRA), Miyazaki, Japan (2018)
10. Hofmann, V., Pierrehumbert, J.B., Schütze, H.: Superbizarre is not superb: Improving Bert's interpretations of complex words with derivational morphology. arXiv preprint arXiv:2101.00403 (2021)
11. Huang, K., Huang, D., Liu, Z., Mo, F.: A joint multiple criteria model in transfer learning for cross-domain chinese word segmentation. In: Proceedings of the 2020 Conference on Empirical Methods in Natural Language Processing (EMNLP), pp. 3873–3882 (2020)

---

[11] https://github.com/google/sentencepiece.

12. Johnson, M., Griffiths, T.L., Goldwater, S.: Adaptor grammars: a framework for specifying compositional nonparametric Bayesian models. In: Advances in Neural Information Processing Systems, pp. 641–648 (2007)
13. Kann, K., Mager, M., Meza-Ruiz, I., Schütze, H.: Fortification of neural morphological segmentation models for polysynthetic minimal-resource languages. arXiv preprint arXiv:1804.06024 (2018)
14. Ke, Z., Shi, L., Meng, E., Wang, B., Qiu, X., Huang, X.: Unified multi-criteria Chinese word segmentation with Bert. arXiv preprint arXiv:2004.05808 (2020)
15. Kudo, T.: Subword regularization: improving neural network translation models with multiple subword candidates. In: Proceedings of the 56th Annual Meeting of the Association for Computational Linguistics (Volume 1: Long Papers), pp. 66–75 (2018)
16. Kudo, T., Richardson, J.: Sentencepiece: a simple and language independent subword tokenizer and detokenizer for neural text processing. In: Proceedings of the 2018 Conference on Empirical Methods in Natural Language Processing: System Demonstrations, pp. 66–71 (2018)
17. Kurimo, M., Virpioja, S., Turunen, V., Lagus, K.: Morpho challenge competition 2005–2010: evaluations and results. In: Proceedings of the 11th Meeting of the ACL Special Interest Group on Computational Morphology and Phonology, pp. 87–95. Association for Computational Linguistics (2010)
18. Mager, M., Maier, E., Medina-Urrea, A., Meza-Ruiz, I., Kann, K.: Lost in translation: analysis of information loss during machine translation between polysynthetic and fusional languages. In: Proceedings of the Workshop on Computational Modeling of Polysynthetic Languages, pp. 73–83 (2018)
19. Mikolov, T., Chen, K., Corrado, G., Dean, J.: Efficient estimation of word representations in vector space. arXiv preprint arXiv:1301.3781 (2013)
20. Provilkov, I., Emelianenko, D., Voita, E.: BPE-dropout: simple and effective subword regularization. In: Proceedings of the 58th Annual Meeting of the Association for Computational Linguistics, pp. 1882–1892 (2020)
21. Ruokolainen, T., Kohonen, O., Virpioja, S., Kurimo, M.: Supervised morphological segmentation in a low-resource learning setting using conditional random fields. In: Proceedings of the Seventeenth Conference on Computational Natural Language Learning, pp. 29–37 (2013)
22. Ruokolainen, T., Kohonen, O., Virpioja, S., et al.: Painless semi-supervised morphological segmentation using conditional random fields. In: Proceedings of the 14th Conference of the European Chapter of the Association for Computational Linguistics, Volume 2: Short Papers, pp. 84–89 (2014)
23. Sennrich, R., Haddow, B., Birch, A.: Neural machine translation of rare words with subword units. In: Proceedings of the 54th Annual Meeting of the Association for Computational Linguistics (Volume 1: Long Papers), pp. 1715–1725 (2016)
24. Shafey, L.E., Soltau, H., Shafran, I.: Joint speech recognition and speaker diarization via sequence transduction. arXiv preprint arXiv:1907.05337 (2019)
25. Sorokin, A.: Convolutional neural networks for low-resource morpheme segmentation: baseline or state-of-the-art? In: Proceedings of the 16th Workshop on Computational Research in Phonetics, Phonology, and Morphology, pp. 154–159 (2019)
26. Sorokin, A., Kravtsova, A.: Deep convolutional networks for supervised morpheme segmentation of Russian language. In: Ustalov, D., Filchenkov, A., Pivovarova, L., Žižka, J. (eds.) AINL 2018. CCIS, vol. 930, pp. 3–10. Springer, Cham (2018). https://doi.org/10.1007/978-3-030-01204-5_1

27. Tian, Y., Song, Y., Xia, F., Zhang, T., Wang, Y.: Improving Chinese word segmentation with wordhood memory networks. In: Proceedings of the 58th Annual Meeting of the Association for Computational Linguistics, pp. 8274–8285 (2020)
28. Ulčar, M., Robnik-Šikonja, M.: Finest Bert and crosloengual Bert: less is more in multilingual models. arXiv preprint arXiv:2006.07890 (2020)
29. Virpioja, S., Smit, P., Grönroos, S.A., Kurimo, M., et al.: Morfessor 2.0: Python implementation and extensions for Morfessor Baseline (2013)
30. Yao, Y., Huang, Z.: Bi-directional LSTM recurrent neural network for Chinese word segmentation. In: Hirose, A., Ozawa, S., Doya, K., Ikeda, K., Lee, M., Liu, D. (eds.) ICONIP 2016. LNCS, vol. 9950, pp. 345–353. Springer, Cham (2016). https://doi.org/10.1007/978-3-319-46681-1_42

Training Dataset and Dictionary Sizes Matter in BERT Models: The Case of Baltic Languages

Matej Ulčar[(✉)] [iD] and Marko Robnik-Šikonja [iD]

Faculty of Computer and Information Science, University of Ljubljana,
Ljubljana, Slovenia
{matej.ulcar,marko.robnik}@fri.uni-lj.si

Abstract. Large pretrained masked language models have become state-of-the-art solutions for many NLP problems. While studies have shown that monolingual models produce better results than multilingual models, the training datasets must be sufficiently large. We trained a trilingual LitLat BERT-like model for Lithuanian, Latvian, and English, and a monolingual Est-RoBERTa model for Estonian. We evaluate their performance on four downstream tasks: named entity recognition, dependency parsing, part-of-speech tagging, and word analogy. To analyze the importance of focusing on a single language and the importance of a large training set, we compare created models with existing monolingual and multilingual BERT models for Estonian, Latvian, and Lithuanian. The results show that the newly created LitLat BERT and Est-RoBERTa models improve the results of existing models on all tested tasks in most situations.

Keywords: Natural language processing · BERT · Transformers · Estonian · Latvian · Lithuanian

1 Introduction

Large pretrained language models, based on transformers [27] present the current state of the art in solving natural language processing tasks. These models are trained on large corpora in a self-supervised fashion and are able to learn basic language principles. Because of this and their ability to quickly and successfully adapt to a wide variety of tasks, they have also been (somewhat impetuously) called foundation models [1,12].

The largest, most complex transformer based models have been trained only for the languages with most resources, mainly English, or in a massive multilingual fashion, covering 100 or more languages in a single model. Two examples of such models are GPT-3 [2], and T5 [19]. For most languages, however, there are neither enough training data nor compute resources to train such complex models. Most monolingual models share the architecture of the BERT-base model

© The Author(s), under exclusive license to Springer Nature Switzerland AG 2022
E. Burnaev et al. (Eds.): AIST 2021, LNCS 13217, pp. 162–172, 2022.
https://doi.org/10.1007/978-3-031-16500-9_14

[5], e.g. RuBERT [8] for Russian, FinBERT [28] for Finnish, KB-BERT [11] for Swedish, or RobeCzech [22] for Czech.

It has been shown [28] that monolingual models outperform massive multilingual models, like multilingual BERT (mBERT) [5] or XLM-RoBERTa (XLM-R) [3]. However, as we show, a good monolingual model needs to be trained on a sufficiently large corpus, and focusing on a single language alone is not sufficient for good performance.

In this work, we introduce two new large models and make them publicly available. Est-RoBERTa[1] is a monolingual Estonian model, while LitLat BERT[2] is a trilingual model, trained on Lithuanian, Latvian, and English corpora. This enables us to compare these models with existing smaller monolingual models. We show that the size of training data matters and that when data is scarce, additional corpora from a similar language can aid in the model's performance.

The paper is divided into five sections. In Sect. 2, we present related Lithuanian, Latvian, and Estonian BERT models. In Sect. 3, we present dataset and training for creation of new models. We evaluate the new models and compare them with existing ones in Sect. 4. We draw conclusions and present ideas for further work in Sect. 6.

2   Existing Models for Estonian, Latvian, and Lithuanian

Latvian and Lithuanian are Baltic languages, belonging in the Indo-European language family. Estonian belongs in the Uralic language family and is not related to the other two. However, all three languages share a few common properties. They are all morphologically complex and they are spoken by a relatively small number of speakers, thus few resources are available.

The massive multilingual models, mBERT [5] and XLM-R [3] support all three languages. Estonian is also covered by the multilingual FinEst BERT [26], which was trained on Estonian, Finnish, and English corpora. Few monolingual models have been trained for Estonian, Latvian, and Lithuanian. EstBERT [24] is a monolingual Estonian model, trained on Estonian National Corpus 2017, which consists mostly of web-crawl texts and includes also newspaper texts and Estonian Wikipedia. LVBERT [29] is a monolingual Latvian model, trained on various Latvian corpora, composed mostly of articles from news portals and Latvian Wikipedia. For Lithuanian we only found a LitBERTa-uncased model[3] for which no details have been published.

In Sect. 3, we present two new models. Est-RoBERTa is a new monolingual model for Estonian, which was trained on a larger corpus than EstBERT. LitLat BERT is a trilingual model, trained on Lithuanian, Latvian, and English. The sizes of training corpora for the existing and new models, supporting Estonian, Latvian, and Lithuanian, are displayed in Table 1.

---

[1] https://doi.org/10.15155/9-00-0000-0000-0000-00226L.

[2] http://hdl.handle.net/20.500.11821/42.

[3] https://huggingface.co/jkeruotis/LitBERTa-uncased.

3 LitLat BERT and Est-RoBERTa

In this section, we present the training of the two new models: Est-RoBERTa and LitLat BERT. Trilingual pre-trained models FinEst BERT and CroSloEngual BERT [26], include two closely related languages and English. They have proved to be successful in both monolingual and cross-lingual applications. They are especially useful for knowledge transfer from a high-resource language (i.e. English), when the training data in one of the two target languages is scarce. With the same motivation, we trained LitLat BERT, combining English, Lithuanian, and Latvian.

The existing Estonian monolingual model Est-BERT was trained on a small dataset and is performing unsatisfactory. We trained a new monolingual Estonian model Est-RoBERTa, on a much larger dataset. We opted not to train a combined Baltic-English model, including Estonian, Latvian, Lithuanian, and English, as Estonian is not related to the other two Baltic languages, and its inclusion would lower the model's ability to encode Lithuanian and Latvian.

We present the datasets used to train the models in Sect. 3.1 and the models' architecture and training parameters in Sect. 3.2.

3.1 Datasets and Preprocessing

The sizes of corpora used for training LitLat BERT and Est-RoBERTa are shown in Table 1, along with dataset sizes for related BERT-like models.

Table 1. Training corpora characteristics of existing and new (at bottom) BERT-like models for Estonian, Lithuanian and Latvian. The training corpora sizes are in billions of tokens per language, and the dictionary sizes are in thousands of tokens. Some numbers are unknown (unk) or do not apply (marked with –). The numbers for mBERT are estimated from current sizes of Wikipedia (marked with *), it's not known which version of Wikipedia was used for training the model. The training size for LitBERTa is extrapolated from the given dataset size in gigabytes (marked with †).

Model	Estonian	Latvian	Lithuanian	Total	Dictionary
mBERT	0.05*	0.03*	0.04*	unk	120
XLM-R	0.84	1.20	1.83	295.09	250
Est-BERT	1.15	–	–	1.15	50
FinEst BERT	0.48	–	–	3.70	75
LVBERT	–	0.50	–	0.50	32
LitBERTa	–	–	0.8†	0.8†	128
LitLat BERT	–	0.53	1.21	4.07	84
Est-RoBERTa	2.51	–	–	2.51	40

Est-RoBERTa was trained on a large Estonian corpus, consisting mostly of news articles from Ekspress Meedia, as well as Estonian part of CoNLL 2017 corpus [6].

LitLat BERT was trained on large corpora from three languages: Lithuanian, Latvian, and English. Lithuanian corpora are composed of Lithuanian Wikipedia from 2018[4] [20], Lithuanian part of DGT corpus[5] [21], and LtTenTen14 corpus [7]. Latvian corpora consist of Latvian parts of CoNLL 2017 corpus [6] and DGT corpus(see footnote 5) [21], Saeima corpus [4], and news articles from Ekspress Meedia. For English corpus, we used English Wikipedia from 2018 (see footnote 4) [20].

The numbers in Table 1 show, that new Est-RoBERTa contains more than twice as much Estonian data compared to other models covering Estonian. The new trilingual LitLat BERT contains approximately the same amount of Latvian data as LVBERT but only half of massively multilingual XLM-R. Concerning Lithuanian data, LitLat BERT contains a third less data compared to XLM-R.

Before training, the corpora need to be preprocessed. First, we deduplicated all the corpora and formatted them to contain one sentence per line. We trained two sentencepiece models[6], one for each BERT-like model, on a subset of the whole corpora. For Est-RoBERTa this subset was randomly sampled, while for LitLat BERT, the subset had equal parts of Lithuanian, Latvian, and English text. The sentencepiece models are used to tokenize the corpora into subword tokens and encode them into numeric representations (i.e. assign each token its unique id). The sentencepiece models contain 40 000 and 84 000 subword tokens for Est-RoBERTa and LitLat BERT, respectively. This sizes coincide with the size of the input layer of the model and represents the dictionary (also referred to as vocabulary) size of the model. The dictionary sizes for related models are shown in the rightmost column of Table 1.

3.2 Architecture and Training

LitLat BERT and Est-RoBERTa both use the same architecture as RoBERTa-base model [10]. They are composed of 12 transformer layers, each layer has the size of 768. We used the input sequence length of 512. The training task for both models was masked token prediction. 15% of input tokens were randomly masked, using whole-word masking, meaning that no word was partially seen and partially masked. Both models were trained using the fairseq toolkit [16] for 40 epochs on the corpora presented in Sect. 3.1. We used the Adam optimizer with parameters $\beta_1 = 0.9$ and $\beta_2 = 0.98$ and dropout value of 0.1. We trained both models on 4 Nvidia Tesla V100 GPUs with the batch size of 20 480 tokens per batch per GPU and gradient accumulation with steps of 32, so a total batch size of 2 621 440 tokens. The training took about 13 h per epoch for Est-RoBERTa and about 35 h per epoch for LitLat BERT (altogether approximately 3 weeks and 2 months, respectively).

4   Evaluation

We evaluated Estonian, Latvian, and Lithuanian mono- and multilingual BERT-like models presented in Table 1 on four tasks: named entity recognition (NER),

---

[4] http://hdl.handle.net/11234/1-2735.

[5] http://hdl.handle.net/11356/1197.

[6] https://github.com/google/sentencepiece.

dependency parsing (DP), part-of-speech (POS) tagging, and word analogy (WA). In Sect. 4.1, we describe the evaluation tasks and procedure, and in Sect. 4.2, we present the results of the evaluation.

4.1 Evaluation Settings

We evaluated the models on the NER and POS-tagging tasks by adding a standard classification head on top of the models and fine-tuned the models on each task. We used the classification code from Huggingface's transformers project[7]. We fine-tuned each model for 3 epochs with the batch size 8. For the NER task, we evaluated models on the Estonian NER corpus [9], TildeNER [18] for Lithuanian, and train data of the LV Tagger [17] for Latvian. We limited the scope of the task to the classification of three common named entity classes: persons, locations, and organizations.

For the POS-tagging and DP tasks we used the datasets from the Universal Dependencies project [15]: EDT [13] for Estonian, LVTB for Latvian, and ALKSNIS for Lithuanian. To solve the DP task, we translated the problem to a sequence labeling task [23], using arc-standard algorithm [14] for encoding the dependency trees. We fine-tuned the models for 10 epochs with a batch size 8, using a modified dep2label-bert tool[8].

Traditional WA task measures the distance between word vectors in a given analogy word1 : word2 ≈ word3 : word4 (e.g., man : king ≈ woman : queen). For contextual embeddings such as BERT, the task is modified by using a boilerplate sentence "If the word [word1] corresponds to the word [word2], then the word [word3] corresponds to the word [word4]." We masked [word2] and attempted to predict it using masked token prediction. As the source of analogies, we used the multilingual culture-independent word analogy dataset [25]. We consider an entry correctly predicted if the correct word is among the top 5 most probable predictions.

4.2 Results

We split results according to the task into four subsections: NER, POS, DP, and WA. In the last subsection we attempt to draw conclusions related to the vocabulary size.

NER Results. The results of the NER task are shown in Table 2. We report macro average F_1 scores of the three named entity classes. In summary, our Est-RoBERTa performs the best for Estonian, and our LitLat BERT performs the best for Lithuanian and Latvian. We present language specific observations below.

While Est-RoBERTa performs best on the Estonian NER, Est-BERT has a larger vocabulary size than Est-RoBERTa, but it was trained on a much smaller

---

[7] https://github.com/huggingface/transformers.
[8] https://github.com/EMBEDDIA/dep2label-transformers.

Table 2. The results of NER evaluation task for various BERT models. The scores are macro average F_1 scores of the three named entity classes.

Model	Estonian	Latvian	Lithuanian
mBERT	0.901	0.849	0.809
XLM-R	0.907	0.867	0.793
Est-BERT	0.870	–	–
FinEst BERT	0.925	–	–
LVBERT	–	0.780	–
LitBERTa	–	–	0.630
LitLat BERT	–	**0.875**	**0.847**
Est-RoBERTa	**0.928**	–	–

corpus. In comparison with multilingual FinEst BERT, Est-BERT was trained on more than twice as large Estonian corpus, but the size of its whole training set was less than the third of the FinEst BERT. FinEst BERT performs significantly better than Est-BERT and almost as good as Est-RoBERTa.

We observe similar behaviour on Latvian. LVBERT and LitLat BERT were both trained on 0.5 billion token large Latvian corpus, but LitLat BERT significantly outperforms LVBERT. LitLat BERT was pre-trained also on Latvian and English corpora, thus on a larger overall training set. XLM-R was trained on an even larger total dataset and used larger Latvian corpus than both LVBERT and LitLat BERT. However, XLM-R performs worse than LitLat BERT. The reason might be that XLM-R was trained on 100 languages, therefore its vocabulary for Latvian has to be significantly smaller compared to trilingual LitLat BERT, which is an overall winner on Latvian. The same relation between XLM-R and FinEst BERT is observed on Estonian: FinEst BERT outperforms XLM-R, despite the latter being trained on a larger corpus with more Estonian data.

Similar relation between massively multilingual, trilingual, and inferior monolingual models can be observed for Lithuanian. The mBERT model performs slightly better than XLM-R, but both lag behind LitLat BERT. Monolingual LitBERTa is significant worse than the other three models, but we do not have enough information about its training to explain its behaviour.

POS-tagging Results. The results of the POS-tagging task (Table 3) show similar behaviour as in the NER task. All the models score well on this task, especially XLM-R closes the gap to other models. On Lithuanian, XLM-R performs the best, narrowly beating LitLat BERT. LitLat BERT performs the best on Latvian, and Est-RoBERTa performs the best on Estonian.

DP Results. The results of the DP task are shown in Table 4. Multilingual models, especially massive multilingual models perform significantly worse on this task. Unlike on the NER and POS-tagging task, all monolingual models

Table 3. The results of the POS-tagging evaluation task for various BERT models. The scores are micro average F_1 scores.

Model	Estonian	Latvian	Lithuanian
mBERT	0.966	0.946	0.934
XLM-R	0.970	0.960	**0.964**
Est-BERT	0.961	–	–
FinEst BERT	0.973	–	–
LVBERT	–	0.945	–
LitBERTa	–	–	0.916
LitLat BERT	–	**0.966**	0.961
Est-RoBERTa	**0.977**	–	–

perform very well here. The gap between monolingual and trilingual models is not large, though, especially between LitBERTa and LitLat BERT. The difference between Est-BERT and Est-RoBERTa suggests that training on more data offers significant improvement in performance, but only if the model is monolingual. XLM-R performs poorly, despite being trained on a larger Latvian and Lithuanian corpora than any other model.

Table 4. The results of DP evaluation task for various BERT models. The results are reported as unlabelled attachment score (UAS) and labelled attachment score (LAS).

Model	Estonian		Latvian		Lithuanian	
	UAS	LAS	UAS	LAS	UAS	LAS
mBERT	66.9	56.3	65.4	54.6	56.1	44.1
XLM-R	76.2	68.9	76.5	69.3	65.0	56.6
Est-BERT	82.1	77.4	–	–	–	–
FinEst BERT	80.9	75.7	–	–	–	–
LVBERT	–	–	**80.8**	**75.7**	–	–
LitBERTa	–	–	–	–	68.6	**62.6**
LitLat BERT	–	–	80.4	74.3	**68.7**	61.7
Est-RoBERTa	**83.2**	**78.6**	–	–	–	–

WA Results. We were solving the word analogy task by predicting masked tokens. Because massively multilingual models split words into several subword tokens more frequently than monolingual models, we do not expect them to perform well here, as they need to be able to correctly predict a few consecutive masked tokens. The results in Table 5 confirm our expectation regarding the mBERT model. However, XLM-R performs well and it achieved the second best score in each language. Est-RoBERTa and LitLat BERT significantly outperform all the other models on this task. Surprisingly, LitBERTa performs poorly, although its vocabulary is large.

Table 5. The results of word analogy task, solved as a masked token prediction task for various BERT models. The results are reported as macro average P@5 score.

Model	Estonian	Latvian	Lithuanian
mBERT	0.093	0.026	0.036
XLM-R	0.251	0.118	0.107
Est-BERT	0.165	–	–
FinEst BERT	0.224	–	–
LVBERT	–	0.118	–
LitBERTa	–	–	0.044
LitLat BERT	–	**0.170**	**0.214**
Est-RoBERTa	**0.393**	–	–

Analysis of Vocabulary Size. In Fig. 1, we plot the relative performance gap of each model to the best model for that language against the model's vocabulary size. We observe, that for monolingual models, increasing the vocabulary size at first improves the performance. Further increasing the vocabulary size decreases the model's performance. The exception is the DP task, where larger vocabulary appears to be particularly important. In multilingual models, the larger vocabulary also improves the performance on all tasks. However, the results are strongly correlated with the training dataset sizes (not shown).

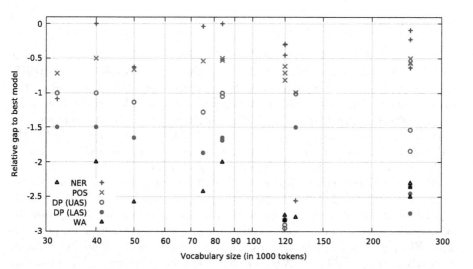

Fig. 1. The results expressed as a relative gap to the best model for each language per task: $1 - \frac{x}{x_{best}}$. The gaps are increased ten-fold for all tasks, except WA, for better readability. The results are offset by task along the y-axis: NER by 0, POS by -0.5, DP (UAS) by -1.0, DP (LAS) by -1.5, WA by -2.0.

To better understand the performance of large pretrained models, we require a joint analysis of both, the training data size and the vocabulary size. As such a study would require building several large models and therefore considerable computational resources, we leave it for further work.

5   Discussion

A performance of large pretrained models, especially multilingual models, depends on several factors, but the impact is not always clear. For example, many models are trained on web-crawl corpora of low quality and this affects the performance. However, such corpora might make the model more robust to the corrupted input (such as misspellings and colloquial language) in an end-task.

A higher number of languages covered by a model negatively correlates with the model's ability to model each language. However, as shown in our study, if models cover more languages, they can be trained on larger corpora, which positively affects the models' performance. It is not clear what is the optimal balance between the two factors. Furthermore, the language similarity must be taken into account when considering these trade-offs.

To study these factors in detail, a large number of different models would have to be trained. The required computing resources exceed our capabilities so we leave these questions for further study.

6   Conclusions

We trained and publicly released two new large pretrained masked language models, a monolingual Estonian Est-RoBERTa, and a trilingual LitLat BERT, trained on Latvian, Lithuanian, and English. Est-RoBERTa improves performance over existing BERT-like models for Estonian on all four evaluation tasks: NER, POS-tagging, DP, and WA. LitLat BERT outperforms related models for Latvian and Lithuanian on three tasks: NER, POS-tagging, and WA. On the DP task it offers comparable performance to the monolingual models, especially on Lithuanian, while outperforming multilingual models.

We have observed that training on more data does not necessarily bring better performance, as XLM-R lags behind monolingual models trained on relatively small corpora. However, a monolingual model's performance can be improved by adding additional training data in other languages. FinEst BERT outperforms Est-BERT, even though it was trained on a much smaller Estonian dataset. Similarly, LitLat BERT and LVBERT were trained on Latvian corpora of similar size, but LitLat BERT significantly outperforms LVBERT on most tasks. The exact benefit of training a model on additional data in other languages over a monolingual model depends on the amount of monolingual data available, and on the evaluation task. On the DP task, a monolingual model is preferred, while on the POS-tagging task, there is little penalty for massively multilingual models.

Despite large requirements for computational resources, it would be interesting to systematically test the main parameters of large pretrained language models such as the number of languages, size of the dictionary, and similarity of included languages.

Acknowledgements. This paper is supported by European Union's Horizon 2020 research and innovation programme under grant agreement No 825153, project EMBEDDIA (Cross-Lingual Embeddings for Less-Represented Languages in European News Media). The results of this publication reflect only the authors' view and the EU Commission is not responsible for any use that may be made of the information it contains. The work was partially supported by the Slovenian Research Agency (ARRS) through the core research programme P6-0411.

References

1. Bommasani, R., et al.: On the opportunities and risks of foundation models. ArXiv preprint 2108.07258 (2021)
2. Brown, T., et al.: Language models are few-shot learners. In: Advances in Neural Information Processing Systems. vol. 33, pp. 1877–1901 (2020)
3. Conneau, A., et al.: Unsupervised cross-lingual representation learning at scale. arXiv preprint arXiv:1911.02116 (2019)
4. Dargis, R., Auziņa, I., Bojārs, U., Paikens, P., Znotiņš, A.: Annotation of the corpus of the Saeima with multilingual standards. In: Proceedings of the Eleventh International Conference on Language Resources and Evaluation (LREC) (2018)
5. Devlin, J., Chang, M.W., Lee, K., Toutanova, K.: BERT: pre-training of deep bidirectional transformers for language understanding. In: Proceedings of the 2019 Conference of the North American Chapter of the Association for Computational Linguistics: Human Language Technologies, Volume 1 (Long and Short Papers), pp. 4171–4186 (2019). https://doi.org/10.18653/v1/N19-1423
6. Ginter, F., Hajič, J., Luotolahti, J., Straka, M., Zeman, D.: CoNLL 2017 shared task - automatically annotated raw texts and word embeddings (2017). http://hdl.handle.net/11234/1-1989, LINDAT/CLARIN digital library
7. Jakubíček, M., Kilgarriff, A., Kovář, V., Rychlý, P., Suchomel, V.: The TenTen corpus family. In: 7th International Corpus Linguistics Conference CL, pp. 125–127 (2013)
8. Kuratov, Y., Arkhipov, M.: Adaptation of deep bidirectional multilingual transformers for Russian language. ArXiv preprint arXiv:1905.07213 (2019)
9. Laur, S.: Nimeüksuste korpus. Center of Estonian Language Resources (2013)
10. Liu, Y., et al.: Roberta: a robustly optimized BERT pretraining approach. ArXiv preprint 1907.11692 (2019)
11. Malmsten, M., Börjeson, L., Haffenden, C.: Playing with Words at the National Library of Sweden - Making a Swedish BERT. ArXiv preprint 2007.01658 (2020)
12. Marcus, G., Davis, E.: Has AI found a new foundation? The Gradient, 11 September 2021 (2021)
13. Muischnek, K., Müürisep, K., Puolakainen, T.: Estonian dependency treebank: from constraint grammar tagset to universal dependencies. In: Proceedings of LREC (2016)
14. Nivre, J.: Algorithms for Deterministic Incremental Dependency Parsing. Comput. Linguist. **34**(4), 513–553 (2008). https://doi.org/10.1162/coli.07-056-R1-07-027

15. Nivre, J., Abrams, M., Agić, Ž.: Universal dependencies 2.3 (2018). http://hdl.handle.net/11234/1-2895
16. Ott, M., et al.: Fairseq: a fast, extensible toolkit for sequence modeling. In: Proceedings of NAACL-HLT 2019: Demonstrations (2019)
17. Paikens, P., Auziņa, I., Garkaje, G., Paegle, M.: Towards named entity annotation of Latvian national library corpus. Front. Artif. Intell. Appl. **247**, 169–175 (2012). https://doi.org/10.3233/978-1-61499-133-5-169
18. Pinnis, M.: Latvian and Lithuanian named entity recognition with TildeNER. In: Proceedings of the 8th International Conference on Language Resources and Evaluation LREC 2012, pp. 1258–1265 (2012)
19. Raffel, C., et al.: Exploring the limits of transfer learning with a unified text-to-text transformer. J. Mach. Learn. Res. **21**, 1–67 (2020)
20. Rosa, R.: Plaintext Wikipedia dump 2018 (2018). http://hdl.handle.net/11234/1-2735, LINDAT/CLARIAH-CZ digital library
21. Steinberger, R., Eisele, A., Klocek, S., Pilos, S., Schlüter, P.: DGT-TM: a freely available translation memory in 22 languages. In: Proceedings of the 8th International Conference on Language Resources and Evaluation LREC (2012)
22. Straka, M., Náplava, J., Straková, J., Samuel, D.: RobeCzech: Czech RoBERTa, a monolingual contextualized language representation model (2021)
23. Strzyz, M., Vilares, D., Gómez-Rodríguez, C.: Viable dependency parsing as sequence labeling. In: Proceedings of the 2019 Conference of the North American Chapter of the Association for Computational Linguistics: Human Language Technologies, Volume 1 (Long and Short Papers), pp. 717–723 (2019). https://doi.org/10.18653/v1/N19-1077
24. Tanvir, H., Kittask, C., Sirts, K.: EstBERT: a pretrained language-specific BERT for Estonian. arXiv preprint 2011.04784 (2020)
25. Ulčar, M., Vaik, K., Lindström, J., Dailidėnaitė, M., Robnik-Šikonja, M.: Multilingual culture-independent word analogy datasets. In: Proceedings of the 12th Language Resources and Evaluation Conference, pp. 4067–4073 (2020)
26. Ulčar, M., Robnik-Šikonja, M.: FinEst BERT and CroSloEngual BERT. In: Sojka, P., Kopeček, I., Pala, K., Horák, A. (eds.) TSD 2020. LNCS (LNAI), vol. 12284, pp. 104–111. Springer, Cham (2020). https://doi.org/10.1007/978-3-030-58323-1_11
27. Vaswani, A., et al.: Attention is all you need. In: Advances in Neural Information Processing Systems, pp. 5998–6008 (2017)
28. Virtanen, A., et al.: Multilingual is not enough: BERT for Finnish. arXiv preprint arXiv:1912.07076 (2019)
29. Znotiņš, A., Barzdiņš, G.: LVBERT: transformer-based model for Latvian language understanding. In: Human Language Technologies-The Baltic Perspective: Proceedings of the Ninth International Conference Baltic HLT 2020, vol. 328, p. 111 (2020)

Computer Vision

Development of a Method for Iris-Based Person Recognition Using Convolutional Neural Networks

Yulia Ganeeva[1]([⊠]) and Evgeny Myasnikov[1,2] [ID]

[1] Samara National Research University, Moskovskoye Shosse 34, 443086 Samara, Russia
jganeeva99@gmail.com
[2] IPSI RAS – Branch of the FSRC "Crystallography and Photonics" RAS, Molodogvardeyskaya 151, 443001 Samara, Russia

Abstract. Recently, convolutional neural networks have become the most commonly used technique for solving many computer vision problems. The classical procedure for the recognition of a person by an iris image consists of the following stages: segmentation, normalization, extraction of indicative representation, and classification. This paper presents the method for iris-based person recognition, in which convolutional neural networks are used at the stages of segmentation and feature extraction. We provide a comparison to alternative iris segmentation approaches and share ground-truth segmentation masks obtained by manual processing of a known open dataset. Besides, we study several feature extraction and classification techniques to achieve the best quality indicators. The results of experimental research have shown that the use of convolutional neural networks allows us to achieve a high quality of irisbased person recognition. In parity on the dataset MMU IrisDatabase recognition accuracy of 99.78% was achieved.

Keywords: Iris segmentation · Person recognition · Convolutional neural networks

1 Introduction

The task of identifying a person on the basis of biometric data is one of the most urgent tasks in the field of information security. The use of biometric data as identification material makes the identification process more reliable. Recently, facial identification [1, 2], hand geometry [3, 4], fingerprints [5, 6], and other biometric data have become common.

Iris identification is one of the most promising identification technologies. The prospective use of the iris as a biometric material is conditioned by the following properties [7]:

1. The structure of the iris is virtually unaffected by time.
2. The iris is present in almost every person.

E. Burnaev et al. (Eds.): AIST 2021, LNCS 13217, pp. 175–189, 2022.
https://doi.org/10.1007/978-3-031-16500-9_15

3. The iris is a unique identifier, meaning that the probability of matching the iris of two different individuals is very small (~10–78) [8].
4. There are methods for extracting and presenting meaningful iris information that allow for further identification with relatively good accuracy.

Traditionally, solving the problem of iris-based person identification, the following stages are distinguished: segmentation, normalization, extraction of a feature representation, classification.

The segmentation phase is one of the most important steps in the considered task. The quality of segmentation directly influences the accuracy of the system as a whole. Since the iris is an annular region between the pupil and sclera, many works reduce the segmentation stage to the approximation of the internal and external iris boundaries by circles. To implement this step, researchers use the Hough transform for circles detection [9–16], the Daugman integral differential operator [17–19], and the boundary point analysis method [20]. However, such approaches do not allow localizing noises caused by eyelashes, eyelids, and the appearance of glare. The interest of the researchers was therefore focused on the development of a segmentation method that would ensure a higher segmentation quality.

The development of neural network technologies has made it possible to have a fresh look at many problems. This included successful attempts to address iris segmentation using convolutional neural networks [21–27].

The normalization stage in many works is realized using Daugman normalization [11–17].

To form a feature representation of the iris, Gabor filters [10–12, 28], log-Gabor filters [29], discrete cosine/sinusoidal transforms [28, 30], wavelet transforms [14, 28, 30], and other techniques have been used. In some works, the downscaling phase [13, 15] follows the feature extraction.

To implement the classification stage in modern works, use is made of the random forest [14], classifier of k-nearest neighbors [11], support vector machine [11, 12, 15], classification algorithm AdaBoost [30], discriminatory analysis [29] and neural network approach [11, 29].

Due to the fact that in recent years neural network methods of deep learning have gained wide popularity in solving problems in the field of computer vision, the application of such methods at various stages of solving the problem under consideration and the comparison of such methods with existing classical approaches becomes an urgent task.

This paper presents a method for iris-based person identification, in which neural networks are used at the stages of segmentation and extraction of a feature representation.

2   Method

2.1   Standard Identification Scheme

The standard scheme for the implementation of the iris personality identification process is shown in Fig. 1.

For the method to work, a digital image of a human iris is required, which can be obtained, for example, using a camera in the infrared range. The image preprocessing stage involves converting an RGB image into a grayscale with subsequent scaling. A preprocessed image is input to the segmentation algorithm, which produces a binary segmentation mask, where a pixel value of 1 indicates that the pixel belongs to the iris, and a value of 0 indicates that the pixel does not belong to the iris. The normalization step following the segmentation transforms a localized region of the iris into a standard form so that iris features have the same spatial location.

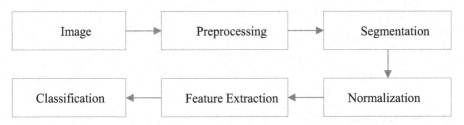

Fig. 1. Schematic representation of the iris-based identification procedure.

The normalized image is redundant in order to feed it to a classification algorithm. In this regard, it is necessary to introduce a separate feature extraction stage. The result of this stage is a feature vector that uniquely describes an instance of a specific class and can be fed to a classification algorithm.

2.2 Segmentation of Iris Image

Earlier in [31], we investigated the method of iris segmentation using the convolutional neural network of the U-Net architecture [32]. To improve the quality of existing methods and achieve high segmentation quality indicators, a series of experimental studies was carried out to find the optimal hyperparameters. After choosing the best hyperparameters, the neural network was trained for 100 epochs using the augmentation mechanism.

In experimental studies, geometric transformations (random rotation, random scaling, random cropping, horizontal and vertical reflection), luminance transformations (changes in brightness across the channel by a random value), and some others were applied to the training set.

In Table 1 we have combined quality estimates of segmentation (accuracy [34]) obtained by known methods [25, 27, 33–38] and proposed method [31]. When comparing specific values in the specified table, one should take into account possible differences in the experiment: the masks of true segmentation used in evaluating the masks of true segmentation, methods of splitting sets, etc. For this reason, we consider it important to provide free access to the segmentation masks [44] used for quality assessment in this paper. These masks were obtained by manual per-pixel markup of images in the known open dataset [43].

Table 1. Comparison of the accuracy of segmentation approaches.

Authors of the approach	Accuracy, %
Pathak, M.P. et al. [25]	97,15
Pathak, M.P. et al. [27]	97,15
Hashim, A.T. et al. [33]	98,78
Chirchi, V. et al. [34]	87,00
Khan, T. et al. [35]	99,01
Jan, F. et al. [36]	98,21
Lin, M. et al. [37]	98,18
Wan, H.L. et al. [38]	97,83
The previously proposed method	**99,09**

2.3 Iris Normalization Step

This work considers two approaches to data normalization. The first approach to normalization was proposed by Daugman in the paper [17]. It helps to combat pupil constriction/dilation. The essence of the approach is to convert a localized iris texture from a Cartesian coordinate system to a polar coordinate system.

As a result of the conversion, the normalized image is a rectangle of size 40×240. The width of the image (240) corresponds to the axis of angles, and the height (40) corresponds to the radial axis.

The second approach is the sequence of the following steps:

1. Cropping of the localized area of the iris on its external borders.
2. Zooms the cropped image to 224×224.

Figure 2 gives an example of normalization for the first and second approaches.

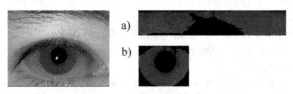

Fig. 2. The original image (left) and the normalization result (right). a) Daugman transform. b) cropping and scaling.

2.4 Feature Extraction

In this work, we studied and compared the following feature systems: Gabor filters [10–12, 28], log-Gabor filters [29] previously trained to recognize the iris of the eye, convolutional neural networks from our earlier work [39].

2.4.1 Gabor 2D Filter

The Gabor filter in image processing is generally used for texture analysis, contour detection, and feature extraction. The use of said method makes it possible to determine whether a component corresponding to a given frequency and direction is present in an image in the neighborhood of a point. In practice, in order to improve the recognition quality, an optimal set of Gabor filters with varying directions and size is used.

In the space domain, the Gabor filter is a harmonic function modulated by the Gaussian function. Let λ be the wavelength of the modulated harmonic function, θ be the orientation of the normal to parallel bands of the Gabor function, ψ be the phase shift of the harmonic function, σ be the standard deviation of the Gaussian function, and γ be the compression factor characterizing the ellipticity of the Gabor function. Then the impulse response of the Gabor filter is constructed according to the following formula [40]:

$$g(x, y; \lambda, \theta, \psi, \sigma, \gamma) = e^{-\left(\frac{x'^2 + \gamma^2 y'^2}{2\sigma^2}\right)} e^{-\left(i\left(2\pi \frac{x'}{\lambda} + \psi\right)\right)}.$$

After constructing a set of Gabor filters, to obtain an indicative representation of the data, the operation of convolution of the normalized image of the iris is carried out with each filter of the set. For further work, the phase component of the result is taken, which for each value is encoded as follows [40]:

1. If the real part is ≥ 0 and the imaginary part is ≥ 0, then the output is "11".
2. If the real part is ≥ 0 and the imaginary part is < 0, then the output is "10".
3. If the real part is < 0 and the imaginary part is ≥ 0, then the output is "01".
4. If the real part is < 0 and the imaginary part is < 0, then the output is "00".

2.4.2 Log-Gabor Filter

The main disadvantage of Gabor filters is the non-zero value of the DC component (the coefficient of the direct Fourier transform with coordinates $(0,0)$). That is, the filter response depends on the average value of the signal. The log-Gabor filter, which is a modification of the Gabor filter, lacks this drawback.

Log-Gabor 1D Filter

The frequency response of a one-dimensional log-Gabor filter is defined as [40]:

$$G(f) = e^{\frac{-\left(\log\left(\frac{f}{f_0}\right)\right)^2}{2\left(\log\left(\frac{g}{f_0}\right)\right)^2}},$$

where f_0 is the central filter frequency, and σ is the width of the filter bandwidth.

The normalized image of the iris is converted to the spectral domain using the Fourier transform and then multiplied line by line by the frequency response of the log Gabor filter. Then the result is translated into the spatial domain using the inverse Fourier transform. The encoding of the resulting representation is performed similarly to the encoding described above in the subsection devoted to the Gabor 2D filter.

Log-Gabor 2D Filter

A 2D log-Gabor filter takes into account not only the frequency component but also the direction component. In the frequency domain, the response of the log-Gabor 2D filter is defined as [40]:

$$G(f, \theta) = e^{\frac{-\left(\log\left(\frac{f}{f_0}\right)\right)^2}{2\left(\log\left(\frac{\sigma_f}{f_0}\right)\right)^2}} e^{\frac{-(\theta-\theta_0)^2}{2\sigma_\theta^2}},$$

where f_0 is the central filter frequency, σ_f is the width of the frequency component, θ_0 is the direction of the center, and σ_θ is the width of the direction component.

The normalized iris image is translated into a spectral domain using the Fourier transform. Then its matching with the two-dimensional log-Gabor filter is performed, and the result translated into a spatial domain using the inverse Fourier transform. The coding of the result is carried out as described earlier.

Pretrained Convolutional Neural Network for Iris Recognition

The procedure for extracting the feature representation consists in "turning off" the last layer in the model trained for iris recognition (which plays the role of a classifier) and passing the data through the modified neural network.

To extract the feature representation from the data, we used the models [39] obtained in the study and showed the best value of the recognition quality for each of the normalization methods described above. Each of the models was trained from scratch. In particular, we used the InceptionV3 model [41] with the Daugman normalization technique, and we used DenseNet121 [42] with the normalization based on the scaling and cropping.

2.5 Dimension Reduction

Solving many complex problems using machine learning algorithms is accompanied by a large number of features for each instance in a dataset. As a rule, the processing of the initial feature representation requires substantial time and computational resources. Also, a large dimension of feature vectors leads to complex procedures for finding a good solution.

In this regard, after extracting the feature representation, a dimension reduction stage is additionally introduced. In this paper, we use the most popular dimensionality reduction technique, i.e. the principal component analysis. The essence of this technique is finding a linear transformation into a subspace of a lower dimension that maximizes the variance of the data, with the subsequent projection of the data into this subspace.

2.6 Classification Step

To solve the classification problem, in this work, three machine learning algorithms are used and compared, namely: the support vector machine (with linear, polynomial, sigmoidal, and radial kernels), the random forest, and the k-nearest neighbors classifier.

3   Experiments

All experimental studies were carried out using the interpreted Python programming language as well as frameworks OpenCV, NumPy, Keras, Scikit-learn, Matplotlib. Training of convolutional neural networks for segmentation and classification was carried out using Google Colab.

3.1   Description of the Dataset

An open dataset MMU Iris Database [43] was chosen to carry out experimental studies. The dataset contains 450 images of irises for 45 people measuring 320 × 240, captured with a near-infrared camera. There are 10 images for each person: the first 5 correspond to the right eye, the second 5 to the left eye.

Due to the fact that the stage of segmentation of the iris was implemented using deep learning methods, namely, convolutional neural networks were trained, it was necessary to obtain ground-truth segmentation masks for the considered dataset. For this reason, manual segmentation of the entire dataset was performed. The prepared set of segmentation masks can be found in the public domain at the link [44].

Figure 3 shows the examples from the dataset as well as the corresponding segmentation masks.

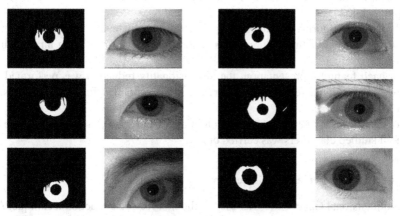

Fig. 3.   An example of images from the MMU Iris Database dataset and prepared true segmentation masks.

3.2   Experimental Research Procedure

The reported study solves the following tasks:

1. Determine the best normalization technique.
2. Choose the best feature extraction method.
3. Estimate the best destination size for the dimensionality reduction stage.

4. Choose the best classification technique.

To estimate the quality of the entire solution, we used the following indicator:

$$Accuracy = \frac{1}{N} \sum_{j=1}^{N} \frac{TP_j + TN_j}{TP_j + TN_j + FP_j + FN_j},$$

Here TP_j is the number of true positive outcomes for the j-th class, TN_j is the number of true negative outcomes for the j-th class, FP_j is the number of false positive outcomes for the j-th class, FN_j is the number of false negative outcomes for the j-th class, and N is the number of classes.

A cross-validation mechanism was used to train and assess the quality of the models. When applying it, a single parameter was set that was responsible for the number of random partitions of the original sample. The dataset was split into five nonoverlapping subsets. During the operation of the cross-validation algorithm, each subset participated in the accuracy testing exactly once. Thus, at each iteration of cross-validation, four subsets (360 images) were submitted for training, and one for testing (90 images). The averaged accuracy estimation was considered as a result of the algorithm.

Experimental studies were carried out as follows:

1. The input image was segmented according to the previously proposed neural network approach [31].
2. The image was normalized depending on the subsequent feature extraction technique.
3. The calculation of the feature representation was performed using all the studied feature systems.
4. For each feature representation, the dimensionality reduction was performed, and lower-dimensional feature representations were generated for a predefined set of output dimensions.
5. For each formed feature representation in a reduced space, the training and testing of all studied classifiers was carried out.

3.3   Results of Experimental Research

A two-dimensional Gabor filter was used as the first feature extraction technique. After normalization by the Daugman method and extraction of the feature representation, the dimensionality reduction, training, and testing of all studied classifiers was done according to the above scheme. One-dimensional and two-dimensional logGabor filters became the next investigated feature extraction techniques. All the subsequent steps for these techniques were the same. The results of the study for these feature systems and use SVM algorithm are shown in Fig. 4.

a)

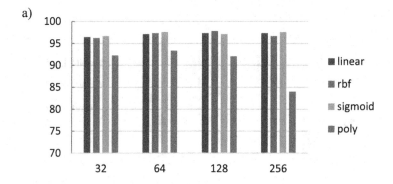

b)

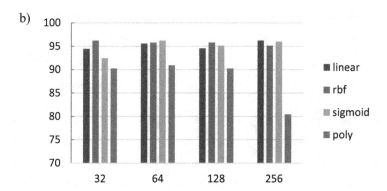

c)

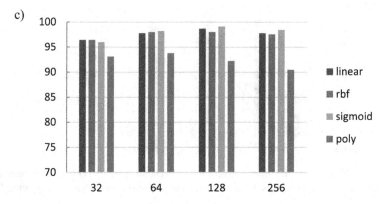

Fig. 4. Study results using for filter extraction a) 2D-Gabor filter b) 1D-logGabor filter c) 2D-logGabor filter. Here the x-axis is the size of the final sequence, the y-axis is the accuracy of the classifier

The results for feature systems Gabor 2D filter, 1D and 2D logGabor filters and use k-NN/Random Forest algorithms are shown in Fig. 5.

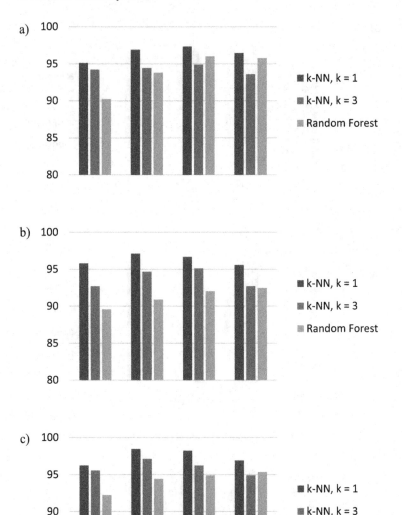

Fig. 5. Study results using for filter extraction a) 2D-Gabor filter b) 1D-logGabor filter c) 2D-logGabor filter. Here the x-axis is the size of the final sequence, the y-axis is the accuracy of the classifier

Considering neural network methods for feature extraction, a modified pre-trained convolutional neural network of the InceptionV3 architecture was studied. The procedure for this technique was as follows: normalization of the original images by the Daugman method, feature extraction using the pre-trained model, dimensionality reduction, and the use of the classifiers.

The final feature extraction technique was a pre-trained convolutional neural network of the DenseNet121 architecture. Normalization was performed by cropping and scaling. All subsequent steps were similar to those when using the Inception V3 network.

The results for neural networks based feature systems and SVM algorithm are shown in Fig. 6.

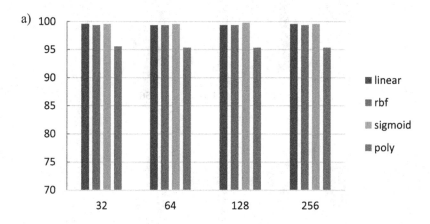

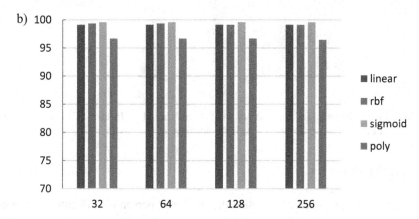

Fig. 6. Study results using for filter extraction a) DenseNet-121, b) InceptionV3. Here the x-axis is the size of the final sequence, the y-axis is the accuracy of the classifier

The results for neural networks based feature systems and k-NN/Random Forest algorithms are shown in Fig. 7.

As can be seen, the best classification quality was provided using the DenseNet neural network and amounted to 99.78%. In this case, the framing and scaling approach was used for normalization. The optimal number of principal components was equal to 128, and the optimal classifier was the support vector machine with a sigmoidal kernel.

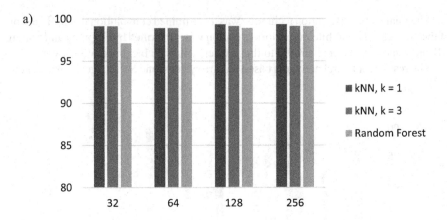

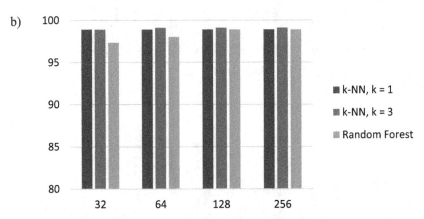

Fig. 7. Study results using for filter extraction a) DenseNet-121, b) InceptionV3. Here the x-axis is the size of the final sequence, the y-axis is the accuracy of the classifier

4 Conclusion

In this paper, we proposed and studied the iris-based person identification technique. The proposed technique is based on the neural network approach for the segmentation and feature extraction stages and uses a classical machine learning technique for classification.

For the segmentation of iris images, we used our previous solution [31], which showed the best segmentation accuracy on the MMU Iris Database. We made the ground-truth segmentation masks obtained by the manual markup of this dataset publicly available [44].

We carried out the comparative analysis of two normalization methods and several feature extraction techniques, including classical approaches based on various modifications of the Gabor filter and neural network approaches. Besides, we studied the influence of the dimensionality reduction stage and compared various classification techniques, including the support vector machines with a number of kernels, the random forest, and k-nearest neighbors classifiers.

The use of the proposed segmentation method as well as the DenseNet121-based feature extraction technique accompanied by the support vector machine classifier with a sigmoid kernel allowed us to achieve 99.78% classification accuracy for 45 classes.

References

1. Nemirovskiy, V.B., Stoyanov, A.K., Goremykina, D.S.: Face recognition based on the proximity measure clustering. Comput. Opt. **40**(5), 740–745 (2016)
2. Vizilter, Y.V., Gorbatsevich, V.S., Vorotnikov, A.V., Kostromov, N.A.: Real-time face identification via CNN and boosted hashing forest. Comput. Opt. **41**(2), 254–265 (2017)
3. Hashemi, J., Fatemizadeh, E.: Biometric identification through hand geometry. In: EUROCON International Conference on Computer as a Tool, vol. 2, pp. 1011–1014 (2005)
4. Prasad, S.M., Govindan, V.K., Sathidevi, P.S.: Bimodal personal recognition using hand images. In: Proceedings of International Conference on Advances in Computing Communication and Control (ICAC3), pp. 403–409 (2009)
5. Yuan, W., Lixiu, Y., Fuqiang, Z.: A real time fingerprint recognition system based on novel fingerprint matching strategy. In: 8th International Conference on Electronic Measurement and Instruments, pp. 1–81 (2007)
6. Kaur, M., Singh, M., Girdhar, A., Parvinder, S.: Fingerprint verification system using minutiae extraction technique. World Acad. Sci. Eng. Technol. **46**, 497–502 (2008)
7. Review of the international market of biometric technologies and their application in the financial sector. https://www.cbr.ru/Content/Document/File/36012/rev_bio.pdf
8. Pavelyeva, E.A., Krylov, A.S., Ushmaev, O.S: Development of information technology of a person's personality on the iris of the eye based on the Hermite transformation. https://elibrary.ru/item.asp?id=13070173
9. Gonzalez, R., Woods, R.: Digital Image Processing, Technosphere, 1072 (2005). (in Russian)
10. Khan, A.A., Kumar, S., Khan, M.: Iris pattern recognition using SVM and ANN. IJIREEICE **2**(12), 2208–2211 (2014)
11. Chen, Y., Liu, Y., Zhu, X., Chen, H., He, F.: Novel approaches to improve iris recognition system performance based on local quality evaluation and feature fusion. Sci. World J. (2014)
12. Firake, S.G., Mahajan, P.M.: Brief review of iris recognition using principal component analysis, independent component analysis and Gabor wavelet. Int. J. Eng. Res. Technol. (IJERT) **3**(3), 1290–1294 (2014)
13. Manisha Nirgude, S.G.: Iris recognition system based on multi-resolution analysis and support vector machine. Int. J. Comput. Appl. (IJCA) **173** (2017)
14. Rana, H.K., Azam, M.D.S., Akhtar, R., Quinn, J.M.W., Moni, M.A.: A fast iris recognition system through optimum feature extraction. https://doi.org/10.7287/peerj.preprints.27363v2
15. Azam, M.D., Rana, H.: Iris recognition using convolutional neural network. Int. J. Comput. Appl. **175**(12), 24–28 (2020)
16. Nguyen, K., Fookes, C., Ross, A., Sridharan, S.: Iris recognition with off-the-shelf CNN features: a deep learning perspective. IEEE Access **6**, 18848–18855 (2018)
17. Daugman, J.G.: How iris recognition works. https://ieeexplore.ieee.org/document/1262028
18. Bakhtiari, A., Shirazi, A., Zahmati, A.: An efficient segmentation method based on local entropy characteristics of iris biometrics. World Acad. Sci. Eng. Technol. **28**, 64–68 (2007)
19. Barzegar, N., Moin, M.S.: A new approach for iris localization in iris recognition systems. In: Proceedings of the 6th IEEE/ACS International Conference on Computer Systems and Applications (AICCSA 2008), pp. 516–523 (2008)

20. Semenov, M.S., Myasnikov, E.V.: A comparison of iris image segmentation techniques. In: CEUR Workshop Proceedings, pp. 163–169 (2018)
21. Liu, N., Li, H., Zhang, M., Liu, J., Sun, Z., Tan, T.: Accurate iris segmentation in non-cooperative environments using fully convolutional networks. In: 2016 International Conference on Biometrics (ICB), pp. 1–8 (2016)
22. Jalilian, E., Uhl, A.: Iris segmentation using fully convolutional encoder–decoder networks. In: Bhanu, B., Kumar, A. (eds.) Deep Learning for Biometrics, pp. 133–155. Springer, Cham (2017). https://doi.org/10.1007/978-3-319-61657-5_6
23. Lozej, J., Meden, B., Štruc, V., Peer, P.: End-to-end iris segmentation using u-net. In: 2018 IEEE International Work Conference on Bioinspired Intelligence (IWOBI), pp. 1–6 (2018)
24. Korobkin, M., Odinokikh, G., Efimov, Y., Solomatin, I., Matveev, I.: Iris segmentation in challenging conditions. Pattern Recognit. Image Anal. **28**(4), 652–657 (2018). https://doi.org/10.1134/S1054661818040193
25. Pathak, M.P., Bairagi, V., Srinivasu, N.: Effective segmentation of sclera, iris and pupil in eye images. TELKOMNIKA (Telecommun. Comput. Electron. Control) **17**, 101–111 (2019)
26. Li, Y.H., Huang, P.J., Juan, Y.: An efficient and robust iris segmentation algorithm using deep learning. https://doi.org/10.1155/2019/4568929
27. Pathak, M.P., Bairagi, V., Srinivasu, N.: Entropy based CNN for segmentation of noisy color eye images using color, texture and brightness contour features. Int. J. Recent Technol. Eng. **8**, 2116–2124 (2019)
28. Poonia, J., Bhurani, P., Gupta, S.K., Agrwal, S.L.: New improved feature extraction approach of iris recognition. Int. J. Comput. Syst. (IJCS) **3**(1), 1–3 (2016)
29. Pathak, M.P., Bairagi, V., Srinivasu, N.: Multimodal eye biometric system based on contour based e-CNN and multi algorithmic feature extraction using SVBF matching. Int. J. Innov. Technol. Explor. Eng. **8**(9), 417–423 (2019)
30. Akbar, S., Ahmad, A., Hayat, M.: Iris detection by discrete sine transform based feature vector using random forest. J. Appl. Environ. Biol. Sci. **4**, 19–23 (2014)
31. Ganeeva, Y., Myasnikov, E.V.: Using convolutional neural networks for segmentation of iris images. In: 2020 International Multi-Conference on Industrial Engineering and Modern Technologies (FarEastCon), pp. 1–4 (2020). https://doi.org/10.1109/FarEastCon50210.2020.9271541
32. Ronneberger, O., Fischer, P., Brox, T.: U-net: convolutional networks for biomedical image segmentation. https://arxiv.org/abs/1505.04597
33. Hashim, A.T., Saleh, Z.A.: Fast iris localization based on image algebra and morphological operations. JUBPAS **27**(2), 143–154 (2019)
34. Chirchi, V., Waghmare, L.M.: Enhanced isocentric segmentor and wavelet rectangular coder to iris segmentation and recognition. Int. J. Intell. Eng. Syst. **10**, 1–10 (2017)
35. Khan, T.M., Bailey, D.G., Khan, M.A.U., Kong, Y.: Real-time iris segmentation and its implementation on FPGA. J. Real-Time Image Proc. **17**(5), 1089–1102 (2019). https://doi.org/10.1007/s11554-019-00859-w
36. Jan, F., Min-Allah, N., Agha, S., Usman, I., Khan, I.: A robust iris localization scheme for the iris recognition. Multimedia Tools Appl. **80**(3), 4579–4605 (2020). https://doi.org/10.1007/s11042-020-09814-5
37. Lin, M., Haifeng, L., Kunpeng, Y.: Fast iris localization algorithm on noisy images based on conformal geometric algebra. Digit. Signal Process. **100**, 102682 (2020)
38. Wan, H.L., Li, Z., Qiao, J.P., Li, B.S.: Non-ideal iris segmentation using anisotropic diffusion. IET Image Proc. **7**, 111–120 (2013)
39. Ganeeva, Y., Myasnikov, E.: Augmentation in neural network training for person identification by iris images. In: 2021 Ural Symposium on Biomedical Engineering, Radioelectronics and Information Technology (USBEREIT) (2021). 0106-0109. https://doi.org/10.1109/USBEREIT51232.2021.9455076

40. Masek, L.: Recognition of human iris patterns for biometric identification. http://www.csse. uwa.edu.au/~pk/studentprojects/libor/
41. Advanced Guide to Inception v3 on Cloud TPU. https://cloud.google.com/tpu/docs/incept ion-v3-avanced
42. Huang, G., Liu, Z., Van der Maaten, L., Weinberger, K.Q.: Densely connected convolutional networks. https://arxiv.org/abs/1608.06993
43. MMU Iris Image Database: Multimedia University. http://pesonna.mmu.edu.my/ccteo/
44. Masks for MMU Iris dataset. https://github.com/jganeeva99/Masks-for-MMU-Iris-dataset

Data Dimensionality Reduction Technique Based on Preservation of Hellinger Divergence

Evgeny Myasnikov$^{(\boxtimes)}$ (iD)

Samara University, Moskovskoe Shosse 34A, Samara 443086, Russia
mevg@geosamara.ru

Abstract. In this paper, we investigate the issue of applying the Hellinger divergence in dimensionality reduction of hyperspectral data. We use two approaches in constructing dimensionality reduction techniques. The first approach is based on approximating the Hellinger divergence by Euclidean distances in a lower-dimensional space, while the second approach is based on preserving the Hellinger divergence when mapping data into a space of reduced dimensionality. In our experiments, we study the efficiency of the proposed techniques using known hyperspectral scenes both in terms of data mapping error and the quality of subsequent classification.

Keywords: Hyperspectral image · Dimensionality reduction · PCA · Nonlinear mapping · SDPM · Spectral information divergence · Hellinger divergence

1 Introduction

The growth in the volume of processed hyperspectral data necessitates an increase in the efficiency of their processing. Elimination of the natural redundancy inherent in such data due to its high spectral resolution is one of the main ways to improve efficiency. To solve the problem of eliminating redundancy, unsupervised dimensionality reduction methods are increasingly used. The main advantage of using unsupervised methods is that there is no need for time-consuming markup of the source data. The use of dimensionality reduction methods conditions the computational efficiency of subsequent processing stages in comparison to data compression techniques, since the use of the latter is associated with preliminary decompression and work with data in the original high dimension.

Despite the variety of nonlinear dimensionality reduction methods [1,3–5] the Principal Components Analysis (PCA) technique [6] holds the leadership in solving the problem under consideration. Although modern nonlinear methods can have advantages when mapping data to low-dimensional spaces, especially in the presence of nonlinear effects and noise, PCA remains the most used method due to its popularity and computational efficiency.

E. Burnaev et al. (Eds.): AIST 2021, LNCS 13217, pp. 190–198, 2022.
https://doi.org/10.1007/978-3-031-16500-9_16

When using the nearest neighbor classifier, excellent classification quality is provided by the nonlinear mapping technique, especially with alternative dissimilarity measures. The most well-known among such measures are spectral angle mapper (SAM) [12] and spectral information divergence (SID) [13].

In fact, we can distinguish two approaches to the use of non-Euclidean dissimilarity measures. The first approach is based on the approximation of the selected dissimilarity measures (calculated in the original space) by the Euclidean distances in the target space [7–10]. This is a more common approach, which does not deeply modify existing data dimensionality reduction routines.

The second approach is based on an attempt to preserve the values of the selected dissimilarity measure in the target space. This approach was previously implemented for spectral angle mapper (SAM) [10] and spectral information divergence (SID) [11]. It was shown that in the latter case, this approach provided significant advantages over the first approach for the SID measure.

The chosen approach determines the dissimilarity measure used in the classifier. If in the first case the nearest neighbors are determined on the basis of the Euclidean distance, then for the second approach, the same dissimilarity measure should be used that was also used in dimensionality reduction technique.

The issue of computational efficiency is worth mentioning here. Acceleration of the query by the nearest neighbor is associated with the use of binary space partitioning trees (BSPTs), such as VP-trees. The construction of such trees is possible only when the used dissimilarity measure is a true metric.

The above requirement is not met for the SID that necessitates the use of slow brute force approach for the search of nearest neighbors in the NN-classifier. For this reason, in a recent paper [17], it was proposed to use the Hellinger divergence instead of the SID when constructing BSPTs and executing NN-queries. It was shown that the use of Hellinger divergence gave a significant increase in the speed of query processing with practically indistinguishable quality of classification in comparison with SID.

In this paper, we investigate the issue of applying the Hellinger divergence for dimensionality reduction of hyperspectral data. When constructing dimensionality reduction methods, we use both the approach based on approximating the Hellinger divergence by Euclidean distances, and the approach based on preserving the Hellinger divergence when mapping data into a space of reduced dimensionality.

The paper structure is as follows. The next section is devoted to the description of the proposed techniques. Section 3 discusses the results of the numerical experiments. The paper ends up with the conclusion and the list of references.

2   Methods

In this section, we assume that the hyperspectral dataset X consists of N vectors (data points) $x_i, i = 1 \ldots N$ in the multidimensional hyperspectral space R^M. The result Y of a dimensionality reduction technique consists of corresponding vectors $y_i, i = 1 \ldots N$ in the lower-dimensional space R^m.

As it was outlined in the introduction, in this paper we use the Hellinger divergence to measure the dissimilarity of data points. The Hellinger divergence for two given vectors x_i and x_j is as follows [18]:

$$HD(x_i, x_j) = \sqrt{1 - \sum_{k}^{M} \sqrt{p_k(x_i)p_k(x_j)}},$$ (1)

where

$$p_k(x_i) = \frac{x_{ik}}{\sum_{l=1}^{M} x_{il}}.$$ (2)

Methods proposed in this paper belong to a class of nonlinear dimensionality reduction techniques operating on the principle of preserving pairwise distances between data points. The basics of this approach appeared in the 1960-s in works by J.B. Kruskal [14] and J.W. Sammon [4]. While previous techniques were focused on preserving Euclidean pairwise distances or approximating pairwise differences by Euclidean distances, in this paper, we introduce both the approximation-based technique and the technique based on preserving pairwise Hellinger divergence. We consider these approaches in the subsections below.

2.1 Approximation of Hellinger Divergence by Euclidean Distances in Lower-Dimension Space

To implement the first approach, let's consider the following data mapping error:

$$\varepsilon_{ED} = \mu \cdot \sum_{i,j=1, i<j}^{N} \rho_{ij}(HD(x_i, x_j) - ED(y_i, y_j))^2,$$ (3)

where $HD(x_i, x_j)$ is the Hellinger divergence (1) calculated for i-th and j-th vectors in a source hyperspectral space, $ED(x_i, x_j)$ is the Euclidean distance between corresponding vectors in the target reduced space, ρ_{ij} is some constant coefficient, and μ is a scale factor. Using the last two coefficients, we can define the particular error function, for example, making small differences more important. In this study, we adopt the most commonly used $\mu = 1/\sum_{i<j} HD^2(x_i, x_j)$, $\rho_{i,j} = 1$.

To minimize the error (3) we use the stochastic gradient descent algorithm as one commonly used in similar cases [10,11]. For brevity of presentation, here we omit the detailed description of the further steps of the method construction since a similar description can be found in the cited works.

In general, the method based on the approximation of Hellinger divergence by the Euclidean distances (hereinafter, HDED) consists of the initialization of low-dimensional coordinates $y_i(t_0), i = 1 \ldots N$ with the subsequent refinement of these coordinates in the reduced space using the following equation:

$$y_i(t+1) = y_i(t) + 2\alpha\mu \cdot \sum_{j \in r(t)} \frac{HD(x_i, x_j) - ED(y_i(t), y_j(t))}{ED(y_i(t), y_j(t))} \cdot (y_i(t) - y_j(t)).$$ (4)

Here t is the number of an iteration, $r(t)$ is a random subset of indices used to approximate the gradient at a particular iteration t so that $i \notin r$, α is the step size of the gradient descent. In our implementation, we used PCA for the initialization and performed error control with step correction. We used another random subsample to estimate the error (3) to avoid long-time calculations.

2.2 Preserving Hellinger Divergence in Lower-Dimensional Space

To implement the second approach, we change the data mapping error, and use the following objective:

$$\varepsilon_{HD} = \mu \sum_{i,j=1(i<j)}^{N} \left(\rho_{i,j} \left(HD(x_i, x_j) - HD(y_i, y_j) \right)^2 \right). \tag{5}$$

According to our problem, we treat the coordinates $Y = (y_1, \ldots y_N)$ in R^m as optimized parameters. The gradient descent technique allows to minimize (5) by the initialization of Y followed by the iterative optimization according to the formula: $Y(t+1) = Y(t) - \alpha \nabla \varepsilon_{HD}$.

To obtain the gradient of the objective function $\nabla \varepsilon_{HD}$, we calculate the partial derivatives of the Hellinger divergence (1):

$$\frac{\partial}{\partial y_{ik}} HD(y_i, y_j) = \frac{1}{4HD(y_i, y_j) \sum\limits_{l=1}^{M} y_{il}} \left(\sum_{l=1}^{M} \sqrt{p_l(y_i)p_l(y_j)} - \sqrt{\frac{p_k(y_j)}{p_k(y_i)}} \right). \tag{6}$$

After transformations, we obtain the partials for the data mapping error (5):

$$\frac{\partial}{\partial y_{ik}} \varepsilon_{HD} = -2\mu \sum_{j=1(j\neq i)}^{N} \left(\frac{HD(x_i,x_j)-HD(y_i,y_j)}{4HD(y_i,y_j) \sum\limits_{l=1}^{M} y_{il}} \cdot \left(\sum_{l=1}^{M} \sqrt{p_l(y_i)p_l(y_j)} - \sqrt{\frac{p_k(y_j)}{p_k(y_i)}} \right) \right). \tag{7}$$

Thus, we can write down the recurrence equation for the optimized parameters Y for the stochastic gradient descent procedure:

$$y_{ik}(t+1) = y_{ik}(t)$$
$$+ 2\alpha\mu \sum_{j\in r(t)} \left(\frac{HD(x_i,x_j)-HD(y_i(t),y_j(t))}{4HD(y_i(t),y_j(t)) \sum\limits_{l=1}^{M} y_{il}(t)} \cdot \left(\sum_{l=1}^{M} \sqrt{p_l(y_i(t))p_l(y_j(t))} - \sqrt{\frac{p_k(y_j(t))}{p_k(y_i(t))}} \right) \right). \tag{8}$$

Here again time t is the number of an iteration, $r(t)$ is a random subset of indices used to approximate the gradient at a particular iteration t so that $i \notin r(t)$, α is the step size of the gradient descent. In our implementation, we used the equally distant components of the source vectors for the initialization, and performed the optimization with the error control as for the first approach.

The Eq. (8) determines the nonlinear mapping algorithm operating on the Hellinger divergence (1) preserving principle, hereinafter, Hellinger divergence preserving mapping (HDPM).

3   Experiments

In this paper, we study the efficiency of proposed techniques both in terms of the data mapping error and the quality of the subsequent classification. To perform the reported study, we used well-known hyperspectral image scenes [16], which were supplied with the ground-truth classification that is necessary to assess the classification quality. In particular, we used the Indian pines scene, registered with the AVIRIS sensor (200 channels) and Kennedy Space Center scene (176 channels).

The proposed techniques were implemented in C++, and the estimation of the classification quality was implemented in MatLab.

In the first experiment, we checked the workability of the approaches. We estimated how the data mapping errors (3) for HDED and (5) for HDPM changed with the number of an iteration. Corresponding dependencies are shown in Fig. 1 for several different output dimensions. As it can be seen, in all the showed cases the error substantially decreased in the optimization process. It is worth noting that the SDPM technique had difficulties in low output dimensionality ($m = 2, 3$, not shown in the figure) caused by bad convergence that resulted in unacceptable values of the error (5). Partly, the reason for this may be the loss of information about the length of the vectors when calculating the Hellinger divergence (1), which means an additional decrease in the internal data dimension by 1 relative to the HDED, which uses Euclidean distance to compare vectors in the target space. The HDED technique, in contrast, showed good convergence for all considered dimensions $dim = 2 \ldots 20$.

In the second experiment, we estimated the embeddings generated by the proposed dimensionality reduction techniques from the viewpoint of the classification quality. We used the Nearest Neighbor (NN-) classifier with the corresponding distance function in our experiment. We split the ground truth pixels into training and testing subsets in the 60:40 proportion and used the overall classification accuracy Acc (defined as the proportion of correctly classified pixels of the test set) as a quality indicator.

In this experiment, we varied the target dimensionality from 2 to 20, performed dimensionality reduction, then trained the classifier on the training subset and estimated the accuracy on the test subset. Besides, we compared the result to the known Principal Component Analysis (PCA) [6], Approximation of Spectral Angles by Euclidean distances (SAED) [10], and Spectral Information Divergence Preserving Mapping (SDPM) [11]. The results of the experimental study for the Indian Pines scene are shown in Fig. 2 (left).

As it can be seen in the figure, the SDPM and proposed HDPM techniques performed poorly for $Dim = 3$, which is consistent with the results of the

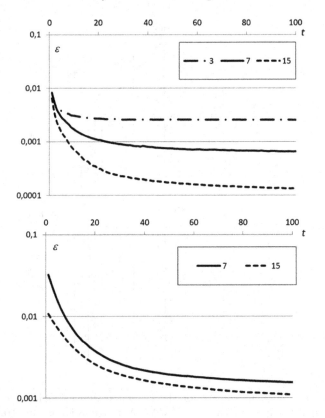

Fig. 1. Dependence of the data mapping error ε on the number of an iteration t for the proposed techniques: HDED (top) and HDPM (bottom). The results for output dimensionality $m = 3$ (dash-dotted line), $m = 7$ (solid line), and $m = 15$ (dashed line) are shown for the Indian Pines scene (200 channels).

first experiment. The PCA technique outperformed SDPM and HDPM for this dimension and for $Dim = 5$, but, PCA gave worse results in all the other cases among all considered techniques.

The SAED and the proposed HDED technique based on the approximation of SAM and HD by Euclidean distances provided the best results for smaller dimensionality $Dim = 3, 5$. But for the higher dimensionality $Dim = 10, 15, 20$ the SDPM and proposed HDPM techniques gave close values, which were the best in this experiment. We cannot choose the best technique between SDPM and HDPM by classification accuracy. However, we recall that the proposed HDPM technique allows building computationally efficient classifiers as the Hellinger divergence, which serves as the basis for the construction of the reduced space, exhibits the metric properties. The latter enables [17] the use of BSPTs for constructing classifiers. In contrast, the previously proposed SDPM technique lacks the described properties.

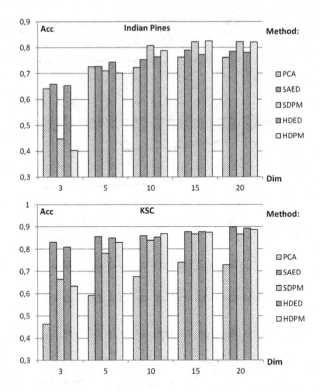

Fig. 2. Classification accuracy *Acc* for the evaluated dimensionality reduction techniques: Principal Component Analysis (PCA), Approximation of Spectral Angles by Euclidean distances (SAED), Spectral Information Divergence Preserving Mapping (SDPM), and the proposed Approximation of Hellinger Divergence by Euclidean Distances (HDED), and Hellinger Divergence Preserving Mapping (HDPM) for the NN classifier: Indian Pines scene (top) and Kennedy Space Center scene (bottom)

Similar results for the Kenedy Space Center (KSC) hyperspectral scene are shown in Fig. 2 (right). Here we observe a slightly different picture. The linear PCA is the worst technique for all the considered target dimensions. For this scene, the leading technique, SAED, is based on the Spectral Angle. It provides the best quality for all the considered dimensions, except $Dim = 10$. In the latter case, the proposed HDPM technique was the best. In almost all the cases, the proposed HDED and HDPM outperformed the previously proposed SDPM. Again, both HDED and HDPM allow for BSPT construction in a target space.

4   Conclusion

In this paper, we proposed two approaches for the construction of nonlinear dimensionality reduction techniques based on the Hellinger divergence. The first technique is based on the approximation of the Hellinger divergence by Euclidean distances in target space, and the second technique is based on the Hellinger divergence preserving principle.

Using open hyperspectral scenes, we studied the efficiency of the proposed techniques in terms of data mapping error and classification quality. We showed that for not too small dimensions, the proposed Hellinger Divergence Preserving Mapping (HDPM) provided competitive classification results compared to other methods.

In contrast to the related Spectral Information Divergence Preserving Mapping (SDPM) technique, the proposed HDPM constructs the target space on the basis of the Hellinger divergence, which exhibits the metric properties. This allows constructing computationally efficient classifiers exploiting binary space partitioning trees to accelerate classification queries.

Acknowledgments. The reported study was funded by RFBR according to the research project no. 18-07-01312-a.

References

1. Tenenbaum, J.B., de Silva, V., Langford, J.C.: A global geometric framework for nonlinear dimensionality reduction. Science **290**, 2319–2323 (2000)
2. Roweis, S.T., Saul, L.K.: Nonlinear dimensionality reduction by locally linear embedding. Science **290**, 2323–2326 (2000)
3. Belkin, M., Niyogi, P.: Laplacian eigenmaps and spectral techniques for embedding and clustering. Adv. Neural. Inf. Process. Syst. **14**, 585–591 (2001)
4. Sammon, J.W.: A nonlinear mapping for data structure analysis. Trans. Comput. **18**, 401–409 (1969)
5. McInnes, L., Healy, J., Melville, J.: Umap: uniform manifold approximation and projection for dimension reduction. arXiv:1802.03426 (2018)
6. Fukunaga, K.: Introduction to Statistical Pattern Recognition, 2nd edn. Academic Press, London (2003)
7. Bachmann, C.M., Ainsworth, L., Fusina, R.A.: Exploiting manifold geometry in hyperspectral imagery. IEEE Trans. Geosci. Remote Sens. **43**(3), 441–454 (2005)
8. Ding, L., Tang, P., Li, H.: Dimensionality reduction and classification for hyperspectral remote sensing data using ISOMAP. Hongwai yu Jiguang Gongcheng/Infrared and Laser Engineering **6**(10), 2707–2711 (2013)
9. Du, P., Wang, X., Tan, K., Xia, J.: Dimensionality reduction and feature extraction from hyperspectral remote sensing imagery based on manifold learning. Wuhan Daxue Xuebao (Xinxi Kexue Ban)/Geomatics and Information Science of Wuhan University **36**(2), 148–152 (2011)
10. Myasnikov, E.: Nonlinear mapping based on spectral angle preserving principle for hyperspectral image analysis. In: Felsberg, M., Heyden, A., Krüger, N. (eds.) CAIP 2017. LNCS, vol. 10425, pp. 416–427. Springer, Cham (2017). https://doi.org/10.1007/978-3-319-64698-5_35
11. Myasnikov, E.: Nonlinear dimensionality reduction of hyperspectral data based on spectral information divergence preserving principle. J. Phys. Conf. Ser. **1368**, 032030 (2019)
12. Kruse, F.A., et al.: The spectral image processing system (SIPS) - interactive visualization and analysis of imaging spectrometer data. Remote Sens. Environ. **44**, 145–163 (1993)

13. Chang, C.-I.: Hyperspectral Data Processing: Algorithm Design and Analysis. Wiley, Hoboken (2013)
14. Kruskal, J.B.: Multidimensional scaling by optimizing goodness of fit to a non-metric hypothesis. Psychometrika **29**, 1–27 (1964)
15. Tenenbaum, J.: Mapping a manifold of perceptual observations. In: Jordan, M., Kearns, M., Solla, S. (eds.) Adv. Neural Inf. Process., vol. 10, pp. 682–688. MIT Press, Cambridge (1998)
16. Hyperspectral Remote Sensing Scenes. http://www.ehu.eus/ccwintco/index.php?title=Hyperspectral_Remote_Sensing_Scenes
17. Myasnikov, E.: Nearest neighbor search in hyperspectral data using binary space partitioning trees. IEEE Xplore Digital Library. In: Proceedings 11th Workshop on Hyperspectral Imaging and Signal Processing: Evolution in Remote Sensing (WHISPERS), pp. 1–4 (2021)
18. Hellinger, E.: Neue Begründung der Theorie quadratischer Formen von unendlichvielenVeränderlichen. Journal für diereine und angewandte Mathematik **136**, 210–271 (1909). (in German)

Group-Level Affect Recognition in Video Using Deviation of Frame Features

Andrey V. Savchenko[1]([✉])(ID), Lyudmila V. Savchenko[1](ID),
and Natalya S. Belova[2](ID)

[1] HSE University, Laboratory of Algorithms and Technologies
for Network Analysis, Nizhny Novgorod, Russia
{avsavchenko,lsavchenko}@hse.ru
[2] HSE University, Moscow, Russia
nbelova@hse.ru

Abstract. In this paper, we propose the novel video-based group-level emotion recognition algorithm. At first, the faces are detected in each video frame, and their features are extracted using a lightweight neural network, e.g., MobileNet pre-trained on large emotional dataset, such as AffectNet. The frame descriptor is defined as a concatenation of STAT features (max, average, standard deviation, etc.). The descriptor of the whole video is computed as a deviation of the frame descriptors, and the resulting video features are fed into a classifier. Experimental results for the VGAF dataset from the EmotiW 2020 challenge demonstrate that the proposed approach has 1% greater accuracy than the best-known single model. It is also at least 5% better than any other facial processing technique. Moreover, a blending of facial expression recognition with a processing of audio features extracted by the OpenSMILE library is comparable with the best-known ensemble for this dataset.

Keywords: Facial analytics · Group-level emotion recognition · Facial expression recognition · VGAF (Video-level Group AFfect) dataset

1 Introduction

Nowadays, one of the most difficult but fast-growing research topics in video-based emotion analysis is group affect recognition [1]. The group-level emotions can be used to predict the overall affect of the students in a classroom or to understand the team performance [2]. In order to facilitate research in audio-visual group emotion recognition, the VGAF (Video-level Group AFfect) dataset was introduced in the EmotiW (Emotion Recognition in the Wild) 2020 challenge [2]. Its participants presented several solutions based on ensembles of deep convolutional neural networks (DCNN). Activity recognition networks were introduced in [3], in which a large labeled video dataset for activity classification was used to pre-train the modified ResNet-50 architecture. An excellent performance was achieved by an ensemble of two K-injection subnetworks [4] with a feature-based

E. Burnaev et al. (Eds.): AIST 2021, LNCS 13217, pp. 199–207, 2022.
https://doi.org/10.1007/978-3-031-16500-9_17

cross attention. The fusion of spatio-temporal and static features (faces, bodies, entire frame, etc.) is discussed in [5]. A privacy-safe model [6] captured the global moods from the whole image without using any individual-based feature, such as pose or facial emotions, but cannot reach the accuracy of the state-of-the-art methods. The winner of this challenge proposed the hybrid networks for audio, faces, video and environmental object statistics [7].

Unfortunately, all these approaches are rather complex due to the usage of many DCNNs, so that their processing time is rather high [8,9]. Indeed, the ablation studies in the above-mentioned papers prove that it is practically impossible to train a single efficient model that can reach a comparable accuracy with the best-known ensembles. However, it is known that it is especially important to develop methods with potential to monitor emotions in real-time to detect the abnormal behavior of a group of people [1]. It is possible that decision should be made at the typical PC/laptop or even mobile or edge device without access to modern expensive graphical cards. Hence, we propose here a novel algorithm by using a single MobileNet-based model pre-trained on the VGGFace2 [10] and AffectNet [11] for facial expression recognition (FER) [12]. The embeddings of individual faces [8] extracted by this DCNN are combined with statistical (STAT) features (maximum, average, standard deviation, etc.), that have been widely used in audio-visual emotion recognition [7,13,14]. The whole video is described by the STAT features applied to the frame descriptors. In this paper, we especially emphasize the need for computing standard deviation of the frame descriptors. It is shown that classification of such video embeddings lead to very accurate results on the validation subset of the VGAF dataset [2]. Moreover, it is possible to improve the overall accuracy if the audio modality is available [15].

The rest of the paper is organized as follows. Section 2 includes detailed description of the proposed approach and training procedures. Section 3 contains experimental results and ablation study for the VGAF dataset. Concluding comments are discussed in Sect. 4.

2   Proposed Approach

In the group affect recognition task, it is required to predict one of the $C > 1$ emotional categories given a sequence of T frames $\{X(t)\}, t = 1, 2, ..., T$. It is assumed that a training set of $N > 1$ example videos $\{X_n(t)\}, t = 1, 2, ..., T_n$ with T_n frames is available, and the class label $c_n \in \{1, ..., C\}$ of the n-th ($n = 1, 2, ..., N$) video is known.

In this paper, we mainly concentrate on facial processing. Hence, $M(t) \geq 0$ facial regions are firstly located in each frame by using an appropriate detector, such as MTCNN. If there are no detected faces in the first frames, then they are completely ignored in further analysis. However, if the MTCNN fails to detect any face at the t-th frame, but $M(t-1) > 0$ for the previous frame, we simply use the bounding boxes from the previous frame to locate the facial region in the t-th frame. Next, the emotional features of each detected face are computed by using the DCNN pre-trained for face identification [8] on large dataset, such as VGGFace2 [10], and further fine-tuned on the emotional dataset with static facial images, e.g., AffectNet [11]. As this paper is focused on low running time, it was decided to use such architectures as MobileNet or EfficientNet as a backbone network. Each m-th facial image ($m \in \{1, ..., M(t)\}$) of the t-th frame is fed into the input of this DCNN, and its D-dimensional embedding $\mathbf{x}_m(t) = [x_{m;1}(t), ..., x_{m;D}(t)]$ is computed output of the last-but-one layer of this network, where D is the number of units in this layer.

In group-level emotion recognition, it is typical to combine features of individual persons into a single descriptor of the whole frame by using simple averaging of facial features. However, many studies in audio-visual emotion classification [13] including the winners of the EmotiW 2020 challenge [7] claimed the success of STAT features (component-wise average, minimum, maximum and standard deviation). The authors of the paper [14] removed the max features, but our experiments demonstrate that they are one of the most important for the VGAF dataset. As a result, we obtain the t-th frame descriptor

$$\mathbf{f}(t) = [f_1(t), ..., f_d(t), ..., f_D(t)], f_d(t) = \max_{m \in \{1, ..., M(t)\}} x_{m;d}(t). \tag{1}$$

It is also possible to compute other STAT features similarly to (1) and concatenate them into high-dimensional frame feature vectors [16], but such approach will increase the feature dimensionality, and, as a result, memory complexity and decision-making time.

In this paper, we used the CNNs from our previous article [12]. The training set for pre-training was created by cropping the facial regions from VGGFace2 dataset [10] using the MTCNN face detector without adding any margins. Thus, the resulted emotional features became rather accurate and do not depend on the background. As a result, our models should be much better than the existing emotional CNNs used even in the hybrid networks [7] though they also used STAT features in one of the classifiers.

Finally, it is required to combine these features into a single video descriptor $\mathbf{v}$ [17]. Nowadays, it is typical to learn frame-wise attention models [4], but we propose to compute the STAT features at this stage. In particular, we believe that the most relevant information about dynamics of emotional state is included in the variance or dissimilarity of the frame-level features. Hence, we decided to mainly concentrate on the usage of standard deviation

$$\mathbf{v} = [v_1, ... v_d, ..., v_D], v_d = \sqrt{\frac{1}{T} \sum_{t=1}^{T} (f_d(t) - \overline{f}_d)^2}, \overline{f}_d = \frac{1}{T} \sum_{t=1}^{T} f_d(t), \tag{2}$$

though other STAT features may be also concatenated to the video descriptor [16]. The same procedure is repeated for every n-th training example to obtain the feature vectors $\mathbf{v}_n$. Finally, an arbitrary classifier $\mathcal{C}$ is learned on the training set $\{(\mathbf{v}_n, c_n)\}, n = 1, 2, ..., N$, and the class of the input descriptor $\mathbf{v}$ is predicted.

Our complete procedure is presented in Algorithm 1. Here optional steps 13–17 were added for the case when the audio modality is available and the

---

Algorithm 1. Proposed approach to video-based group-level affect recognition

---

Require: N training videos, observed video frames $\{X(t)\}$
Ensure: Emotion label of the input video
1: **for** each frame $t \in \{1, ..., T\}$ **do**
2:     Detect $M(t) \geq 0$ facial regions
3:     **if** $M(t) = 0$ AND $t > 0$ AND $M(t-1) > 0$ **then**
4:         Use facial regions from the previous frame and assign $M(t) := M(t-1)$.
5:     **end if**
6:     **for** each facial region $m \in \{1, ..., M(t)\}$ **do**
7:         Extract embeddings $\mathbf{x}_m(t)$ using the emotional CNN
8:     **end for**
9:     Compute frame feature vector $\mathbf{f}(t)$ using STAT features (1)
10: **end for**
11: Compute the video descriptor $\mathbf{v}$ as a standard deviation of $\{\mathbf{f}(t), t = 1, ..., T\}$
12: Feed $\mathbf{v}$ into classifier $\mathcal{C}$ to obtain the confidence scores $s_c, c = 1, 2, ..., C$
13: **if** audio modality is available **then**
14:     Compute emotional embeddings $\mathbf{a}$ of the input audio
15:     Feed $\mathbf{a}$ into the audio classifier to obtain the confidence scores $s_c^{(a)}, c = 1, 2, ..., C$
16:     Assign $s_c := \alpha \cdot s_c + (1 - \alpha) \cdot s_c^{(a)}$
17: **end if**
18: **return** class label with the maximal score $\max s_c$

---

requirements to the real-time processing are not too strong. Though deep learning works incredibly well for various speech processing tasks [18,19], emotions in audio are still better recognized [2,15] by using feature extraction from the OpenSMILE library [20]. In this paper, we used the ComParE feature set ("IS13_ComParE" from is09-13) [21] that contains 6373 features such as MFCC, LSP and pitch [22]. A final decision is made by blending, i.e., the linear weighting with hyper-parameter α of the confidence scores returned by classifiers of visual and audio modalities. The algorithm was implemented in a Python application using OpenCV, scikit-learn, PyTorch and TensorFlow 2 frameworks. The source code and all pre-trained DCNNs are made publicly available[1].

3   Experimental Results

The proposed approach was examined on the VGAF dataset. The training set consists of 2661 video clips associated with one of $C = 3$ classes (Positive, Neutral and Negative). The labels of the testing set are unknown, so that we measure the accuracy on 766 video files from the validation set. Sample frames from this set are shown in Fig. 1. The videos have rather low resolution, so that 42 training and 25 validation files do not have any detected face in all frames. As a result, the classifier $\mathcal{C}$ was learned on the remaining training set. However, in order to be comparable with the reported results for this dataset, we did not remove the validation files without faces, but associate them with zero descriptors $\mathbf{v} = \mathbf{0}$.

---

[1] https://github.com/HSE-asavchenko/face-emotion-recognition.

Fig. 1. Class-wise sample frames from the VGAF validation set: Positive (first row), Neutral (second row) and Negative (third row).

At first, the ablation study of the proposed Algorithm 1 was performed for two DCNNs, namely, MobileNet v1 and EfficientNet-B0, described in our previous work [12]. All the details about training including loss functions and training parameters are available in the latter paper. We used features extracted by these neural networks with various STAT features and their concatenation for faces in a frame at step 9 and for frames in a video at step 11. The validation accuracy of the SVM classifier with RBF kernel and regularization parameter 1 is presented in Table 1.

Surprisingly, the MobileNet architecture here is 1–3% more accurate than much deeper EfficientNet-B2. Our additional experiments with other architectures (RexFace, EfficientNet-B0, ResNet-50, etc.) demonstrated that they are even worse than EfficientNet-B2. Secondly, concatenation of average and standard deviation features for aggregation of both facial embeddings and frame features provides the best possible quality. However, the usage of only maximum pooling for face descriptors (1) and standard deviation for frames together with MobileNet has only 0.13% lower accuracy. Moreover, if we examine various classifiers and preprocessing techniques (Table 2), the latter approach reaches the best validation accuracy 69.58%.

Comparison of our best result with existing state-of-the-art models is shown in Table 3. Here the proposed approach has 1% greater accuracy than the best-known single classifier [5]. It is also at least 5% better than any other facial processing model [7]. If we combine our facial processing technique with conventional OpenSMILE-based audio emotion classification with 55.3% accuracy (steps 13–17 of Algorithm 1), we can improve the overall accuracy up to 71.95%.

Table 1. Validation accuracy (%) for various STAT features, SVM RBF.

DCNN	STAT features for faces in a frame	STAT features for frames in a video					
		Avg	Max	Std	Min	(Max, Std)	(Avg, Std)
MobileNet	Avg	62.66	64.75	65.93	59.66	65.53	64.75
	Max	67.36	68.15	68.80	55.22	68.67	67.89
	Std	63.31	66.57	66.19	39.43	66.84	65.40
	Min	61.23	59.92	60.97	54.18	60.05	61.49
	(Max, Std)	66.71	67.62	68.93	55.09	68.28	68.14
	(Avg, Std)	65.80	67.10	67.49	60.05	67.10	68.93
EfficientNet	Avg	65.14	64.62	63.32	62.27	65.01	65.27
	Max	65.40	65.80	65.58	57.05	66.32	65.80
	Std	61.62	64.62	62.66	36.55	64.23	65.14
	(Max, Std)	65.93	66.19	66.06	57.18	66.84	66.58
	(Avg, Std)	66.32	68.02	64.75	59.27	66.55	67.75

Table 2. Validation accuracy (%) for various classifiers, MobileNet feature extractor.

Classifier	Max/Std	(Avg, Std)/(Avg, Std)
SVM RBF, original features	68.80	68.93
SVM RBF, L_2-normed features	69.58	67.89
Linear SVM, L_2-normed features	64.49	67.10
Random Forest	65.01	67.62
XGBoost	65.80	67.10
GRU	61.40	64.25
Frame-level attention	65.34	67.23

Here the hyper-parameter $\alpha = 0.9$ was chosen by the cross-validation on the given training set. The confusion matrices of three classifiers for faces, audio and the result of a blending are presented in Fig. 2. The latter is only slightly worse than the results of the best ensemble (74.2%) [7], but better than the accuracy (71.93%) of the second place [5] that combines 14 different models.

As we mentioned in the introduction, another important challenge for real-time systems is the frame processing time. The performance of CNNs are shown in Table 4. The running time to predict emotions was measured on the MSI GP63 8RE laptop (CPU Intel Core i7-8750H 2.2 GHz, GPU NVidia GTX1060, RAM 16Gb). Here it was assumed based on the statistics for the validation set that each frame contains approximately 3.5 faces found by MTCNN. As one can notice, the group-level emotion recognition is quite time-consuming operation even if only faces are processed, so that only 9 frames per second (FPS) can be processed even if GPU of a laptop is utilized. The usage of the state-of-the-art

Table 3. Validation accuracy (%) of existing single models for VGAF dataset.

Method	Modality	Accuracy, %
Hybrid networks [7]	two faces + audio + whole video + environmental object statistics + fighting detector	74.28
Our MobileNet-v1	faces + audio	71.95
Fusion of 14 models [5]	face (emotions/keypoints) + body + audio + whole video + optical flow	71.93
K-injection networks [4]	audio (acoustic/linguistic) + video	66.19
VGAFNet [2]	face + holistic + audio	61.61
Our MobileNet-v1	faces	69.58
Slowfast [5]	entire video	68.57
Self-attention from K-injection network [4]	text (video situation)	65.01
DenseNet-121, Hybrid Networks [7]	faces	64.75
Temporal shift model, Hybrid Networks [7]	entire video	63.71
ResNet-50, Activity Recognition Networks [3]	entire video	62.40
VGAFNet [2]	faces	60.18
Frame attention network [5]	faces	55.88
VGG19 privacy-safe [6]	entire video	52.36

Table 4. Performance of our visual models.

Metric	Hardware	MobileNet	EfficientNet
Avg inference time (ms)	CPU	40.67	80.11
per one face	GPU	9.29	13.84
Avg frame processing	CPU	219.104	352.78
time (ms)	GPU	112.24	127.98

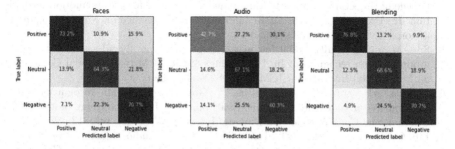

Fig. 2. Confusion matrix of the proposed approach, VGAF validation set.

ensembles with complex features extracted for each subject [5,7] can potentially lead to performance lower than 1 FPS. Hence, the improvement of the running time in this task is very acute.

4 Conclusion

In this paper, we introduced the novel Algorithm 1 for the group-level emotion recognition in videos based on facial analysis. Our approach is based on the MobileNet architecture, so that it can be applied even in embedded systems, e.g., in mobile applications. It is emphasized that excellent accuracy can be obtained by classifying only standard deviation of the frame-level features. Our technique is much better when compared to existing facial processing techniques and outperforms all existing single models (Table 3) for the group affect prediction in videos. If it is combined with conventional OpenSMILE-based audio emotion recognition, the accuracy will become higher than the very difficult ensemble [5] that took the second place in the EmotiW 2020 challenge.

Unfortunately, our best accuracy is 2% lower when compared to the best-known hybrid networks [7]. Hence, in future it is necessary to improve the quality by using better algorithms for audio emotion classification, video analysis (Slowfast or Temporal Binding Network [5]) or processing of situation textual descriptions that have been automatically extracted from a video [4].

Acknowledgements. The work is supported by RSF (Russian Science Foundation) grant 20-71-10010.

References

1. Veltmeijer, E.A., Gerritsen, C., Hindriks, K.: Automatic emotion recognition for groups: a review. IEEE Trans. Affect. Comput. (2021)
2. Sharma, G., Dhall, A., Cai, J.: Audio-visual automatic group affect analysis. IEEE Trans. Affect. Comput. (2021)
3. Pinto, J.R., et al.: Audiovisual classification of group emotion valence using activity recognition networks. In: Proceedings of the 4th International Conference on Image Processing, Applications and Systems (IPAS), pp. 114–119. IEEE (2020)
4. Wang, Y., Wu, J., Heracleous, P., Wada, S., Kimura, R., Kurihara, S.: Implicit knowledge injectable cross attention audiovisual model for group emotion recognition. In: Proceedings of the ACM International Conference on Multimodal Interaction (ICMI), pp. 827–834 (2020)
5. Sun, M., et al.: Multi-modal fusion using spatio-temporal and static features for group emotion recognition. In: Proceedings of the ACM International Conference on Multimodal Interaction (ICMI), pp. 835–840 (2020)
6. Petrova, A., Vaufreydaz, D., Dessus, P.: Group-level emotion recognition using a unimodal privacy-safe non-individual approach. In: Proceedings of the ACM International Conference on Multimodal Interaction (ICMI), pp. 813–820 (2020)
7. Liu, C., Jiang, W., Wang, M., Tang, T.: Group level audio-video emotion recognition using hybrid networks. In: Proceedings of the ACM International Conference on Multimodal Interaction (ICMI), pp. 807–812 (2020)
8. Savchenko, A.V.: Maximum-likelihood dissimilarities in image recognition with deep neural networks. Comput. Opt. **41**(3), 422–430 (2017)
9. Savchenko, A.V.: Probabilistic neural network with complex exponential activation functions in image recognition. IEEE Trans. Neural Netw. Learn. Syst. **31**(2), 651–660 (2020)

10. Cao, Q., Shen, L., Xie, W., Parkhi, O.M., Zisserman, A.: Vggface2: a dataset for recognising faces across pose and age. In: Proceedings of International Conference on Automatic Face & Gesture Recognition (FG), pp. 67–74. IEEE (2018)
11. Mollahosseini, A., Hasani, B., Mahoor, M.H.: AffectNet: a database for facial expression, valence, and arousal computing in the wild. IEEE Trans. Affect. Comput. **10**(1), 18–31 (2017)
12. Savchenko, A.V.: Facial expression and attributes recognition based on multi-task learning of lightweight neural networks. In: Proceedings of the 19th International Symposium on Intelligent Systems and Informatics (SISY), pp. 119–124. IEEE (2021)
13. Bargal, S.A., Barsoum, E., Ferrer, C.C., Zhang, C.: Emotion recognition in the wild from videos using images. In: Proceedings of the ACM International Conference on Multimodal Interaction (ICMI), pp. 433–436 (2016)
14. Knyazev, B., Shvetsov, R., Efremova, N., Kuharenko, A.: Convolutional neural networks pretrained on large face recognition datasets for emotion classification from video. arXiv preprint arXiv:1711.04598 (2017)
15. Savchenko, L., V. Savchenko, A.: Speaker-aware training of speech emotion classifier with speaker recognition. In: Karpov, A., Potapova, R. (eds.) SPECOM 2021. LNCS (LNAI), vol. 12997, pp. 614–625. Springer, Cham (2021). https://doi.org/10.1007/978-3-030-87802-3_55
16. Demochkina, P., Savchenko, A.V.: MobileEmotiFace: efficient facial image representations in video-based emotion recognition on mobile devices. In: Del Bimbo, A., et al. (eds.) ICPR 2021. LNCS, vol. 12665, pp. 266–274. Springer, Cham (2021). https://doi.org/10.1007/978-3-030-68821-9_25
17. Lomotin, K., Makarov, I.: Automated image and video quality assessment for computational video editing. In: van der Aalst, W.M.P., et al. (eds.) AIST 2020. LNCS, vol. 12602, pp. 243–256. Springer, Cham (2021). https://doi.org/10.1007/978-3-030-72610-2_18
18. Zuenko, D., Makarov, I.: Style-transfer autoencoder for efficient deep voice conversation. In: Proceedings of the International Symposium on Computational Intelligence and Informatics (CINTI), pp. 41–6. IEEE (2021)
19. Savchenko, A.V.: Phonetic words decoding software in the problem of Russian speech recognition. Autom. Remote. Control. **74**(7), 1225–1232 (2013)
20. Eyben, F., Wöllmer, M., Schuller, B.: OpenSMILE: the Munich versatile and fast open-source audio feature extractor. In: Proceedings of the 18th ACM International Conference on Multimedia, pp. 1459–1462 (2010)
21. Schuller, B., et al.: The INTERSPEECH 2013 computational paralinguistics challenge: Social signals, conflict, emotion, autism. In: Proceedings of 14th Annual Conference of the International Speech Communication Association (INTERSPEECH) (2013)
22. Savchenko, A.V., Savchenko, V.V.: A method for measuring the pitch frequency of speech signals for the systems of acoustic speech analysis. Meas. Tech. **62**(3), 282–288 (2019)

Outfit Recommendation using Graph Neural Networks via Visual Similarity

Diana Zagidullina[1] and Ilya Makarov[1,2(✉)]

[1] HSE University, Moscow, Russia
drzagidullina@edu.hse.ru, iamakarov@hse.ru
[2] Artificial Intelligence Research Institute (AIRI), Moscow, Russia

Abstract. Computer vision plays an important role in the development of the fashion industry. There has been a lot of research done on various fashion recommendations, and determining the compatibility of clothing is a key factor in most of them. Solving this problem can help users buy items that go well with their current wardrobe, and help stores sell multiple clothing items at once. Previous research has mainly focused on learning compatibility between two clothing elements. There are several approaches that take into account the outfit as a whole but they require rich textual data. In this work, we only use images of clothing from the Polyvore dataset to extract visual features. By representing outfits in the form of a graph, we train node embeddings based on graph structure and node features. We then train multi-layer perceptron to classify the set of embeddings representing the outfit. We compare our method with the relevant works in two tasks: outfit compatibility prediction and fill-in-the-blank. Our approach showed the best result among approaches that use only images on the first task and showed state-of-the-art result on the second task.

Keywords: Fashion recommendation · Outfit recommendation · Fashion outfit compatibility · Graph neural networks · graphSAGE · Node embeddings

1 Introduction

The fashion industry offers several interesting challenges for computer vision. New challenges are emerging along with the rapid evolution of the fashion industry towards online business, which is becoming more personalized. Clothing parsing [25,54], clothing style recognition [47], trend forecasting [28,55], different types of fashion recommendations [18,24,27] - a lot of research has been done using computational methods to solve these problems in fashion. Research in this area directly influences what people buy, how they shop, and how they dress.

The purpose of this work is to learn how to answer the question: "What item of clothing should be chosen from the proposed ones so that it complements the outfit well?". This task is called "Fill in the blank". The answer to this question is based on determining the compatibility of a set of clothing items that are our second task. Both tasks are illustrated in Fig. 1.

© The Author(s), under exclusive license to Springer Nature Switzerland AG 2022
E. Burnaev et al. (Eds.): AIST 2021, LNCS 13217, pp. 208–222, 2022.
https://doi.org/10.1007/978-3-031-16500-9_18

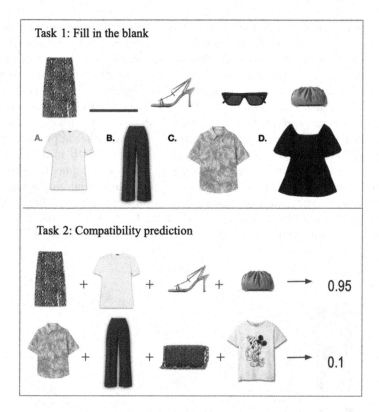

Fig. 1. Outfit recommendation

The outfit compatibility prediction task allows answering many user questions, for example:"What pair of shoes to choose for this dress?", "Which T-shirt should I buy to make my wardrobe more functional?", "How to make a beautiful outfit?", "How to create a capsule wardrobe?". Determining the compatibility of clothing items is also the basis of many recommendation systems [17,18,27]. Also more and more online stores started to provide customers with ready-made collections of outfits to attract new customers and make their life easier, as well as to sell not only one unit of goods, but several at once.

Previous works [6,16,40,46,51] on the problem of outfit compatibility prediction mainly focused on the compatibility of a pair of items and failed to show good results on outfit compatibility. In such models, we get the same prediction score for any given pair of garments every time, regardless of the outfit. Some works represent outfit as a sequence [14] of elements where a certain order of the elements is required. The problem with such approaches is that the elements in the outfit have relations not only with the neighbors but also with all the elements. Moreover, there is no fixed order of elements in the outfit. Other works [8,14,24] also have limitations, they need rich item contextual information such as category labels (e.g., top, dress) and descriptions (e.g., "dark blue denim shorts", "leather knotted saddle bag") to train models. This requirement reduces the applicability of the model since often not all information is available.

To remedy these limitations, in this paper, we use only images of clothing items to extract visual features (see Sect. 4.2) and no other data is required. We represent the outfit in the form of a graph (see Fig. 2), where the nodes correspond to the garments, and two nodes are connected if these elements occur in the same outfit. Graph representation helps to fully exploit the complex relationships between garments bypassing the problems that arise in other approaches. There are different graph neural networks that help to work with graph-structured data. The key idea behind them is to generate node embeddings based on local neighborhoods. Nodes aggregate information from their neighbors using neural networks. As a result, "similar" nodes in the graph have embeddings that are close together. Such representation should work for outfits because the embedding of one clothing item stores information about items from the whole outfit.

Fig. 2. Graph representation of an outfit.

We chose GraphSAGE model [13] to learn node embeddings, and multi-layer perceptron to classify embeddings of outfit items. We perform experiments on Polyvore dataset [14] and through two tasks (1) outfit compatibility prediction (see Sect. 5.1) and (2) fill-in-the-blank (see Sect. 5.2). We compare our model with other methods and obtain state-of-the-art results on the FITB task.

This paper is organized as follows. The "Literature Review" section contains existing approaches to solving relevant problems. The "Dataset" section describes the dataset that is used to train and test our model. In the "Proposed Method" section we first formulate the problem, then we describe our approach. The "Experiments" section presents the metrics that are used to evaluate performance, and also contains a comparison of our approach with previous methods on the described metrics. The "Discussion" section provides examples of how our model works. The "Conclusion" section describes the results of our work and plans for future work.

2   Literature Review

2.1   Outfit Compatibility Prediction

Many articles explore different approaches to determining the compatibility of garments in an outfit. This growing interest is because outfit compatibility is the basis of many fashion recommendations. Wei-Lin et al. [17] solved the problem of outfit compatibility as part of the automatic creation of capsule wardrobes. They proposed an algorithm that should collect the minimum set of items from the given inventory, which will ensure the maximum possible mix-and-match outfits.

The first works on this topic were mainly devoted to pair-wise compatibility. McAuley et al. [40] learned style space where compatible garments fall close to each other. They used Low-rank Mahalanobis Transformation to represent clothing items in this space and calculated distances there. Similar to that, Veit et al. [51] applied an end-to-end Siamese convolutional neural network to transform images of co-purchased items into a similar metric distance space that determines the compatibility of garments. Song et al. [46] showed that the integration of visual and contextual information can improve the compatibility prediction. They learned a style space by using a dual autoencoder network and a Bayesian Personalized Ranking (BPR) [15] framework. Some other works [6,16,45] have also used similar latent spaces to determine pair-wise compatibility of elements in various aspects.

Some work, in addition to determining the outfit compatibility, has also been focused on creating additional comments. Lin et al. [26] developed a model that creates a sentence for each recommendation explaining why the top and bottom match together. Wang et al. [52] also focused on diagnosing the outfit, that is, identifying incompatible factors in a fashionable outfit, such factors are indicated as an explanation.

However, all of these works focused only on pair-wise compatibility, which does not allow us to fully take into account the complex relationships of elements in the outfit. To address this limitation, some studies represented outfits as a sequence, in which the elements are in a special order (top, bottom, accessories). Li et al. [24] used a multimodal approach to get the most out of the extracted information from items and trained a Recurrent Neural Network to evaluate the outfit compatibility. Han et al. [14] used a bidirectional LSTM model to predict the next element sequentially based on the previously seen items in both directions.

The sequence representation of the outfit outperforms pair representation but still has a disadvantage because the relations among garments in an outfit are not ordered since a garment has relations with not only its neighbors in sequence. To remedy this, Cui et al. [8] presented a graph representation of the outfit, they built a graph, where each node corresponds to a specific category, and each edge contains information about the interaction between the two categories. They used Node-wise Graph Neural Networks and the attention mechanism to calculate outfit compatibility. This approach requires extensive information about the garment, such as the category label and description.

There have been many studies on compatibility prediction, but the majority of them focused on pair-wise compatibility. Some works considered an entire outfit, but they either used a fixed order of elements in the outfit or required rich contextual data. In this work, we address these limitations by using only images of garments and using a graph representation of the outfit.

2.2 Graph Embeddings

There exists a large variety of graph embeddings approaches that learn low-dimensional embeddings. The first graph embedding methods were based on matrix-factorization, they use a matrix describing graph and factorize it to create the embedding of the network. Some methods directly factorize the adjacency matrix [9,22,53], other use proximity matrix of the graph [43,49] or Laplacian matrices [4,5]. These methods have a weakness, they are computationally expensive for large graphs.

Other methods transform data into a sequence-structured format based on random walks or diffusion, the main idea is to maximize the probability of observing the neighborhood of a node taking into account its embedding. Perozzi et al. first proposed a sequence-based algorithm called Deep-Walk [4], inspired by word2vec [41]. After that, Grover et al. improved this algorithm by using second-order random walks in the node2vec [12] algorithm. Sequence-based models only work with structural information and they are not able to capture additional node attributes that make them inherently transductive and they can not generate embeddings for unseen nodes.

With the development of deep learning, some works started to apply neural networks to graph-structured data directly. These algorithms are able to capture node features to generate node embeddings. Kipf et al. proposed Graph Convolutional Network [20], the idea is to apply convolutional neural network on graphs under the transductive semi-supervised learning setting. Hamilton et al. offered the inductive GraphSAGE [13] algorithm that is an extension of the GCN framework.

Different network representations via manual feature engineering and network embeddings can be used for social network recommendations [3,29,30,33–36], node attributes inference [1,44] and link prediction [11,21,31,32,38]. Surveys on network embedding applications for attributed networks can be found in [37,39].

In this work, we use GraphSAGE algorithm since it allows us to efficiently generate embeddings for previously unseen nodes. This helps us predict outfit compatibility for data that was not in the training dataset.

3   Dataset

We chose the Polyvore dataset that was released by Han et al. [14] and additionally filtered by Cui et al. [8]. Polyvore is a popular fashion site that has been heavily used by stylists and people interested in fashion until it was bought by multi-brand retailer SSENSE and closed in 2018. The site allows the user to create personal collages of clothes, shoes, accessories, jewelry, cosmetics, household, and interior items. Polyvore has its large database of such items, and the user can also use their photos and pictures to compose collages. It contained a large number of different items of clothing and people could upload their photos of clothing items as well. Users could also collect outfits from these images right on the site.

The advantage of this dataset is that compatibility ratings are based on the opinions of people who are professional stylists or fashion influencers. Such data is more accurate since it can be more difficult for ordinary users to evaluate the combination of elements in a complex outfit. The dataset has been used in many papers on a similar topic [7,8,14,50,52] which makes it possible to compare the results. An example of an outfit in the dataset is illustrated in Fig. 3 which contains rich information about the outfit such as clear images of clothing items, categories, and descriptions. We only use garments images in our work.

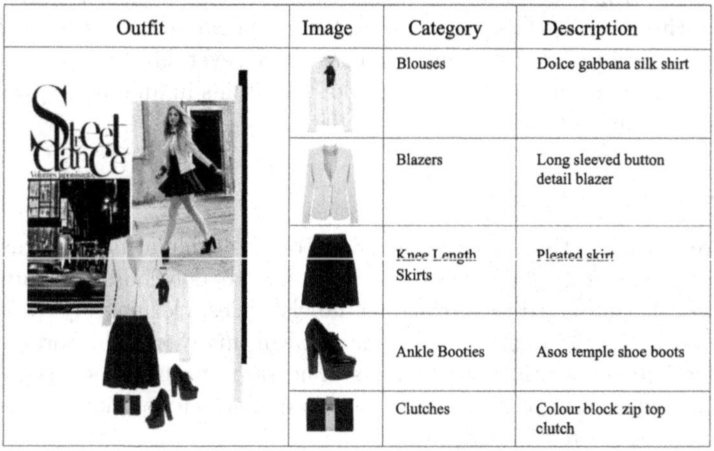

Outfit	Image	Category	Description
		Blouses	Dolce gabbana silk shirt
		Blazers	Long sleeved button detail blazer
		Knee Length Skirts	Pleated skirt
		Ankle Booties	Asos temple shoe boots
		Clutches	Colour block zip top clutch

Fig. 3. Example of outfit collage on Polyvore.com.

The dataset contains a total of 164,379 items covering 380 categories, of which 21,899 different outfits are composed. The outfits are divided into 3 non-overlapping sets. 17,316 outfits in the training set, 1,497 in validation and 3,076 in testing. The dataset has been cleaned of categories that are not garments or that are rare (less than 100 times, similar to [8]). Also, only outfits containing more than 3 elements were left in the dataset. As a result, there are 16983 outfits left in the training set, 1497 and 2697 in the validation and testing sets, respectively, a total of 126 054 garments covering 120 categories. For simplicity, the maximum number of clothing items in an outfit has been reduced to 8, the average number is 6.2.

The FITB task contains 3,076 questions and the outfit compatibility task has 3,076 valid and 4,000 invalid outfits. The training set in the original dataset includes only compatible outfits. To create negative samples, we randomly replace one item in each compatible garment with a random clothing item of the same category. Perhaps the resulting outfits are also compatible, but we expect that these outfits are less compatible than those collected by the experts at Polyvore. Thus, the size of the training set is doubled from 16983 to 33966.

4 Proposed Method

In this section, we first formulate the outfit compatibility problem. We then discuss feature extraction. We also explain how we build a graph from outfits. After that, we describe how we produce node embeddings. Finally, we introduce how we classify outfits into compatible and non-compatible based on node features.

4.1 Problem Formulation

We aim to determine if a set of clothing items is compatible. We have a set of positive outfits $S_+ = \{s_1^+, s_2^+, s_3^+, ...\}$ and set of negative outfits $S_- = \{s_1^-, s_2^-, s_3^-, ...\}$ as a training data. The outfit s_i consists of several fashion items, for each of which we have an image. Having the clothing items in an outfit s, we need to count the compatibility score p_s.

4.2 Features

Fashion recommendations are usually based on multimodal inputs (images and text) from users [2,10,23,42]. Most approaches to determining compatibility require rich textual attributes (category labels, titles, clothing styles) and they will not work if only images are available. But many online stores may not have a detailed description of garments. The same problem occurs in fashion applications, where it is easier for the user to upload only a photo of the clothes than a photo and a text description.

To remove this limitation, we use only visual information of items to determine fashion compatibility. To extract feature vectors from images we use GoogleNet InceptionV3 [48] model that is a convolutional neural network for assisting in image analysis and object detection. The model has been shown to attain greater than 78.1% accuracy on the ImageNet dataset. InceptionV3 generates visual feature vectors of dimension 2048.

4.3 Graph Construction

We construct undirected graph $\mathcal{G} = (\mathcal{N}, \mathcal{E})$ using the training dataset. Each node $n_i \in \mathcal{N}$ represents the clothing item, two nodes (n_i, n_j) are connected by an edge $(i, j) \in \mathcal{E}$ if the corresponding garments are present in the same outfit. Each node in the graph corresponds to a vector of features $f_i \in \mathbb{R}^{2048}$ obtained from the image of clothing item.

4.4 Node Embeddings

We want to produce node embeddings with the following intuition: map each node to a low-dimensional vector so that nearby nodes in the graph have embeddings that are close together. This means that the embeddings of clothing items that can be combined in one outfit will be close since each node corresponds to a garment.

To generate node embeddings we use GraphSAGE [13] model that is an inductive framework for constructing embeddings of nodes using graph structure and nodes feature inputs. We use visual features extracted from images (see Sect. 4.2) as feature inputs. The advantage of this approach comparing to transductive approaches is that it allows to efficiently generate embeddings for previously unseen nodes. Thus, we do not have to retrain the model to determine the compatibility of new clothing items that expands the application of this model.

The GraphSAGE idea is to aggregate "messages" from a node's local neighborhood. At the beginning for each node $n_i \in \mathcal{N}$, its embedding is initialized with feature input f_i^0. Then the node embedding is updated K ("search depth") times using the following formulas:

$$
\begin{aligned}
f_{N(i)}^k &= AGG(\{f_j^{k-1}, \forall j \in N(i)\}) \\
f_i^k &= \sigma(W_k \cdot CONCAT(f_i^{k-1}, f_{N(i)}^k))
\end{aligned}
\tag{1}
$$

where $N(i)$ - neighborhood function, AGG - any differentiable function that maps set of vectors to a single vector, σ - non-linearity function, W_k - weight matrices. $CONCAT$ function concatenates self embedding and neighbor embeddings from the previous step. As aggregation function (AGG) we use mean operator which takes the elementwise mean of the vectors in $\{f_j^{k-1}, \forall j \in \forall N(i)\}$:

$$
AGG = \sum_{\forall j \in N(i)} \frac{f_j^{k-1}}{|N(i)|}
\tag{2}
$$

To learn the parameters in a fully unsupervised setting, a graph-based loss function is used that makes nearby nodes have similar embeddings, and disparate nodes have different embeddings:

$$
J(f_i) = -\log(\sigma(f_i^\top f_j)) - S \cdot \mathbb{E}_{j_n \sim NEG_n(i)} \log(\sigma(-f_i^\top f_{j_n}))
\tag{3}
$$

where σ is the sigmoid function, j is a node co-occurring near node j on fixed-length random walk, NEG_n is a negative sampling distribution, and S defines the number of negative samples.

4.5 Classifier

Given embeddings for items in outfit $f_s = \{f_1, f_2, f_3, ... | f_i \in \mathbb{R}^{256}\}$ we need to predict the compatibility score of outfit. We use multi-layer perceptron (MLP) to classify outfits into 2 classes: positive and negative. We concatenate $f_i \in f_s$ in one vector $\overline{f_s}$ to feed this vector to our model. If outfit has less than 8 clothing items we append 256 dimensional zero vectors to $\overline{f_s}$. As a result each outfit is represented as a 2048-dimensional vector.

MLP consists of 6 layers of nodes: an input layer, 4 hidden layers, and an output layer. Hidden layers have sizes 512, 512, 256, and 128. Each node in the hidden layers is a neuron that uses ReLU activation function that is defined as $f(x) = max(0, x)$. To learn the parameters, we use Adam optimizer [19]. As a loss function, we use Cross-Entropy that is defined as follows:

$$Loss(\hat{y}, y, W) = -y\ln\hat{y} - (1 - y)\ln(1 - \hat{y}) + \alpha\|W\|_2^2 \tag{4}$$

where W - model weights, y - ground truth label, $\hat{y}$ - model predicted label, $\alpha\|W\|_2^2$ - is an L2-regularization term that penalizes the model and $\alpha > 0$ is a hyperparameter that controls the magnitude of the penalty.

5   Experiments

To evaluate the performance of our model, we run a set of experiments on "Outfit Compatibility Prediction" and "Fill-in-the-blank Fashion Recommendation" tasks, and then we compare the results with alternative methods.

5.1   Outfit Compatibility Prediction

In this task, we need to predict an outfit compatibility score that corresponds to the overall compatibility of the set of garments that make up the outfit. This can be useful in real life, for example, a user might create an outfit and want a more professional opinion on the outfit's compatibility. If the compatibility score from the model is greater than 0.5, we consider the outfit to be compatible, otherwise, incompatible. We estimate performance using the area under the receiver operating characteristic curve (AUC), this metric is widely used in various recommendation systems. We show the results of comparison with other models in Table 1.

Table 1. Evaluation of our model on the compatibility prediction task

Method	Compatibility AUC
Random	0.5104
SiameseNet [51]	0.7087
LMT [40]	0.6782
Bi-LSTM [14]	0.8427
NGNN [8]	**0.9722**
Ours	0.9653

5.2   Fill-in-the-Blank Fashion Recommendation

The problem of filling-in-the-blank (FITB) questions was first presented by Han et al. [14]. In this task, we have a sequence of garments, we need to choose the most compatible item from the proposed elements to complete the outfit. People often have to solve this problem in real life, for example, a user wants to choose a jacket that matches his T-shirt, trousers, and shoes. Fill-in-the-blank questions were created from the test dataset. For each positive outfit, we replace a randomly selected item with a blank. We then randomly select 3 negative items from the dataset and use them as suggested items along with the correct item. Random elements are expected to be less compatible than the ones chosen by users on Polyvore. To get an answer, we supplement the outfit with 4 options and count compatibility for each. The item with the highest compatibility is selected as the answer. We measure performance by the accuracy of choosing the correct answer from four candidates (FITB accuracy). We show the results of comparison with other models in Table 2.

Table 2. Evaluation of our model on the FITB task

Method	FITB accuracy
Random	24.07%
SiameseNet [51]	67.01%
LMT [40]	50.91%
Bi-LSTM [14]	67.01%
NGNN [8]	78.13%
Ours	**87.67%**

6   Discussion

In this section we visualize some examples of our model on outfit compatibility prediction task (see Fig. 4).

Our model gave high compatibility score to the first two examples. Indeed, in the first outfit, items go well with each other, as they are all black/white and share a smart-casual style. Items in the second row have high compatibility score because they share casual/safari style and have well-matching colors. We can see that the model gave a lower rating to the next two outfits. We suppose that this may be since these outfits have repeated clothing images (two bodysuits on both sides in the third outfit and two shoes in the fourth outfit). The fifth outfit received a low score since among the elements of clothing there was also an image of the inscription, which is not an element of clothing. The last outfit has two shirts and our network has given a low compatibility score to this outfit since the outfit has redundant items.

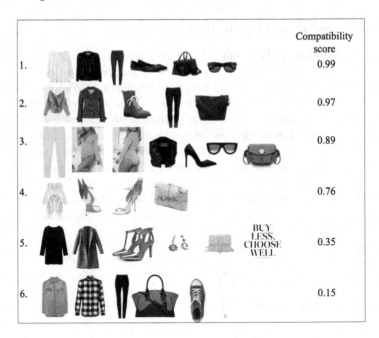

Fig. 4. Examples of our method on the fashion outfit compatibility prediction task.

7   Conclusion

In this paper, we propose an approach to predict outfit compatibility score. We use a graph representation of the outfit because the complex combination of garments affects the outfit compatibility and the graph structure allows us to express this. We generate node embeddings using GraphSAGE based on graph structure and visual features from images of garments. We then train multi-layer perceptron to classify the set of embeddings representing the outfit into 2 classes: compatible and incompatible outfits. Our model showed the best result in the outfit compatibility task (Table 5.1) among approaches that use only images and state-of-the-art performance in the fill-in-the-blank task (Table 5.2) based on the Polyvore dataset. In addition, we use only images of clothing items to determine the outfit compatibility, which can significantly expand the application of this model.

As to future work, we would like to make a personalized outfit recommendation system that takes into account individual aspects such as style preference and human body shape. We would also like to train our model on a dataset collected from images from recent fashion shows. Thus, the model will give a fashion evaluation of the outfit, reflecting its modernity, style, and presence of trends in the outfit.

Acknowledgement. The work of Ilya Makarov was supported by the Russian Science Foundation under grant 22-21-00227 and performed at HSE University, Moscow, Russia.

References

1. Ananyeva, M., Makarov, I., Pendiukhov, M.: GSM: inductive learning on dynamic graph embeddings. In: Bychkov, I., Kalyagin, V.A., Pardalos, P.M., Prokopyev, O. (eds.) NET 2018. SPMS, vol. 315, pp. 85–99. Springer, Cham (2020). https://doi. org/10.1007/978-3-030-37157-9_6
2. Andreeva, E., Ignatov, D.I., Grachev, A., Savchenko, A.V.: Extraction of visual features for recommendation of products via deep learning. In: van der Aalst, W.M.P., et al. (eds.) AIST 2018. LNCS, vol. 11179, pp. 201–210. Springer, Cham (2018). https://doi.org/10.1007/978-3-030-11027-7_20
3. Averchenkova, A., et al.: Collaborator recommender system. In: Bychkov, I., Kalyagin, V.A., Pardalos, P.M., Prokopyev, O. (eds.) NET 2018. SPMS, vol. 315, pp. 101–119. Springer, Cham (2020). https://doi.org/10.1007/978-3-030-37157-9_7
4. Belkin, M., Niyogi, P.: Laplacian eigenmaps and spectral techniques for embedding and clustering. In: Nips, vol. 14, pp. 585–591 (2001)
5. Brand, M.: Continuous nonlinear dimensionality reduction by kernel eigenmaps. In: IJCAI, pp. 547–554. Citeseer (2003)
6. Chen, L., He, Y.: Dress fashionably: learn fashion collocation with deep mixed-category metric learning. In: Proceedings of the AAAI Conference on Artificial Intelligence, vol. 32, no. 1 (2018)
7. Cucurull, G., Taslakian, P., Vazquez, D.: Context-aware visual compatibility prediction. In: Proceedings of the IEEE/CVF Conference on Computer Vision and Pattern Recognition, pp. 12617–12626 (2019)
8. Cui, Z., Li, Z., Wu, S., Zhang, X.Y., Wang, L.: Dressing as a whole: outfit compatibility learning based on node-wise graph neural networks. In: The World Wide Web Conference, pp. 307–317 (2019)
9. Deerwester, S., Dumais, S.T., Furnas, G.W., Landauer, T.K., Harshman, R.: Indexing by latent semantic analysis. J. Am. Soc. Inf. Sci. **41**(6), 391–407 (1990)
10. Demochkin, K., Savchenko, A.V.: Multi-label image set recognition in visually-aware recommender systems. In: van der Aalst, W.M.P., et al. (eds.) AIST 2019. LNCS, vol. 11832, pp. 291–297. Springer, Cham (2019). https://doi.org/10.1007/978-3-030-37334-4_26
11. Gerasimova, O., Makarov, I.: Higher school of economics co-authorship network study. In: 2019 2nd International Conference on Computer Applications & Information Security (ICCAIS), pp. 1–4. IEEE (2019)
12. Grover, A., Leskovec, J.: node2vec: scalable feature learning for networks. In: Proceedings of the 22nd ACM SIGKDD International Conference on Knowledge Discovery and Data Mining, pp. 855–864 (2016)
13. Hamilton, W.L., Ying, R., Leskovec, J.: Inductive representation learning on large graphs. In: Proceedings of the 31st International Conference on Neural Information Processing Systems, pp. 1025–1035 (2017)
14. Han, X., Wu, Z., Jiang, Y.G., Davis, L.S.: Learning fashion compatibility with bidirectional lstms. In: Proceedings of the 25th ACM International Conference on Multimedia, pp. 1078–1086 (2017)

15. He, R., McAuley, J.: VBPR: visual bayesian personalized ranking from implicit feedback. In: Proceedings of the AAAI Conference on Artificial Intelligence, vol. 30, no. 1 (2016)
16. He, R., Packer, C., McAuley, J.: Learning compatibility across categories for heterogeneous item recommendation. In: 2016 IEEE 16th International Conference on Data Mining (ICDM), pp. 937–942. IEEE (2016)
17. Hsiao, W.L., Grauman, K.: Creating capsule wardrobes from fashion images. In: Proceedings of the IEEE Conference on Computer Vision and Pattern Recognition, pp. 7161–7170 (2018)
18. Hu, Y., Yi, X., Davis, L.S.: Collaborative fashion recommendation: a functional tensor factorization approach. In: Proceedings of the 23rd ACM International Conference on Multimedia, pp. 129–138 (2015)
19. Kingma, D.P., Ba, J.: Adam: a method for stochastic optimization. arXiv preprint arXiv:1412.6980 (2014)
20. Kipf, T.N., Welling, M.: Semi-supervised classification with graph convolutional networks. arXiv preprint arXiv:1609.02907 (2016)
21. Kiselev, D., Makarov, I.: Prediction of new itinerary markets for airlines via network embedding. In: van der Aalst, W.M.P., et al. (eds.) AIST 2019. CCIS, vol. 1086, pp. 315–325. Springer, Cham (2020). https://doi.org/10.1007/978-3-030-39575-9_32
22. Kruskal, J.B.: Multidimensional scaling. Sage (1978)
23. Li, X., Wang, X., He, X., Chen, L., Xiao, J., Chua, T.S.: Hierarchical fashion graph network for personalized outfit recommendation. In: Proceedings of the 43rd International ACM SIGIR Conference on Research and Development in Information Retrieval, pp. 159–168 (2020)
24. Li, Y., Cao, L., Zhu, J., Luo, J.: Mining fashion outfit composition using an end-to-end deep learning approach on set data. IEEE Trans. Multimedia $19(8)$, 1946–1955 (2017)
25. Liang, X., Lin, L., Yang, W., Luo, P., Huang, J., Yan, S.: Clothes co-parsing via joint image segmentation and labeling with application to clothing retrieval. IEEE Trans. Multimedia $18(6)$, 1175–1186 (2016)
26. Lin, Y., Ren, P., Chen, Z., Ren, Z., Ma, J., De Rijke, M.: Explainable outfit recommendation with joint outfit matching and comment generation. IEEE Trans. Knowl. Data Eng. $32(8)$, 1502–1516 (2019)
27. Liu, S., et al.: Hi, magic closet, tell me what to wear! In: Proceedings of the 20th ACM International Conference on Multimedia, pp. 619–628 (2012)
28. Ma, Y., Ding, Y., Yang, X., Liao, L., Wong, W.K., Chua, T.S.: Knowledge enhanced neural fashion trend forecasting. In: Proceedings of the 2020 International Conference on Multimedia Retrieval, pp. 82–90 (2020)
29. Makarov, I., Bulanov, O., Gerasimova, O., Meshcheryakova, N., Karpov, I., Zhukov, L.E.: Scientific matchmaker: collaborator recommender system. In: van der Aalst, W.M.P., et al. (eds.) AIST 2017. LNCS, vol. 10716, pp. 404–410. Springer, Cham (2018). https://doi.org/10.1007/978-3-319-73013-4_37
30. Makarov, I., Bulanov, O., Zhukov, L.E.: Co-author recommender system. In: Kalyagin, V.A., Nikolaev, A.I., Pardalos, P.M., Prokopyev, O.A. (eds.) NET 2016. SPMS, vol. 197, pp. 251–257. Springer, Cham (2017). https://doi.org/10.1007/978-3-319-56829-4_18
31. Makarov, I., Gerasimova, O.: Link prediction regression for weighted co-authorship networks. In: Rojas, I., Joya, G., Catala, A. (eds.) IWANN 2019. LNCS, vol. 11507, pp. 667–677. Springer, Cham (2019). https://doi.org/10.1007/978-3-030-20518-8_55

32. Makarov, I., Gerasimova, O.: Predicting collaborations in co-authorship network. In: 2019 14th International Workshop on Semantic and Social Media Adaptation and Personalization (SMAP), pp. 1–6. IEEE (2019)
33. Makarov, I., Gerasimova, O., Sulimov, P., Korovina, K., Zhukov, L.E.: Joint node-edge network embedding for link prediction. In: van der Aalst, W.M.P., et al. (eds.) AIST 2018. LNCS, vol. 11179, pp. 20–31. Springer, Cham (2018). https://doi.org/10.1007/978-3-030-11027-7_3
34. Makarov, I., Gerasimova, O., Sulimov, P., Zhukov, L.E.: Co-authorship network embedding and recommending collaborators via network embedding. In: van der Aalst, W.M.P., et al. (eds.) AIST 2018. LNCS, vol. 11179, pp. 32–38. Springer, Cham (2018). https://doi.org/10.1007/978-3-030-11027-7_4
35. Makarov, I., Gerasimova, O., Sulimov, P., Zhukov, L.E.: Recommending co-authorship via network embeddings and feature engineering: the case of national research university higher school of economics. In: Proceedings of the 18th ACM/IEEE on Joint Conference on Digital Libraries, pp. 365–366. ACM (2018)
36. Makarov, I., Gerasimova, O., Sulimov, P., Zhukov, L.E.: Dual network embedding for representing research interests in the link prediction problem on co-authorship networks. PeerJ Comput. Sci. **5**, e172 (2019)
37. Makarov, I., Kiselev, D., Nikitinsky, N., Subelj, L.: Survey on graph embeddings and their applications to machine learning problems on graphs. PeerJ Comput. Sci. **7**, e357 (2021)
38. Makarov, I., Korovina, K., Kiselev, D.: Jonnee: joint network nodes and edges embedding. IEEE Access **9**, 144646–144659 (2021)
39. Makarov, I., Makarov, M., Kiselev, D.: Fusion of text and graph information for machine learning problems on networks. PeerJ Comput. Sci. **7**, e526 (2021)
40. McAuley, J., Targett, C., Shi, Q., van Den Hengel, A.: Image-based recommendations on styles and substitutes. In: Proceedings of the 38th International ACM SIGIR Conference on Research and Development in Information Retrieval, pp. 43–52 (2015)
41. Mikolov, T., Sutskever, I., Chen, K., Corrado, G.S., Dean, J.: Distributed representations of words and phrases and their compositionality. Adv. Neural. Inf. Process. Syst. **26**, 3111–3119 (2013)
42. Revanur, A., Kumar, V., Sharma, D.: Semi-supervised visual representation learning for fashion compatibility. In: Fifteenth ACM Conference on Recommender Systems, pp. 463–472 (2021)
43. Roweis, S.T., Saul, L.K.: Nonlinear dimensionality reduction by locally linear embedding. Science **290**(5500), 2323–2326 (2000)
44. Rustem, M.K., Makarov, I., Zhukov, L.E.: Predicting psychology attributes of a social network user. In: Proceedings of the Fourth Workshop on Experimental Economics and Machine Learning (EEML 2017), Dresden, Germany, 17–18 September 2017, pp. 1–7. CEUR WP (2017)
45. Shih, Y.S., Chang, K.Y., Lin, H.T., Sun, M.: Compatibility family learning for item recommendation and generation. In: Proceedings of the AAAI Conference on Artificial Intelligence. vol. 32, no. 1 (2018)
46. Song, X., Feng, F., Liu, J., Li, Z., Nie, L., Ma, J.: Neurostylist: neural compatibility modeling for clothing matching. In: Proceedings of the 25th ACM International Conference on Multimedia, pp. 753–761 (2017)
47. Sun, G.L., Wu, X., Chen, H.H., Peng, Q.: Clothing style recognition using fashion attribute detection. In: Proceedings of the 8th International Conference on Mobile Multimedia Communications, pp. 145–148 (2015)

48. Szegedy, C., Vanhoucke, V., Ioffe, S., Shlens, J., Wojna, Z.: Rethinking the inception architecture for computer vision. In: Proceedings of the IEEE Conference on Computer Vision and Pattern Recognition, pp. 2818–2826 (2016)

49. Tenenbaum, J.B., De Silva, V., Langford, J.C.: A global geometric framework for nonlinear dimensionality reduction. Science **290**(5500), 2319–2323 (2000)

50. Vasileva, M.I., Plummer, B.A., Dusad, K., Rajpal, S., Kumar, R., Forsyth, D.: Learning type-aware embeddings for fashion compatibility. In: Proceedings of the European Conference on Computer Vision (ECCV), pp. 390–405 (2018)

51. Veit, A., Kovacs, B., Bell, S., McAuley, J., Bala, K., Belongie, S.: Learning visual clothing style with heterogeneous dyadic co-occurrences. In: Proceedings of the IEEE International Conference on Computer Vision, pp. 4642–4650 (2015)

52. Wang, X., Wu, B., Zhong, Y.: Outfit compatibility prediction and diagnosis with multi-layered comparison network. In: Proceedings of the 27th ACM International Conference on Multimedia, pp. 329–337 (2019)

53. Wold, S., Esbensen, K., Geladi, P.: Principal component analysis. Chemom. Intell. Lab. Syst. **2**(1–3), 37–52 (1987)

54. Yamaguchi, K., Kiapour, M.H., Ortiz, L.E., Berg, T.L.: Retrieving similar styles to parse clothing. IEEE Trans. Pattern Anal. Mach. Intell. **37**(5), 1028–1040 (2014)

55. Yu, Y., Hui, C.L., Choi, T.M.: An empirical study of intelligent expert systems on forecasting of fashion color trend. Expert Syst. Appl. **39**(4), 4383–4389 (2012)

Data Analysis and Machine Learning

Scalable Computation of Prediction Intervals for Neural Networks via Matrix Sketching

Alexander Fishkov[✉] and Maxim Panov[✉]

Skolkovo Institute of Science and Technology,
Bolshoy Boulevard 30, bld. 1, Moscow 121205, Russia
{alexander.fishkov,m.panov}@skoltech.ru

Abstract. Accounting for the uncertainty in the predictions of modern neural networks is a challenging and important task in many domains. Existing algorithms for uncertainty estimation require modifying the model architecture and training procedure (e.g., Bayesian neural networks) or dramatically increase the computational cost of predictions such as approaches based on ensembling. This work proposes a new algorithm that can be applied to a given trained neural network and produces approximate prediction intervals. The method is based on the classical delta method in statistics but achieves computational efficiency by using matrix sketching to approximate the Jacobian matrix. The resulting algorithm is competitive with state-of-the-art approaches for constructing predictive intervals on various regression datasets from the UCI repository.

Keywords: Uncertainty estimation · Matrix sketching · Confidence intervals

1 Introduction and Related Work

Modern neural networks achieve great results on various predictive modelling tasks. In the fields like computer vision and natural language processing neural-network based approaches are ubiquitous. When the task at hand has high cost of wrong decisions one has to assess the reliability of the model's predictions. There exist different ways to proceed with this task for neural networks.

The standard approach for uncertainty estimation is ensembling which proposes to train several neural networks from different starting parameter values. Next, one can compute uncertainty estimates based on the discrepancies between the predictions of ensemble members [10,14]. Ensembling usually gives high quality uncertainty estimates but for the price of large computational overhead both on training and inference stages. The different direction is to make weights of the network stochastic via Bayesian approach which results in the concept of Bayesian neural networks [11]. One can then quantify predictive uncertainty using the posterior distribution of the weights conditioned on the training

© The Author(s), under exclusive license to Springer Nature Switzerland AG 2022
E. Burnaev et al. (Eds.): AIST 2021, LNCS 13217, pp. 225–238, 2022.
https://doi.org/10.1007/978-3-031-16500-9_19

dataset. Bayesian inference is intractable for practical neural networks and thus approximate methods like variational inference [1,8,18] and stochastic gradient MCMC [3] are in use. These methods have a higher computational cost compared to training regular neural networks which limits their practical applicability. The other popular alternative is dropout which adds stochasticity to a standard neural network via randomly setting some of the weights to zero. This technique leads to the regularization of training [22] and can provide uncertainty estimates if applied at prediction time [6,21,23,24].

All the mentioned methods require either a modification of network architecture or training multiple models. As there are many existing pre-trained models for various tasks, it is beneficial to have a method that can be applied to a given neural network in order to obtain high-quality uncertainty estimates.

In this work we will focus on the classical form of uncertainty estimates: prediction intervals [2]. We propose a general method to construct approximate prediction intervals for a trained neural network model in the non-linear regression problem. Our method is based on the classical ideas of linearization (or first-order Taylor expansion) and so-called *delta method* (see e.g. [4]). While the original method requires inverting the estimated parameter covariance matrix of the model, we propose an efficient computational scheme based on SVD and matrix sketching techniques. These techniques make the proposed *Important Directions* method viable for modern deep neural networks.

The rest of the work is organized as follows. Section 2 is devoted to the formal problem statement and introduces the proposed method. Section 3 describes the implementation and experiments. After that Sect. 4 points to possible extensions of the considered method and Sect. 5 concludes the work.

2 Approximate Prediction Intervals for Neural Networks

2.1 Prediction Intervals for Neural Networks

We will be working in a general regression setting, where we are given a training dataset $\mathcal{D} = \{\mathbf{x}_i, y_i\}_{i=1}^n$, $\mathbf{x}_i \in \mathbb{R}^m$, $y_i \in \mathbb{R}$. The task is to estimate the hypothesised relationship $y_i \approx f(\mathbf{x}_i, \mathbf{w})$ by providing an estimate $\hat{\mathbf{w}}$ of the parameter vector $\mathbf{w} \in \mathbb{R}^p$. To quantify the approximate nature of the statement, an unobserved noise random variable is introduced:

$$y_i = f(\mathbf{x}_i, \mathbf{w}) + \varepsilon_i, \quad \varepsilon_i \sim \mathcal{N}(0, \sigma^2), \quad i = 1, \dots, n. \tag{1}$$

A level $(1 - \alpha)$ prediction interval for the function value $y_0 = f(\mathbf{x}_0, \mathbf{w})$ is a random interval $[L_\alpha(\mathbf{x}_0, \mathcal{D}), U_\alpha(\mathbf{x}_0, \mathcal{D})]$ with the following property:

$$\mathbb{P}\Big(y_0 \in \big[L_\alpha(\mathbf{x}_0, \mathcal{D}), U_\alpha(\mathbf{x}_0, \mathcal{D})\big]\Big) \geq 1 - \alpha.$$

Linear Regression. In the simplest case of a linear relationship f there is a closed-form expression for the prediction interval. For ease of exposition need to slightly modify the training set here $\mathbf{x}_i \in \mathbb{R}^{m+1}$ with a dummy first dimension equal to 1 and $\mathbf{X}$ is a matrix with the vectors $\mathbf{x}_i$ as rows. Using the fact that the noise is normally distributed, the following formula for interval endpoints can be derived [20]:

$$\widehat{y}_0 \pm t_{n-m-1}^{(\alpha/2)} \widehat{\sigma} \sqrt{1 + \mathbf{x}_0^T \left(\mathbf{X}^T \mathbf{X}\right)^{-1} \mathbf{x}_0},$$

where $\widehat{\sigma}$ is an estimate of the noise standard deviation, $\widehat{y}_0$ is a linear regression prediction at point $\mathbf{x}_0$ and $t_{n-m-1}^{(\alpha/2)}$ is a level $\alpha/2$ quantile of the Student's t-distribution with $n - m - 1$ degrees of freedom.

From the expression under the square root we see that even in this simple case the required computation scales quadratically with the size of the whole dataset and cubically with the feature dimension. Overall, computational complexity is $O(n^2 m + m^3)$.

Nonlinear Regression. Now we move to the general setting of (1) with f being a neural network with parameters $\mathbf{w}$. In [25] authors provide a method of constructing a prediction interval for a trained neural network based on delta-method. We briefly state main results here while the derivation and discussion can be found in the original paper.

Let us assume that parameter estimates $\widehat{\mathbf{w}}$ for the model are obtained by minimizing sum-of-squares loss with L_2 regularization with parameter λ based on the training dataset $\mathcal{D} = \{\mathbf{x}_i, y_i\}_{i=1}^n$. We introduce some additional notations. Matrix J consists of the gradients of the output of the neural network, computed at the final parameter estimates $\widehat{\mathbf{w}}$:

$$J = \{J_{ij}\}_{i,j=1}^{n,m}, \quad J_{i,j} = \left.\frac{\partial f}{\partial w_j}\right|_{\mathbf{x}=\mathbf{x}_i, \mathbf{w}=\widehat{\mathbf{w}}},$$

and matrix $\boldsymbol{\Sigma}$ is the approximate parameter covariance matrix:

$$\boldsymbol{\Sigma}^{-1} = (J^T J + \lambda I)^{-1}(J^T J)(J^T J + \lambda I)^{-1}. \tag{2}$$

Let $\mathbf{x}_0$ be a new unseen point where we aim to predict the response value y_0. Additionally, let $g_0 = \left.\frac{\partial f}{\partial w_j}\right|_{\mathbf{x}=\mathbf{x}_0, \mathbf{w}=\widehat{\mathbf{w}}}$ be the gradient of the output of the neural network at the new point $\mathbf{x}_0$. The approximate level $(1 - \alpha)$ prediction interval at a new point $\mathbf{x}_0$ is given by [25]:

$$\widehat{y}_0 \pm t_{n-p^*}^{\alpha/2} \widehat{s} \sqrt{1 + g_0^T \boldsymbol{\Sigma}^{-1} g_0}. \tag{3}$$

Here, $t_{n-p^*}^{\alpha/2}$ is the $\alpha/2$-level quantile of Student's t-distribution with $n - p^*$ degrees of freedom. Quantity p^* is the so-called *effective number of parameters*. Authors of [25] propose the following approximation for it:

$$p^* = \mathrm{Tr}(2H - H^2) \quad \text{with} \quad H = J(J^T J + \lambda I)^{-1} J^T.$$

The last component we have to cover is $\widehat{s}$, which is an estimate of the noise variance σ. This quantity can be found from the sum of squared residuals of the model on the training data:

$$\widehat{s}^2 = \frac{\sum_{i=1}^{n}(y_i - \widehat{y}_i)^2}{n - p^*} = \frac{\sum_{i=1}^{n}\left(y_i - f(\mathbf{x}_i, \widehat{\mathbf{w}})\right)^2}{n - p^*}.$$

From the applied point of view, we can divide this method into two parts: calculating Σ and calculating p^*. Computational complexity of both parts is $O(nm^2 + m^3)$, not including the cost of computing gradients. Space complexity is at least $O(nm + m^2)$ since we need to store the full covariance matrix and the matrix of gradients.

Unfortunately, for most modern deep neural networks this approach can not be applied as is. For example, in contemporary computer vision, a neural network can have tens of millions of parameters (m) and millions of examples in the training set (n). This will make the matrices involved in the formulas above prohibitively large to carry out computations.

In this work we propose a set of approximation techniques to reduce the computational burden of the formulas above and obtain a practical algorithm to construct approximate prediction intervals for deep neural networks.

2.2 Simplification Using SVD

If we assume that matrix J admits the following singular value decomposition:

$$J = UDV^T, \quad D = \mathrm{diag}(d_1, d_2, \ldots, d_m), \tag{4}$$

we can then simplify formula (2) using the fact that U and V are unitary matrices:

$$J^T J = VDU^T UDV^T = VD^2 V^T,$$
$$(J^T J + \lambda I)^{-1} = (VD^2 V^T + \lambda I)^{-1} = (V(D^2 + \lambda I)V^T)^{-1} = V(D^2 + \lambda I)^{-1}V^T$$

and obtain

$$\Sigma^{-1} = V(D^2 + \lambda I)^{-1}V^T VD^2 V^T V(D^2 + \lambda I)^{-1}V^T = VD^2(D^2 + \lambda I)^{-2}V^T.$$

We can also observe that

$$D^2(D^2 + \lambda I)^{-2} = \mathrm{diag}\left(\left\{\frac{d_j^2}{(d_j^2 + \lambda)^2}\right\}_{j=1}^{m}\right). \tag{5}$$

Using this form we can greatly simplify the computation of the formula (2) if the SVD decomposition is precomputed.

2.3 Proposed Approximation

While obtaining the full decomposition (4) may still be intractable, we can use a truncated rank-k SVD. However, most existing implementations assume that the target matrix is stored and accessed in a sparse format.

Since J is a matrix of gradients of a neural network, it is dense, and, thus, we have a slightly different setup. First, the target matrix is accessible row-by-row, because modern automatic differentiation packages allow computation of the gradients of a function that outputs a single scalar. Second, we can only store a limited number of rows because of the large number of trainable parameters in the neural network.

Based on these two requirements, we need an online algorithm that computes a low rank approximation of a matrix by reading through its rows one at a time. Low rank matrix approximation in the presence of constraints on access to the matrix elements is sometimes called matrix sketching. We refer to [19] and [15] for further reading.

In this work, we consider an online low rank matrix approximation algorithm Robust Frequent Directions (RFD) [16]. RFD is a modification of Frequent Directions (FD) algorithm [7,15]. Given a matrix A, the algorithm approximates $A^T A$ by $B^T B + \lambda' I$, where matrix B has only $k << n$ rows. Reading matrix A row-by-row, RFD updates the approximation B and the regularization parameter λ' simultaneously[1].

Parameter Covariance. We propose a new algorithm to approximate formula (2) using a modification of the RFD that we call Robust Important Directions, presented in Algorithm 1. The key element of the algorithm is the use of a scoring function to select a pivotal singular value (line 6) instead of just using the minimal one. This scoring function is chosen based on a higher-level problem at hand. In our case the end goal is estimating the parameter covariance matrix and (5) can be directly used here:

$$score(D, \lambda) = \left[\frac{d_j^2}{(d_j^2 + \lambda)^2} \right]_{j=1}^m. \tag{6}$$

Since we want to preserve as much information about Σ^{-1} as possible, this score function will help the algorithm retain the largest singular values.

After going through our training dataset one time we store the final SVD decomposition of the matrix B. This approximation of matrix J is used later for computation of the intervals.

Computational complexity of this procedure is the same as for Frequent Directions [7]: $O(nmk)$, where k is the rank of the approximation. We again ignore the cost of computing gradients since it is done only once and the resulting computational complexity will depend on the specific architecture. Space complexity is only $O(km)$ as we store $B \in \mathbb{R}^{2k,m}$.

---

[1] When $A^T A + \lambda I$ is the target matrix the final approximation becomes $B^T B + (\lambda + \lambda')I$.

Algorithm 1: Update Approximation

   Input : $B \in \mathbb{R}^{2k \times m}$ – current approximation to the target matrix, λ –
                regularization parameter used for network training, λ' – current
                additional regularization parameter, R – next row of the target matrix
   Output: B' – updated low-rank approximation, λ'' – updated additional
                regularization coefficient

1 $r \leftarrow$ index of the first zero row of B
2 $B_r \leftarrow R$
3   **if** $r = 2k$ **then**
4 $U, D, V \leftarrow \texttt{SVD}(B)$
5 $S, \text{idx} \leftarrow$ top k entries of D based on $score(D, \lambda)$ together with indices
6 $\delta \leftarrow \min S$
7 $B' \leftarrow 0^{2k \times m}$
8 $B'_{1:k} \leftarrow \sqrt{\max(S^2 - \delta^2 I, 0)}V^T_{\text{idx}}$ `# use singular vectors indexed with`
            `idx and scale corresponding singular values`
9 $\lambda'' \leftarrow \lambda' + \delta^2/2$
10 **else**
11 $B' \leftarrow B$
12 $\lambda'' \leftarrow \lambda'$
13 **return** B', λ''

Noise Variance. So far uncovered ingredient of the formula (3) is the estimation of the noise variance $\widehat{s}^2$. Using our SVD decomposition of J we can represent H in the following form:

$$H = J(J^T J + \lambda I)^{-1} J = UDV^T V(D^2 + \lambda I)^{-1} V^T V D U^T = U D^2 (D^2 + \lambda I)^{-1} U^T$$

and $D^2(D^2 + \lambda I)^{-1}$ can be computed by (5).

To compute the unknown quantity p^* we need to find the trace of the equation depending on H. Using the properties of SVD decomposition and trace we get:

$$p^* = \text{Tr}(2H - H^2) = 2\text{Tr}(H) - \text{Tr}(H^2) = \sum_{j=1}^{m} \frac{2d_j^2}{(d_j^2 + \lambda)} - \frac{d_j^4}{(d_j^2 + \lambda)^2}.$$

Having obtained this simple expression, we can evaluate it using our truncated SVD approximation.

Algorithms. We collect the steps described above in the following subprograms.

Algorithm 1 computes an update to existing low rank approximation. It runs when a new row of the target matrix is available: if the buffer (matrix B) is full, we perform SVD and truncate our approximation. We use this later as a subroutine.

Algorithm 2: Robust Important Directions

Input : k – rank of the approximation, $A \in \mathbb{R}^{n \times m}$ – target matrix, λ –
regularization parameter

Output: B – low-rank approximation of A, λ_n – corrected regularization
coefficient

1 $B \leftarrow 0^{2k \times m}$

2 $\lambda_0 \leftarrow 0$

3 **foreach** *row* $A_i \in A$ **do**

4 $\quad \mid \quad B, \lambda_i \leftarrow \text{UpdateApproximation}(B, \lambda, \lambda_{i-1}, A_i)$

5 **end**

6 **return** $B_{1:k}, \lambda_n$

Algorithm 3: Estimate the covariance matrix of the NN parameters and
effective number of parameters

Input : $f(\mathbf{x}, \widehat{\mathbf{w}})$ - trained neural network, λ - L_2-regularization parameter that
was used to train the network, $\mathcal{D} = \{\mathbf{x}_i, y_i\}_{i=1,\dots,n}$ - training dataset,
k – rank of the approximation

Output: $D_{\Sigma^{-1}}, V$ - low-rank approximation to the parameter covariance
matrix, p^* - estimate of the effective number of parameters

1 $B \leftarrow 0^{2k \times m}$

2 $\lambda_0 \leftarrow 0$

3 **for** $(\mathbf{x}_i, y_i)$ *in* $\mathcal{D}$ **do**

4 $\quad \mid \quad \mathbf{g}_i \leftarrow \nabla f_i = \frac{\partial f}{\partial w}\big|_{x=x_i, w=\widehat{w}}$

5 $\quad \mid \quad B_i, \lambda_i \leftarrow \text{UpdateApproximation}(B, \lambda, \lambda_{i-1}, \mathbf{g}_i)$

6 **end**

7 $U, D, V \leftarrow \text{SVD}(B_n)$

8 $D_{\Sigma^{-1}} \leftarrow (D^2 + \lambda_n)(D^2 + (\lambda_n + \lambda)I)^{-2}$

9 $D_H \leftarrow (D^2 + \lambda_n)(D^2 + (\lambda_n + \lambda)I)^{-1}$

10 $p^* \leftarrow \text{Tr}(2D_H - D_H^2)$

11 **return** $D_{\Sigma^{-1}}, V, p^*$

Algorithm 2 is the procedure for computing low-rank approximation of a
matrix for the case when a *score* function is provided (implicitly used in Algorithm 1). This algorithm can be used as a drop-in replacement for other low-rank
approximation algorithms when more information about the problem is available. The end user will select the *score* function based on the task at hand.

Algorithm 3 summarizes all the steps that are required to compute prediction
intervals of a trained neural network for the proposed *Important Directions* approach. It precomputes a compact low rank approximation to the inverse covariance matrix of the network parameters and effective number of parameters. To
construct the intervals for new data points the end user can employ Eq. (3)
together with these approximations.

2.4   Motivation Behind the Approximation

The proposed modification of the low-rank approximation procedure introduces the score function (6) in order to capture the most important eigenvalues and eigenvectors of the parameter covariance matrix. We demonstrate the effectiveness of this heuristic on small-scale regression datasets from the UCI repository [5]. For a sufficiently small neural network it is feasible to apply formula (3) directly.

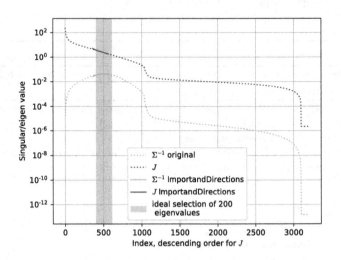

Fig. 1. Approximation quality of Important Directions (rank $k = 200$) for a $[50, 50]$ fully-connected network trained on the Wine Quality dataset. Dotted lines show the true specters of the matrices and solid lines show their respective approximations, shifted horizontally for visual clarity. (Color figure online)

Figure 1 shows comparison of the Important Directions approximation with exact computations: singular values of the Jacobian matrix J and eigenvalues of the covariance Σ^{-1} are plotted on the same axes. Using Robust Frequent Directions corresponds to selecting largest singular values of J, which will correspond to much smaller values for Σ^{-1} (left-most part of the graph). At the same time, the largest eigenvalues of the covariance (shown in pink) will have more influence on the length of the resulting prediction intervals in (3). The presented graph shows that Important Directions approach achieves the goal of preserving the more informative part of the specter.

2.5   Hyperparameters

The proposed method essentially has a single hyperparameter: the rank of the approximation k. Our experiments have shown diminishing returns with its increase.

One advantage of this hyperparameter is that it can be chosen without the use of a hold-out dataset: based on the relative magnitude of the recovered singular values of the Jacobian. If the smallest and largest singular values of the approximation differ by several orders, additional singular values and singular vectors will contribute very little to the final length of the interval. An example of a sharp decline like this can be seen in Fig. 1 around index 1000.

Since the size of the additional memory required for the approximation is a multiple of the total number of parameters of the neural network, it is reasonable to expect significant memory constraints playing a bigger role in the choice of this parameter. Thus we recommend to set it as high as your computational budget allows if you do not observe the behavior described in the previous paragraph.

3   Experiments

We evaluate our method against a number of existing approaches on publicly available regression datasets from UCI repository [5]. For comparison we have chosen methods designed specifically for neural networks and compared performance of the resulting prediction intervals.

Every dataset was split 20 times into training and test sets and each method was independently applied on each split. Average performance metrics of these 20 runs are reported.

3.1   Methods

MC-Dropout. We have based our implementation on the reference implementation[2] of [6]. Dropout probability is tuned via grid search on a validation set (taken out of training set) and the then the final model is trained from scratch on the whole training set. Resulting prediction interval is obtained using 10 000 stochastic forward passes.

Deep Ensembles. We train the neural network multiple times from different initializations. Probabilistic output is a Gaussian with mean and variance obtained by aggregating their predictions.

Important Directions. For our method we set rank of the approximation to a fixed value of 500 for most datasets with some exceptions: 200 for "Wine" and 300 for "Boston" due to their small size and 1000 for "Year". Number of training epochs is tuned on a validation set similar and then the network is retrained from scratch (similar to Dropout).

For all methods we have used the training of a neural network as a black-box building block. The same fully connected neural network with two hidden layers of size 50 was used. We used Adam optimizer [12] with the default learning rate of 0.001.

Neural Network Prediction intervals [25]. We have implemented the base method using the fact that our neural networks and some datasets are rather

---

[2] https://github.com/yaringal/DropoutUncertaintyExps.

small. All the data was stored in GPU memory to speed up gradient computation. Jacobian product matrix $J^T J$ was also pre-computed on GPU which will be impossible for larger networks.

3.2   Metrics and Results

While many existing papers on probabilistic regression use log-likelihood on the test set as the target metric, the quality of the resulting interval estimates is rarely explored [13]. We argue that for a practitioner working on a regression problem a prediction interval with a specified level is much more useful than a simple ranking of the test sample by an uncertainty estimate. With this in mind we have chosen the following metrics to measure the quality of the obtained prediction intervals:

- Probability of coverage p_{cov} – proportion of the test examples for which estimated intervals cover the true response values.
- Pearson's correlation coefficient r between width of the estimated interval and true absolute error.
- Width of the prediction interval in terms of the standard deviation of the true response values w_{sd}. Naturally one would prefer narrower intervals while still providing the coverage guaranties.

Comparison with the Exact Method. We have compared our method with the exact application of formula (3), setting the rank at 500 for all of the datasets using a powerful workstation. While the computation time was impractical even for these simple datasets, it will allow us to assess whether our approximation is adequate. Results are summarised in Table 1. Column Δt corresponds to wall clock time in seconds that it took to run the method (excluding training of the neural network, etc.).

As we can see the results for Important Directions are comparable with the exact computation. Rank of the approximation of 500 was enough even for larger datasets where p^* is underestimated. Our hypothesis is that the original formula often produces over-conservative intervals – their coverage is often greater than the nominal 95%, so the lower p^* actually improves the metrics. See also the relevant discussion in [25].

In some cases running time of the exact method is slightly better. This might be due to different implementations of SVD on CPU and GPU: we could not fit J and/or Σ^{-1} in the GPU memory.

Comparison with Other Methods. We have also preformed a comparison with other popular methods of uncertainty estimation for neural networks: MC-dropout and Deep Ensembles. Results of these experiments are presented in Table 2. For p_{cov} the closest value to the nominal 95% coverage is shown in bold. For r the highest correlation coefficient is shown in bold.

Overall it is a tie between ensembles and Important Directions: both methods are very close to the nominal coverage probability and provide intervals that

Table 1. Comparison of the low-rank approximation with the exact method

Metric	p_{cov}		r		w_{sd}		Δt	
Method	Exact	ID	Exact	ID	Exact	ID	Exact	ID
Dataset								
Ailerons	0.94	0.95	0.22	0.25	1.55	1.55	25	6
Boston	0.98	0.96	0.36	0.34	1.61	1.33	7	1
CT	0.96	0.96	0.36	0.33	0.38	0.37	2398	24
Concrete	0.96	0.96	0.21	0.21	1.41	1.37	5	1
Energy	0.96	0.96	0.42	0.42	1.75	1.75	6	1
Protein	0.95	0.95	0.10	0.10	2.95	2.95	17	19
SGEMM	0.96	0.96	0.41	0.41	0.69	0.69	71	99
Superconduct	0.94	0.94	0.10	0.12	1.50	1.52	76	9
Wine	0.98	0.95	0.18	0.14	3.85	3.08	6	1
Yacht	0.98	0.95	0.17	0.13	3.86	3.20	6	1
Year	0.94	0.93	0.22	0.22	3.10	3.07	5357	214

correlate with the true error. Interesting observation is that methods are close in the three metrics at the same time, so ensemble does not for example provide universally tighter intervals.

Our proposed explanation for the poor performance of the dropout is the following. Authors of [6] propose to tune both drop-out rate and τ parameter (related to noise precision) at the same time using grid search over predefined ranges. Even with standardizing the datasets it proves to be a challenging scenario given the observed coverage statistics.

Given all the above we conclude that Important Directions method achieves results on par with more established but computationally heavy methods without the need to retrain the network and tune multiple hyper-parameters.

3.3 Implementation

We have implemented our method in Python 3 language. Neural network training and gradient computations were done in PyTorch [17]. Subsequent matrix and vector calculations used NumPy library [9]. Full source code for the method and experiments is made publicly available[3].

4 Discussion and Future Work

While we introduce Important Directions as method for regression tasks, it can also be applied in the classification setting. For a given class index the network's logits can be combined in a one-vs-rest fashion to produce a single logit value. For this we will need to run our method independently for each class (although

---

[3] https://github.com/stat-ml/id_prediction_intervals.

Table 2. Probability of coverage and other metrics for different methods.

Metric	p_{cov}			r			w_{sd}		
Method	D-out	Ens.	ID	D-out	Ens.	ID	D-out	Ens.	ID
Dataset									
Ailerons	0.965	**0.948**	0.945	0.008	0.158	**0.245**	1.960	1.573	1.550
Boston	**0.947**	0.957	0.962	0.032	0.261	**0.344**	1.519	1.388	1.330
CT	0.999	0.978	**0.956**	0.005	**0.372**	0.331	0.657	0.411	0.368
Concrete	0.908	**0.950**	0.960	-0.062	**0.211**	0.206	1.029	1.265	1.372
Energy	0.963	**0.952**	0.959	0.022	0.419	**0.421**	1.862	1.715	1.748
Protein	0.870	0.953	**0.952**	-0.086	0.072	**0.097**	1.960	2.965	2.949
SGEMM	0.990	0.967	**0.963**	-0.069	**0.503**	0.411	0.392	0.709	0.693
Sup.con.	0.903	**0.944**	0.943	-0.097	**0.271**	0.121	1.029	1.523	1.524
Wine	0.801	**0.946**	**0.946**	0.032	0.114	**0.144**	1.960	3.135	3.082
Yacht	0.802	0.946	**0.952**	0.041	**0.139**	0.133	1.960	3.135	3.202
Year	0.847	**0.936**	0.934	0.006	0.186	**0.216**	1.960	3.116	3.070

some gradient computations could be reused). Other ways to extend our method to classification task is the topic for future research.

The described increase in computational cost urges us to find other ways to improve performance. One way is to limit the number of the parameters of the neural network that are considered in the gradient computation and related matrix operations. We can restrict our focus only to top layers or any other subset. This is similar in spirit to performing Bayesian inference on the last layer only. A direction of possible future research is the choice of higher rank vs larger number of parameters.

5   Conclusions

This work presents a new method to construct prediction intervals for a trained neural network based on matrix sketching. The method was developed for regression tasks but can also be applied in a classification setting targeting class logits. We demonstrate its advantages compared to other interval estimation methods for neural networks on a range of benchmark regression datasets.

Acknowledgements. The research was supported by the Russian Science Foundation grant 20-71-10135.

References

1. Blundell, C., Cornebise, J., Kavukcuoglu, K., Wierstra, D.: Weight uncertainty in neural networks. In: ICML, pp. 1613–1622. PMLR (2015)
2. Casella, G., Berger, R.L.: Statistical Inference. Duxbury. Pacific Grove (2002)
3. Chen, T., Fox, E., Guestrin, C.: Stochastic gradient Hamiltonian Monte Carlo. In: International Conference on Machine Learning, pp. 1683–1691. PMLR (2014)

4. Doob, J.L.: The limiting distributions of certain statistics. Ann. Math. Stat. **6**(3), 160–169 (1935)
5. Dua, D., Graff, C.: UCI machine learning repository (2017). http://archive.ics.uci.edu/ml
6. Gal, Y., Ghahramani, Z.: Dropout as a Bayesian approximation: representing model uncertainty in deep learning. In: ICML, pp. 1050–1059. PMLR (2016)
7. Ghashami, M., Liberty, E., Phillips, J.M., Woodruff, D.: Frequent directions: simple and deterministic matrix sketching. SIAM J. Comput. **45**, 1762–1792 (2016)
8. Graves, A.: Practical variational inference for neural networks. In: Advances in Neural Information Processing Systems, vol. 24 (2011)
9. Harris, C.R., et al.: Array programming with NumPy. Nature **585**(7825), 357–362 (2020). https://doi.org/10.1038/s41586-020-2649-2
10. Heskes, T., Wiegerinck, W., Kappen, H.: Practical confidence and prediction intervals for prediction tasks. In: Progress in Neural Processing, pp. 128–135 (1997)
11. Jospin, L.V., Buntine, W., Boussaid, F., Laga, H., Bennamoun, M.: Hands-on Bayesian neural networks-a tutorial for deep learning users. arXiv preprint arXiv:2007.06823 (2020)
12. Kingma, D.P., Ba, J.: Adam: a method for stochastic optimization. arXiv preprint arXiv:1412.6980 (2014)
13. Kompa, B., Snoek, J., Beam, A.: Empirical frequentist coverage of deep learning uncertainty quantification procedures (2021)
14. Lakshminarayanan, B., Pritzel, A., Blundell, C.: Simple and scalable predictive uncertainty estimation using deep ensembles. arXiv preprint arXiv:1612.01474 (2016)
15. Liberty, E.: Simple and deterministic matrix sketching. In: Proceedings of the 19th ACM SIGKDD International Conference on Knowledge Discovery and Data Mining (2013)
16. Luo, L., Chen, C., Zhang, Z., Li, W.J., Zhang, T.: Robust frequent directions with application in online learning. J. Mach. Learn. Res. **20**(1), 1697–1737 (2019)
17. Paszke, A., et al.: Pytorch: an imperative style, high-performance deep learning library. In: Wallach, H.M., Larochelle, H., Beygelzimer, A., d'Alché-Buc, F., Fox, E.B., Garnett, R. (eds.) Advances in Neural Information Processing Systems 32: Annual Conference on Neural Information Processing Systems 2019, NeurIPS 2019, Vancouver, BC, Canada, 8–14 December 2019, pp. 8024–8035 (2019). https://proceedings.neurips.cc/paper/2019/hash/bdbca288fee7f92f2bfa9f7012727740-Abstract.html
18. Rezende, D., Mohamed, S.: Variational inference with normalizing flows. In: International Conference on Machine Learning, pp. 1530–1538. PMLR (2015)
19. Sarkar, B., Bhattacharyya, M.: Spectral algorithms for streaming graph analysis: a survey. Ann. Data Sci. **8**, 667–681 (2020)
20. Seber, G., Lee, A.: Linear Regression Analysis. Wiley Series in Probability and Statistics. Wiley (2003)
21. Shelmanov, A., Tsymbalov, E., Puzyrev, D., Fedyanin, K., Panchenko, A., Panov, M.: How certain is your Transformer? In: Proceedings of the 16th Conference of the European Chapter of the Association for Computational Linguistics: Main Volume, pp. 1833–1840 (2021). https://doi.org/10.18653/v1/2021.eacl-main.157
22. Srivastava, N., Hinton, G., Krizhevsky, A., Sutskever, I., Salakhutdinov, R.: Dropout: a simple way to prevent neural networks from overfitting. J. Mach. Learn. Res. **15**(1), 1929–1958 (2014)

23. Tsymbalov, E., Fedyanin, K., Panov, M.: Dropout strikes back: improved uncertainty estimation via diversity sampled implicit ensembles. In: International Conference on Analysis of Images, Social Networks and Texts (2021). https://arxiv.org/abs/2003.03274

24. Tsymbalov, E., Panov, M., Shapeev, A.: Dropout-based active learning for regression. In: van der Aalst, W.M.P., et al. (eds.) AIST 2018. LNCS, vol. 11179, pp. 247–258. Springer, Cham (2018). https://doi.org/10.1007/978-3-030-11027-7_24

25. Veaux, R.D., Schweinsberg, J., Schumi, J., Ungar, L.H.: Prediction intervals for neural networks via nonlinear regression. Technometrics 40, 273–282 (1998)

Application of Data Analysis Methods for Optimizing the Multifunctional Service Center Operation

Ekaterina Kasatkina$^{(\boxtimes)}$ (ID) and Daiana Vavilova (ID)

Kalashnikov Izhevsk State Technical University, Izhevsk, Russia
kasatkina@istu.ru

Abstract. The article presents the work of the multifunctional center for provision of state and municipal services to citizens in the Udmurt Republic. The analysis was carried out according to the work of the multifunctional center branch office in Pervomaisky district of Izhevsk city for the period of 2017–2021. During the data analysis, problems in the MFC operation were defined and recommendations for their solving were developed. As a result of the recommendations implement, it was possible to increase the remote registration availability by 22%, reduce the standard time of customer service by 6% and increase the number of customers' talons served by 30% and the efficiency of employees up to 96%.

Keywords: Multifunctional Service Center (MFC) · Product analytics · Queuing theory · Machine learning · Optimization · Forecasting

1 Introduction

In terms of the information society development, the formation of modern public administration is becoming an important issue. Today, a person is a client and a service consumer, provided by government agents (institutions, departments), and the activities of government agencies are viewed through the prism of meeting specific requests and needs of clients. Need satisfaction of society and improving the population life quality directly depend on the implemented information and communication technologies, the digitalization level and scientific-technological progress [1,2].

Obviously, the problem of digitalization is urgent in the Russian Federation, both for the automated collection, storage and processing of information, and for the development of new digital management business models and the digital services for the population. Currently, the multifunctional center for the provision of state and municipal services to citizens (MFC) is one of the main elements of digital transformation in the country.

© The Author(s), under exclusive license to Springer Nature Switzerland AG 2022
E. Burnaev et al. (Eds.): AIST 2021, LNCS 13217, pp. 239–249, 2022.
https://doi.org/10.1007/978-3-031-16500-9_20

The MFC is a unified place for receiving, registering and obtaining the necessary documentation to citizens and legal entities in providing state and municipal services. The quality and timeliness of the MFC services provision to the population are important criteria for assessing the digitalization level in the country. In addition, the relevance of the study and the MFC work optimization is explained by their insufficient bandwidth for reservation through remote services (the Regional Portal of State and Municipal Services (RPS) and the Central Service Center (CSC)), leading to unnecessary wait for submitting an electronic application for services. Consequently, there is a problem of the accessibility to state and municipal services through the MFC, which requires a solution. The purpose of this research is analyzing the MFC operation data to optimize its functioning.

Further, the second part of the research presents the description of approaches used to analyze, predict and optimize of the MFC operation. The third part describes the results of product analytics methods application and machine learning methods to forecast the MFC services demand in the UR. The fourth part provides with the results of solving the optimization problem for the MFC considered.

2   Approaches of Analyzing, Forecasting and Optimizing the MFC Operation

The general analysis methodology of optimize the MFC's operation can be presented in the form of the following steps:

1. Conduct an FRM analysis of services and form a list of MFC's TOP-services.
2. Conduct an XYZ analysis of services and identify the categories of services based on the demand variation.
3. Execute a joint FMR-XYZ analysis and identify 9 service categories based on trends and demand variations.
4. Forecast the demand for each MFC service using different time series forecasting methods for each service category.
5. Calculate the main characteristics and parameters of the MFC system functioning as of a queuing system.
6. Optimize the MFC work by redistributing service delivery windows for a service category.

2.1   FMR-XYZ Analysis

When analyzing data on the service volatility, FMR analysis (Fastest, Medium, Rare) and XYZ analysis are mostly carried out [4,5,10,12]. They allow us to group and classify services depending on the stability of demand.

FMR analysis shows the service demand it is performed on the basis of the calculated frequency request rate factor [10]:

$$W_i = \frac{K_i}{\sum\limits_{j=1}^{n} K_j} 100\%, \tag{1}$$

where K_i is the number of the i-th provided service, $\sum_{j=1}^{n} K_j$ is the total number of provided services.

As the FMR analysis results, all MFC services are divided into three categories:

- Category F is the most frequently requested services (accounting for 80% of the MFC's work);
- Category M is less frequently requested services (15%);
- Category R is rarely requested services (5%).

XYZ analysis allows us to predict changes in service needs within a specific time frame; it is based on the coefficient of variation evaluation according to the formula [5]:

$$V_i = \frac{\sigma_{x_i}}{\overline{x_i}} 100\%, \qquad (2)$$

where σ_{x_i} is standard deviation of the provided service, $\overline{x_i}$ - arithmetic mean of the service provided for the period.

According to the variation coefficient, all services are divided into three categories:

- Category X is a stable demand for the service, minor fluctuations and high forecast accuracy (variation ranges from 0 to 10%);
- Category Y is known trends in determining the service need, seasonal fluctuations and average predictive capabilities (variation ranges from 10 to 25%);
- Category Z is the irregular service demand, trends are not available, low forecasting accuracy (the variation is above 25%).

In practice, organizations use a joint FMR-XYZ classification to optimize their work, dividing the provided services range into a larger number of groups and selecting a management and/or forecasting methodology for each of them.

2.2 Forecasting Methods

A forecast is a scientifically based judgment about the possible states of an object in the future. The methods used for forecasting various indicators, processes and phenomena make it possible to forecast with certain accuracy when based on retrospective statistical data about the forecast object, as well as using newly obtained data. Only mathematical models and forecasting methods are accurate.

Currently, the main mathematical forecasting methods include regression analysis statistical models: trend models, trend-seasonal models, ARIMA or SARIMA models, etc. [6–8].

The baseline time series is divided into training and test sets. The Mean Absolute Percent Error [16] metric was calculated to assess the quality of forecasts and select the best predictive model.

2.3   Optimization Task

The MFC's operation with talons received remotely can be considered as a multiple queue system (MQS) with a limited electronic queue [3,13,14].

The MQS properties for the MFC:

- The input receives a Poisson flow of MFC services requests with the distribution parameter λ, which is determined from the forecast estimate of the MFC services demand.
- The n number of service channels (windows), which can be registered via remote services, is a variable and can range from 1 to N (the maximum number of service windows).
- Each window can serve only one talon at a time.
- If there are no free talons on the current date, then the talon is queued (electronic entry).
- The number of places in the electronic queue m is limited and is determined from the number of service windows n and the MFC operating mode.
- If the application for the MFC services founds all the windows occupied and there are no free queue talons, then the application is denied.
- The talon service time is a random variable that obeys the exponential distribution law with the parameter μ. Average service time of one talon $t_s = 1/\mu$.
- The reduced intensity of the request flow is $\rho = \lambda/\mu$.
- Possible states of the MQS: S_0 (all windows are free), S_1 (one window busy, the rest are free), S_2 (two windows busy/the rest are free),..., S_n (all windows busy), S_{n+1} (all windows busy, there is one customer in the electronic queue), ..., S_{n+m} (all windows occupied, there are m customers in the electronic queue).

The number of the MQS possible states is finite, therefore, there are limited probabilities of these states and the operation parameters of this system can be estimated. Probability if all windows are free:

$$P_0 = \left(\sum_{i=1}^{n} \frac{\rho^i}{i!} + \frac{\rho^{n+1}}{n!n} * \frac{1 - (\frac{\rho}{n})^m}{1 - \frac{\rho}{n}} \right)^{-1}. \tag{3}$$

Probability of denial (all servicing windows busy, there are no free places in the queue):

$$P_f = P_{n+m} = \frac{\rho^{n+m}}{n!n^m} P_0. \tag{4}$$

Average wait time for a queue request:

$$T_r = \frac{\rho^{n+1} \left(1 - (\frac{\rho}{n})^m (m + 1 - m\frac{\rho}{n}) \right)}{n!n(1 - \frac{\rho}{n})^2 (1 - P_f)\lambda} P_0. \tag{5}$$

Window work load factor:

$$K_{load} = \frac{\rho}{n} \left(1 - \frac{\rho^{n+m}}{n!n^m} P_0 \right) 100\%. \tag{6}$$

The MFC operation is customer-focused, for that reason the optimization goal is to minimize the service denial:

$$P_f \to min. \tag{7}$$

There are some controlled variables in this case:

- The number of windows for servicing talons through RPS and CSC: n.
- The share of talons allocated for registration through RPS and CSC for service windows: v.

According to our formulation the problem of optimizing the considered QS work is a two-parameter and is solved by cyclic coordinate wise descent [9].

3 Product Analytics of the MFC Services

The paper analyses the work of one of the MFC in the UR, located in the Pervomaisky district of Izhevsk city. The data set for analysis consists of the talon information, issued through various booking methods (through a manager, service provision console, RPS and CSC) in the first quarter of 2021. Additionally, we use monthly data on the MFC operation in Pervomaisky District within the period of 2017–2020, presented as reports on the official website of the institution [15].

The MFC operation data initial analysis includes the analysis of services by request frequency (Fastest Medium Rare (FMR analysis). Analysis of the MFC data in Pervomaisky district within the period of 2017–2021 showed that 80% of the 178 services provided are TOP-8 basic services (see Fig. 1).

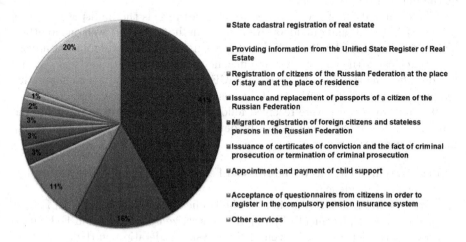

Fig. 1. Result of FMR analysis of the MFC services in Pervomaisky district (category F)

Figure 1 shows that 41% of all services provided in the MFC is the State cadastral registration of real estate service. According to the Pareto principle [11], for a deeper analysis of the MFC work, it is quite effective to work out TOP-8 dedicated services provision to obtain much of the planned result in optimizing the MFC work.

Further analysis of the MFC work data assumes the classification of services, depending on the service consumption pattern (XYZ-analysis). The MFC services in Pervomaisky district of the Udmurt Republic, included in X category, are presented in Table 1.

Table 1. The most demanded MFC services in Pervomaisky district (X category)

Name of service	Variation, %
Registration of citizens of the Russian Federation	3.4
Informing citizens about pre-retirement age	6.1
Providing information contained in Register of Real Estate	7.7
Acceptance of questionnaires for registration in the pension insurance system	8.6
Payment pensions	9.5
Issuance and replacement of passports of a citizen of the RF	9.6
Other services (172 services)	More 10

Table 1 demonstrates that the most stable service in the citizen request is the Registration of citizens of the Russian Federation at the place of stay and at the place of residence service.

The combination of FMR and XYZ analyzes reveals the undisputed service leaders (FX group) and outsiders (RZ). Both methods support well one another. The use of combined FMR and XYZ analyzes has a number of significant advantages: increasing the control system efficiency; identification of key services and reasons affecting the overall dynamics of MFC services demand; staff effort redistribution in accordance with qualifications and experience.

9 groups of services were received following the multivariate combined analysis. The FX service group is highly demanded by MFC clients. This group includes such services as:

– Registration of citizens of the Russian Federation at the place of stay and at the place of residence.
– Providing information contained in the Unified State Register of Real Estate.
– Issuance and replacement of passports of a citizen of the Russian Federation proving the identity of a citizen of the Russian Federation on the territory of the Russian Federation.

– Acceptance of questionnaires from citizens for the purpose of registration in the compulsory pension insurance system, including receipt of applications from insured persons for exchange or for issuance of a duplicate insurance certificate.

It is necessary to bring the stability of its receipt and the constant availability of specialists providing these services at the MFC. Client demand for this service is well predicted. Since the demand for these services has stable dynamics, trend models can be used for forecasting.

Figure 2 shows an example of service forecast from the FX group - Issuance and replacement of passports of a citizen of the Russian Federation using a trend power-law model.

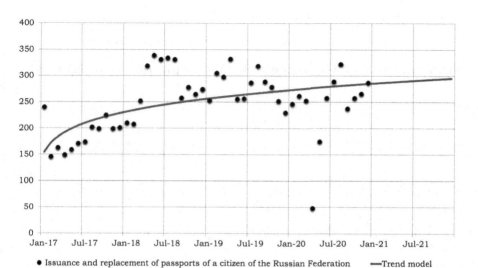

Fig. 2. Forecasting the service demand Issuance and replacement of passports of a citizen of the Russian Federation

The results of this analysis can be used to predict the service demand and optimize the MFC work.

The FY service group is distinguished by the average seasonal demand among the MFC clients. The FY group includes such services as:

– Migration registration of foreign citizens and stateless persons in the Russian Federation.
– Issuance of certificates of conviction and the fact of criminal prosecution or termination of criminal prosecution.
– Allocation and payment of child support.

Figure 3 presents an example of the service forecast from the FY group - Allocation and payment of child support using a multiplicative trend and seasonality model.

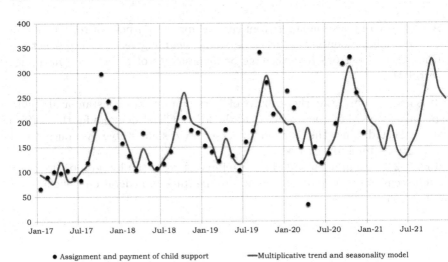

Fig. 3. Forecasting the demand for the Allocation and payment of child support service

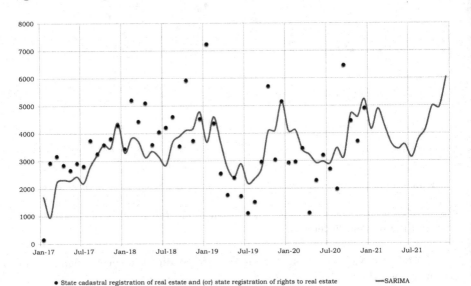

Fig. 4. Forecasting the service demand of State cadastral registration of real estate and (or) state registration of rights to real estate

The FZ service group with high demand is characterized with low predictability. This category of services includes only one service - State cadastral registration of real estate and (or) state registration of rights to real estate. Seasonal autoregressive integrated moving average (SARIMA) models are used to predict FZ services (see Fig. 4).

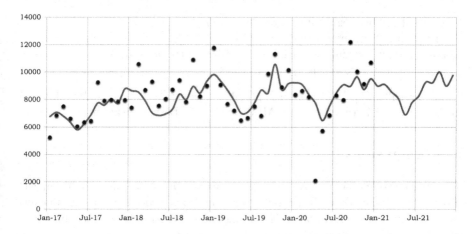

Fig. 5. Modeling and forecasting demand for all MFC services in Pervomaisky district

The use of various forecasting methods to assess the need for MFC services helped to obtain an accurate summary forecast for all MFC services in the short term (until the end of 2021), shown in Fig. 5.

Figure 5 displays that the dynamics of the demand indicator for all MFC services is well described by the combined model with account taken of seasonality and trend. Average relative approximation error is 8.5%. It can also be seen that in the period of April-May 2020, there is a sharp decline in the MFC services received, which is associated with the COVID-19 pandemic.

4 The MFC's Operation Optimization

The research is aimed at increasing the availability of MFC services remote registration through the remote services - the Regional Portal of State and Municipal Services (RPS) and the Central Service Center (CSC), while maintaining the high efficiency of the MFC employees. When optimizing the MFC work, the following restrictions were taken into account:

– The maximum wait time in the electronic queue is up to 7 calendar days.
– The minimum window work load should be 90%.

The average talon servicing time is calculated as the weighted average standard time of MFC services provision and is equal to $t_s = 24.7$ minutes. With the average service time and the MFC operating mode by weekdays, the talon service average intensity by one window per day was calculated $\mu_{avg} = 16.2$ talons/day.

The results of solving the optimization problem for the MFC of the Pervomaisky District in the Udmurt Republic are as follows:

– the probability of denial is $P_f = 8 * 1e - 6$;
– number of service windows for remote registration $n = 16$;

- share of remote registration talons $v = 0.67$;
- maximum electronic queue $m = 1056$;
- one window servicing intensity is $\mu = 16.2$ talons per day;
- employee utilization rate is $K_{load} = 96\%$.

As a result of optimization, it is possible to construct an optimal weekly plan for talons provision for the main MFC services in Pervomaisky district.

5 Conclusions

The article provides a Data Analysis of one of the MFC offices in the Pervomaisky district of Izhevsk city with the use of these methods: product analytics, machine learning (within the framework of solving the forecasting time series problem), modeling methods and queuing system optimization.

As part of the research, we have developed a toolkit for analyzing and optimizing the MFC work. It simulates the MFC operation using the queuing theory. This model assumes that the MFC is a multi-channel service center with a limited queue. Using a mathematical model, the service window optimal number is determined with a breakdown into various registration forms: preliminary registration, preliminary registration for the current day and with a breakdown into service types.

By using the FMR analysis, we found that the TOP-8 services account for more than 80% of the MFC requests in Pervomaisky District. The largest number of windows began to provide these services. The F category services are assigned to certain MFC employees, the rest of the services are allocated to general-duty MFC employees. Several forecasting methods were used various service categories and they let obtain a fairly accurate forecast for the MFC services.

In the process of modeling the multi-channel queuing system and its optimization, it was found that in order to keep the work load of windows servicing talons from the RPS (Regional Portal of State and Municipal Services) and the CSC (Central Service Center) at the level of not less than 90% and the registration possibility for no more than a week, it is necessary to allocate 11 windows (out of 17) for servicing talons received through the preliminary registration forward by one week. To provide the basic Russian Register service - State cadastral registration of real estate and (or) state registration of rights to real estate, it is necessary to allocate at least 5 windows for preliminary registration with the registration possibility one week in advance. It is needed to assign two specialists and 40% of another specialist's working time for a preliminary registration for the current day.

In the course of the recommendations implementing to optimize the MFC work, it was possible to increase the employee efficiency up to 96%, raise the number of served talons by 30%, reduce the electronic queue and thereby improve the remote registration availability by 22%, as well as reduce the electronic queue wait time up to 7 days.

References

1. Ketova, K.V., Saburova, E.A.: Addressing a problem of regional socio-economic system control with growth in the social and engineering fields using an index method for building a transitional period. In: Silhavy, R., Silhavy, P., Prokopova, Z. (eds.) CoMeSySo 2020. AISC, vol. 1295, pp. 385–396. Springer, Cham (2020). https://doi.org/10.1007/978-3-030-63319-6_35
2. Kasatkina, E.V., Vavilova, D.D.: Information-analytical system to forecast the factors of regional development. Control Sci. **4**, 25–34 (2015)
3. Sutyagina, N.I.: Modeling the activity of a multifunctional center as a queuing system. Karelian Sci. J. **1**(10), 199–203 (2015)
4. Konikov, A., Konikov, G.: Methodology of construction site marketing analysis. Procedia Eng. **165**, 1052–1056 (2016). https://doi.org/10.1016/j.proeng.2016.11.819
5. Devarajan, D., Jayamohan, M.S.: Stock control in a chemical firm: combined FSN and XYZ analysis. Procedia Technol. **24**, 562–567 (2016). https://doi.org/10.1016/j.protcy.2016.05.111
6. Wang, Z., Liu, K., Li, J., Zhu, Y., Zhang, Y.: Various frameworks and libraries of machine learning and deep learning: a survey. Arch. Comput. Methods Eng. (2019). https://doi.org/10.1007/s11831-018-09312-w
7. Hošovský, A., Piteľ, J., Adámek, M., Mižáková, J., Židek, K.: Comparative study of week-ahead forecasting of daily gas consumption in buildings using regression ARMA/SARMA and genetic-algorithm-optimized regression wavelet neural network models. J. Build. Eng. **34**(9), 101955 (2021). https://doi.org/10.1016/j.jobe.2020.101955
8. Dubey, A.K., Kumar, A., García-Díaz, V., Sharma, A.K., Kanhaiya, K.: Study and analysis of SARIMA and LSTM in forecasting time series data. Sustain. Energy Technol. Assess **47**, 101474 (2021). https://doi.org/10.1016/j.seta.2021.101474
9. Wright, S.J.: Coordinate descent algorithms. Math. Program. **151**(1), 3–34 (2015). https://doi.org/10.1007/s10107-015-0892-3
10. Zenkova, Z., Kabanova, T.: The ABC-XYZ analysis modified for data with outliers. In: Proceedings - GOL 2018: 4th IEEE International Conference on Logistics Operations Management, 137114 (2018). https://doi.org/10.1109/GOL.2018.8378073
11. Liu, J., Xiong, L., Pei, J., Zhang, H., Yu, W.: Group-based skyline for pareto optimal groups. In: Proceedings - 2020 5th International Conference on Logistics Operations Management, GOL, 931473 (2020). https://doi.org/10.1109/TKDE.2019.2960347
12. Zenkova, Z., Musoni, W., Tarima, S.: Accounting for deficit in ABC-XYZ analysis. IEEE Trans. Knowl. Data Eng. **33**(7), 8935189, 2914–2929 (2021). https://doi.org/10.1109/GOL49479.2020.9314731
13. Lavrenchenko, S.A., Zgonnik, L.V., Gladskaya, I.G.: Statistical approach to the management of Moscow multifunctional centers. Serv. Russia Abroad **6**(67), 36–49 (2016). https://doi.org/10.12737/21207
14. Dhoka, D., Choudary, L.Y.: Challenges with multi-dimensional inventory classifications and optimization. Asian Soc. Sci. **11**(4), 365–370 (2015). https://doi.org/10.5539/ass.v11n4p365
15. AI "MFC UR". http://mfcur.ru. Accessed 30 July 2021
16. Mean absolute percentage error regression loss. https://scikit-learn.org. Accessed 30 July 2021

Depression Detection by Person's Voice

Evgeniya Zavorina[1] and Ilya Makarov[1,2(✉)] (iD)

[1] HSE University, Moscow, Russia
iamakarov@hse.ru
[2] Artificial Intelligence Research Institute (AIRI), Moscow, Russia

Abstract. In this work, a machine learning algorithm is proposed to detect depression. The Transformer encoder network is considered and compared with top baseline approaches. Low-level features are extracted from audio recordings and then are augmented to overcome the problem of the small size of available dataset. The Transformer network achieves recognition accuracy of 73.51% on DAIC-WOZ database, which compare favourably to the accuracy of 65.85% and 66.35% obtained by traditional approaches.

Keywords: Deep learning · Speech recognition · Depression detection

1 Introduction

Depression is common health problems in society. In contrast to emotions, which are of a short-term nature, mood can remain unchanged for a sufficiently long time. Clinical depression is a disorder that can last for weeks, months, or even years. It differs from normal mood changes and short-term emotional responses to the challenges of everyday life.

From 76% to 85% of people in low- and middle-income countries are not receiving treatment for their disorder due to lack of resources [46]. The consequences of severe depression can be avoided if patients seek professional help on time and if healthcare professionals would be provided with the appropriate technology to detect and diagnose depression [35]. One of such technologies could be a model that relies on the extraction of voice markers that are reliable indicators of depression. According to DSM-5, individuals with depression experience depressed mood, loss of interest in the activities and feeling of worthlessness [4]. Studies have shown that depressed people have a slow, monotonous, and expressionless voice [28,48]. This work focuses on patterns extracted from human speech, which have been observed to change depending on the mental and emotional state of the patient. The most commonly used voice characteristics are cepstral characteristics [9,11,17,27,29,32].

Different machine learning tasks including speech emotion recognition and mental disorder recognition have known significant improvements over the last years with the advent of Deep Neural Networks (DNNs). The rise of deep learning techniques had led researchers to explore different deep architectures and methods for depression assessment [31].

E. Burnaev et al. (Eds.): AIST 2021, LNCS 13217, pp. 250–262, 2022.
https://doi.org/10.1007/978-3-031-16500-9_21

Machine learning methods have been utilized to extract high-level feature representations [1,13]. Yang et al. [47] presented a Convolutional Neural Network for the assessment of the level of depression which outperformed all the existing approaches on DAIC-WOZ dataset. Bhargava et al. [7] researched stacked bottleneck DNNs trained on windowed speech waveforms and achieved results that could be compared with the same architecture but trained on MFCC.

In this paper, a deep neural networks architecture based on Transformer encoder is proposed and evaluated on audio data. The remainder of this article is organised as follows. Section 2 introduces related works devoted depression recognition from speech. Section 3 introduces description of data and details of experiments. Section 4 describes the details of the results of each experiment. The analysis of mistakes of proposed methods is done in Sect. 5. Finally, the conclusion is presented in Sect. 6.

2   Related Work

As noted before, typically depressed individuals tend to change their expressions at a very slow rate and pronounce flat sentences with stretched pauses. Therefore, to assess depression audio features are frequently used for their ability and consistency to reveal signs of depression. Extracted from speech patterns are then fed to the classifier.

The depression recognition task has known significant improvements over the past years with the advent of DNNs. The most popular end-to-end speech recognition approaches typically include a combination of recurrent neural network (RNN) and Convolutional Neural Network (CNN), Graph Convolutional Recurrent Network [33] and attention-based encoder-decoder architecture [6,8,10,16,34].

Convolutional Neural Networks (CNNs) were applied on the speech recognition task (SER) and reached remarkable results on various datasets by learning affective-salient features [26]. Recurrent neural networks (RNNs) have also been introduced for SER purpose with a deep Bidirectional Long Short-Term Memory (Bi-LTSM) [15]. Several papers have presented CNNs in combination with LSTM cells to improve speech emotion recognition [14,43].

The self-attention based methods also called transformers have demonstrated promising results in various deep learning tasks recently [18]. The transformer model exploits the short/long range context by connecting arbitrary pairs of positions in the input sequence directly. Furthermore the model can be trained in a parallel way, which is much more efficient than conventional recurrent neural networks.

2.1   Acoustic Features

There are several categories of voice characteristics. Such patterns can carry linguistic information, as well as information about a person's gender, age, dialect, the presence of stress or nervousness, etc. Patterns are divided into: prosodic features (pitch, energy, fundamental frequency), spectral and cepstral features, voice quality features (jitter, shimmer, amplitude). In this work the Mel-frequency cepstral coefficients (MFCC) were selected as main features. MFCC features describe the audio cepstrum energies in a non-linear scale.

At first step of MFCC extraction the original audio signal is split into separate fragments of 20–40 ms duration, then each frame is multiplied by the Hamming window function to reduce leakage effect on the ends of frames. At next step the discrete Fourier transform (FFT) is applied to the result. The speech signal is analyzed on short segments, because it is simple enough on these segments and it can be described using coefficients. The received energies are logarithmized because the human ear perceives loudness as non-linear. Finally, the cosine transform is used, which results are the mel-cepstral coefficients.

This approach is quite popular in speech recognition problems, since it is relatively simple to implement and focuses on the part of the signal that is usually the most informative.

2.2   Classifiers in Speech Processing

Deep neural networks (DNNs) have proven very effective in traditional approaches on tasks like Automatic Speech Recognition (ASR) and speaker identification, the gains observed on emotion recognition and mood disorders recognition are limited, likely due to the small size of the data. In this section different approaches to emotional and depression recognition using Deep neural networks would be considered.

Convolutional Neural Networks (CNNs) and Long-Short-Term-Memory Recurrent Neural Networks (LSTM RNNs) are excellent DNN candidates for audio data classification. LSTM RNNs are successfully used for SER purposes due to the ability to interpret sequential data such as features of the audio waveform represented as a time series. The implementation of CNNs could be justified by audio spectrogram resemblance to images, in which CNNs excel at recognizing and discriminating between distinct patterns. For applying this strategy the audio signal transforms to a spectrogram which is then used as an input to convolutional layers, followed by recurrent ones. Such an approach of treating SER problems has recently demonstrated very competitive performance [2,38,39].

Deep learning on graph data has emerged as a major topic in the past few years. This is because graphs provide a natural and convenient way to deal with large data. The graph-based network is applied in this work to detect depression by speech. The proposed architecture consists of graph CNN for feature extraction and an LSTM for sequence learning (Fig. 1).

A graph construction approach is a frame-to-node transformation, where each node v_i corresponds to a short windowed segment of the audio signal with a feature vector x_i [41]. A feature vector contains MFCCs extracted from the corresponding audio segment. In this work a graph convolution layer was applied to learn higher abstraction levels for the node features. To learn not spatial features but also dynamic patters a LSTM layer was applied [40]. Details on application of graph embeddings and graph neural networks can be found in our previous studies [23–25], however, we mostly apply graph embeddings for

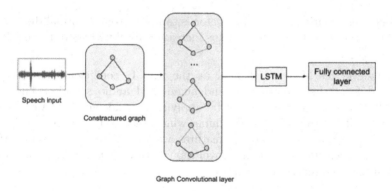

Fig. 1. The graph-based architecture consists of a convolution layer and a LSTM to learn graph embedding from node embeddings to facilitate depression detection.

recommender systems by extracting joint features from discrete structures [3, 5, 19, 20, 22] or texts [21, 37, 42], while in this paper we consider graph appearing from feature similarity of consequent audio segments representations.

A new attention technique based on an encoder-decoder structure that uses weighted correlations between the elements of the input sequence was introduced [44]. This algorithm, which is able to model more complex patterns within speech utterances, has led to more robust models for emotional recognition.

The role of the encoder part is to map the input sequence into several attention matrices. The decoder uses those matrices to generate a new token. In this work only the encoder layer is used since the network could predict frequency distributions of different emotions as it had shown in Transformer-CNN based architecture.[1]

A commonly used dataset within depression detection is the DAIC-WOZ corpus consisting of 50 h of data collected from 189 clinical interviews from a total of 142 patients [12, 36]. Each record is manually labeled with Patient Health Questionnaire score (PHQ-8). The PHQ-8 binary defines whether the participant is depressed or not. The training set contains audio records of 107 patients and the development set contains samples of 35 patients. There is no personal information about patient. The dataset curators removed mentions of personal names and locations from the audio samples.

Data Augmentation
Depression datasets available to assess depression from speech are relatively small. To overcome this challenge and avoid overfitting, data augmentation is performed. Generating more real samples is a difficult process, so instead the white noise is added to the audio data. It is a common method that allows the creation of pseudo-new samples and helps train robust model. Additive White Gaussian Noise (AWGN) was used in this paper, because after the transformation the white noise will add power to an audio signal uniformly across the

---

[1] https://github.com/IliaZenkov/transformer-cnn-emotion-recognition.

frequency distribution. The noise vector is sampled from a normal distribution. Thus, generated syntactic audio helps increase the robustness of the model.

Features Extraction
The low-level features are defined as the MFCC coefficients and they are extracted from the preprocessed and augmented audio samples. The speech signal is first divided into frames by applying a windowing function of 8s at fixed intervals of 500 ms. The Hamming window is used as a window function to remove edge effects. 40 cepstral features are obtained for each frame. MFCC were used as features in Graph Convolutional Recurrent Network and in CNN Transformer architecture.

3   Experiment Setup

Experiment 1. Convolutional neural network and long short-term memory model (CNN LSTM)
2D CNN LSTM network, were constructed to learn local and global emotion-related features from speech and log-mel spectrograms respectively. Network is consisting of four Convolutional feature learning blocks and one long short-term memory (LSTM) layer. Convolutional blocks are applied for learning local correlations along with extracting hierarchical correlations. LSTM layer is used to learn long-term dependencies from the learned local features. The number of channels of the convolutional and pooling layers are both 128. While their filter size is 3×3. RELU is used as an activation function for all the layers.

Experiment 2. Graph Convolutional Recurrent Network
A set of Mel-Frequency Cepstral Coefficients is extracted from the raw speech. Each audio frame produces a graph of 120 nodes, where each node corresponds to a (overlapping) speech segment of length 10 ms. Padding is used to make the samples of equal length. The dimension of the graph embedding is set to $Q = 64$.

Experiment 3. Transformer encoder and 2 parallel CNN.
The proposed model consists of three parallel blocks: two Convolutional networks and one Transformer-based network (Fig. 2).
The transformer encoder layer consists of 4 multi-head self-attention layers and a feedforward network. The Relu activation layer was used in Feed Forward Network to tame gradients and reduce compute time.
Parallel with Transformer the two similar Convolutional Neural networks are trained. The 3×3 kernels were applied in all Convolutional layers in both CNNs.
The first layer has a single input channel creating a $1 \times 3 \times 3$ filter with 16 outputs channels. The next layer has 16 input and 32 output channels therefore producing 32 unique filters of size $16 \times 3 \times 3$. The last convolutional layer has 32 input channels and 64 output channels, and after 4 max pooling (stride 4×4) produces an output feature map of $64 \times 1 \times 8$ size. The outputs of both CNN and Transformer are concatenated and fed to the final linear softmax layer.

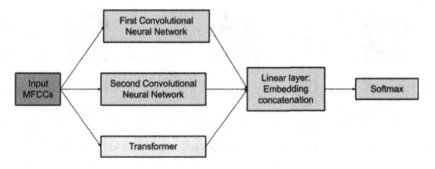

Fig. 2. Top-level representation of the Transformer architecture.

4 Results

The experimental results of the three proposed architectures on depression detecting are presented in Table 1 and Fig. 3. Transformer outperforms CNN-LSTM and Graph Convolutional RNN with an accuracy of 73.5%. The accuracy of the CNN-LSTM and Graph-based network is around 65.8% and 66.3% respectively. The precision of the Transformer approach is higher by 8.07% and 6.91% as compared to LSTM-based and Graph-based, respectively. There is also an improvement in quality for recall and F1-score. 87% of the samples are correctly labeled with non-depression, whereas, only 39% are correctly diagnosed with depression. The low rate of correct classification of depression can be explained by the small amount of the participants labeled as depressed. The number of samples labeled with depression is twice smaller then the number of non-depressed samples. Thus, collecting more samples of depressed participants would significantly increase the model's ability to recognize depression.

Table 1. Comparison of the performances of the proposed deep neural networks with Transformer method for predicting the PHQ-8 binary.

Experiments	Accuracy	Precision	Recall	F1-score
CNN-LSTM	65.85	78.94	32.50	46.04
Graph Convolutional RNN	66.35	80.10	31.31	45.03
Transformer	**73.51**	**87.01**	**39.14**	**53.99**

The Transformer network reached the higher Recall value. It is shown in Fig. 3 that at the first two experiments only 32% and 31% of depressed people were detected correctly.

Table 2 compares the performance of the given Transformer approach with recent depression recognition research in terms of Accuracy, Precision and Recall. The accuracy of the proposed Transformer model is slightly lower as compared

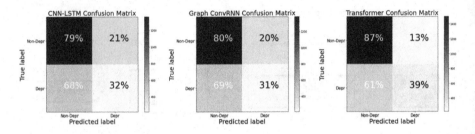

Fig. 3. Confusion Matrices on the test set of three experiments.

Table 2. Comparison of proposed network with resent methods for PHQ-8 binary in terms of Accuracy, Precision and Recall.

Methods	Accuracy	Precision	Recall
1D-CBBG [45]	71.43	50.00	50.00
Transformer (Our)	73.51	87.01	39.14
3D-CBHGA [45]	74.29	58.33	41.67
MFCC-AU LSTM [30]	77.16	83	44

to 3D-CBHGA model [45], yet a notable increment has been found in precision. It means that Transformer architecture better deals with I type error when non-depressed patient is incorrectly classified as a depressed person. According to research of multi-modal depression recognition in [30], it has been found that the visual features allows to reach higher accuracy score, thus accuracy of MFCC-AU LSTM model is noticeably higher than accuracy of our approach.

5　Discussion

The several examples of correct predictions and mistakes are considered for a qualitative assessment of the model's performance. The MFCC features for each example are introduced to represent the difference in speech in terms of loudness, the spread of emotions, and frequency ranges.

Example #1. True positive.
The proposed model based on Transformer correctly predicted the label of depressed participant, where the main pattern could be long pauses in speech (Fig. 4). The wide dark regions indicate there is low sound level. Therefore it can be judged that the model has learned such logical patterns as pauses. The rest of two models also classified the example correctly.

Example #2. True positive.
The participant's speech is monotonous, without prolonged pauses. It is an example of a mistake for the CNN-LSTM and Graph models. Probably convolutions at these methods could learn patterns associated with contrast speech when small pauses are followed by stressed pronounce and classify this example as an

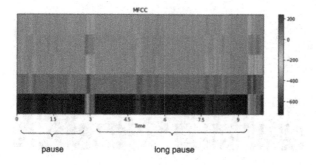

Fig. 4. The example of depressed patient. Several pauses in speech could be found in the time domain of intensity.

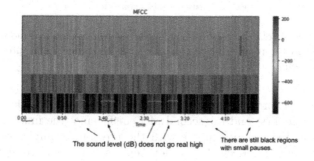

Fig. 5. The example of depressed patient.

enthusiastic. Convolution layers could incorrectly learn such pattern as "contrast stripes" as pattern of absence of depression. However the Transformer marked the inexpressive speech as depressed. The attention mechanism allows learn enough information about context and detect monotonous speech.

Example 3. True Negative.

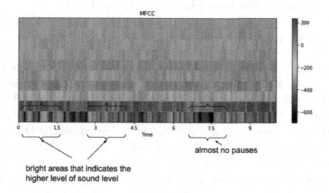

Fig. 6. The speech of a non-depressed person is more contrasting.

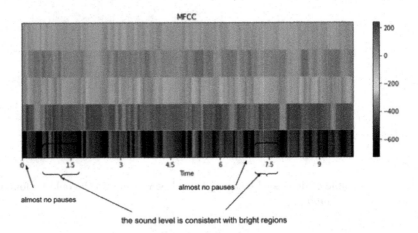

Fig. 7. The example of depressed patient. Speech is less excited and more monotonous than speech from the previous example. The sound level in the brighter areas is still lower than the sound level of non-depressed speakers.

The example of expressive speech without pauses is also considered. Taking a closer look at the first and the second Mel coefficients (the first two lines from the bottom) in Fig. 6 and compare with the previous examples in Fig. 5, it is clear that the values of decibel for a non-depressive patient are higher than for a depressive one. The whole first MFCC line looks like a wide and bright area without dark regions. Thus the model learned the patterns of expressive speech of non-depressed persons.

Example 4. False Negative. Error II type.
The speech is more expressive and almost without pauses in contrast to previous examples with depression. Most likely because of the emotional color and presence of bright areas, all of the models classified such example incorrectly. Also perhaps it is hard to predict that here is an example of a depressed person without looking at context and semantic features. Therefore the linguistic or visual patterns could solve these kinds of mistakes (Fig. 7).

6 Conclusion

In this study, an MFCC-based CNN Transformer Neural Network is proposed to detect depression by speech. The audio samples are preprocessed and augmented and the MFCC features are then extracted and normalized. The MFCC features are used as inputs for two parallel CNN networks and as an input for a Transformer. The proposed architecture is evaluated on the DAIC-WOZ corpus and results are compared with two baseline models: CNN-LSTM Network and Graph Convolutional Recurrent network. Transformer model outperforms existing state-of-the-art approaches in depression recognition. These results show that the proposed self-attention method is highly reliable in automatic diagnosis of clinical depression.

7   Future Work

In future work several approaches to improve the proposed model will be considered. As it may be observed, depressed patients tend to speak slowly and use simple phrases in speech. Therefore methods of automatic speech transcription and linguistic feature extraction could improve accuracy of Deep Neural Network for clinical depression recognition. The other possible solution which will be investigated is to transfer knowledge from an independent, yet related, previously learned task. For that, first need to pretrain the model on a related task. The emotions recognition from speech could be chosen as the related task. Then fine-tune a pretrained model on the DIAC-WOZ dataset for depression recognition. The main aim is to create in future an AI-based application designed to automatically detect clinical depression without any medical assistance.

Acknowledgments. The work of Ilya Makarov was supported by the Russian Science Foundation under grant 22-11-00323 and performed at HSE University, Moscow, Russia.

References

1. Al Hanai, T., Ghassemi, M.M., Glass, J.R.: Detecting depression with audio/text sequence modeling of interviews. In: Interspeech, pp. 1716–1720 (2018)
2. Amodei, D., et al.: Deep speech 2: end-to-end speech recognition in English and mandarin. In: International Conference on Machine Learning, pp. 173–182. PMLR (2016)
3. Ananyeva, M., Makarov, I., Pendiukhov, M.: GSM: inductive learning on dynamic graph embeddings. In: Bychkov, I., Kalyagin, V.A., Pardalos, P.M., Prokopyev, O. (eds.) NET 2018. SPMS, vol. 315, pp. 85–99. Springer, Cham (2020). https://doi.org/10.1007/978-3-030-37157-9_6
4. American Psychiatric Association et al.: Diagnostic and Statistical Manual of Mental Disorders: DSM-5. Arlington (2013)
5. Averchenkova, A., et al.: Collaborator recommender system. In: Bychkov, I., Kalyagin, V.A., Pardalos, P.M., Prokopyev, O. (eds.) NET 2018. SPMS, vol. 315, pp. 101–119. Springer, Cham (2020). https://doi.org/10.1007/978-3-030-37157-9_7
6. Bahdanau, D., Chorowski, J., Serdyuk, D., Brakel, P., Bengio, Y.: End-to-end attention-based large vocabulary speech recognition. In: 2016 IEEE International Conference on Acoustics, Speech and Signal Processing (ICASSP), pp. 4945–4949. IEEE (2016)
7. Bhargava, M., Rose, R.: Architectures for deep neural network based acoustic models defined over windowed speech waveforms. In: Sixteenth Annual Conference of the International Speech Communication Association (2015)
8. Chan, W., Jaitly, N., Le, Q., Vinyals, O.: Listen, attend and spell: a neural network for large vocabulary conversational speech recognition. In: 2016 IEEE International Conference on Acoustics, Speech and Signal Processing (ICASSP), pp. 4960–4964. IEEE (2016)
9. Cohn, J.F., et al.: Detecting depression from facial actions and vocal prosody. In: 2009 3rd International Conference on Affective Computing and Intelligent Interaction and Workshops, pp. 1–7. IEEE (2009)

10. Dong, L., Xu, S., Xu, B.: Speech-transformer: a no-recurrence sequence-to-sequence model for speech recognition. In: 2018 IEEE International Conference on Acoustics, Speech and Signal Processing (ICASSP), pp. 5884–5888. IEEE (2018)
11. France, D.J., Shiavi, R.G., Silverman, S., Silverman, M., Wilkes, M.: Acoustical properties of speech as indicators of depression and suicidal risk. IEEE Trans. Biomed. Eng. **47**(7), 829–837 (2000)
12. Gratch, J., et al.: The distress analysis interview corpus of human and computer interviews. In: Proceedings of the Ninth International Conference on Language Resources and Evaluation (LREC 2014), pp. 3123–3128 (2014)
13. Haque, A., Guo, M., Miner, A.S., Fei-Fei, L.: Measuring depression symptom severity from spoken language and 3d facial expressions. arXiv preprint arXiv:1811.08592 (2018)
14. Keren, G., Schuller, B.: Convolutional RNN: an enhanced model for extracting features from sequential data. In: 2016 International Joint Conference on Neural Networks (IJCNN), pp. 3412–3419. IEEE (2016)
15. Lee, J., Tashev, I.: High-level feature representation using recurrent neural network for speech emotion recognition. In: Interspeech 2015 (2015)
16. Li, S., Raj, D., Lu, X., Shen, P., Kawahara, T., Kawai, H.: Improving transformer-based speech recognition systems with compressed structure and speech attributes augmentation. In: Interspeech, pp. 4400–4404 (2019)
17. Low, L.S.A., Maddage, N.C., Lech, M., Sheeber, L., Allen, N.: Influence of acoustic low-level descriptors in the detection of clinical depression in adolescents. In: 2010 IEEE International Conference on Acoustics, Speech and Signal Processing, pp. 5154–5157. IEEE (2010)
18. Makarov, I., Borisenko, G.: Depth inpainting via vision transformer. In: 2021 IEEE International Symposium on Mixed and Augmented Reality Adjunct (ISMAR-Adjunct), pp. 286–291. IEEE (2021)
19. Makarov, I., Gerasimova, O.: Link prediction regression for weighted co-authorship networks. In: Rojas, I., Joya, G., Catala, A. (eds.) IWANN 2019, Part II. LNCS, vol. 11507, pp. 667–677. Springer, Cham (2019). https://doi.org/10.1007/978-3-030-20518-8_55
20. Makarov, I., Gerasimova, O.: Predicting collaborations in co-authorship network. In: 2019 14th International Workshop on Semantic and Social Media Adaptation and Personalization (SMAP), pp. 1–6. IEEE (2019)
21. Makarov, I., Gerasimova, O., Sulimov, P., Zhukov, L.E.: Co-authorship network embedding and recommending collaborators via network embedding. In: van der Aalst, W.M.P., et al. (eds.) AIST 2018. LNCS, vol. 11179, pp. 32–38. Springer, Cham (2018). https://doi.org/10.1007/978-3-030-11027-7_4
22. Makarov, I., Gerasimova, O., Sulimov, P., Zhukov, L.E.: Dual network embedding for representing research interests in the link prediction problem on co-authorship networks. PeerJ Comput. Sci. **5**, e172 (2019)
23. Makarov, I., Kiselev, D., Nikitinsky, N., Subelj, L.: Survey on graph embeddings and their applications to machine learning problems on graphs. PeerJ Comput. Sci. **7**, e357 (2021)
24. Makarov, I., Korovina, K., Kiselev, D.: JONNEE: joint network nodes and edges embedding. IEEE Access **9**, 144646–144659 (2021)
25. Makarov, I., Makarov, M., Kiselev, D.: Fusion of text and graph information for machine learning problems on networks. PeerJ Comput. Sci. **7**, e526 (2021)
26. Mao, Q., Dong, M., Huang, Z., Zhan, Y.: Learning salient features for speech emotion recognition using convolutional neural networks. IEEE Trans. Multimed. **16**(8), 2203–2213 (2014)

27. Moore, E., Clements, M., Peifer, J., Weisser, L.: Analysis of prosodic variation in speech for clinical depression. In: Proceedings of the 25th Annual International Conference of the IEEE Engineering in Medicine and Biology Society (IEEE Cat. No. 03CH37439), vol. 3, pp. 2925–2928. IEEE (2003)

28. Moore, E., II., Clements, M.A., Peifer, J.W., Weisser, L.: Critical analysis of the impact of glottal features in the classification of clinical depression in speech. IEEE Trans. Biomed. Eng. **55**(1), 96–107 (2007)

29. Mundt, J.C., Snyder, P.J., Cannizzaro, M.S., Chappie, K., Geralts, D.S.: Voice acoustic measures of depression severity and treatment response collected via interactive voice response (IVR) technology. J. Neurolinguistics **20**(1), 50–64 (2007)

30. Muzammel, M., Salam, H., Othmani, A.: End-to-end multimodal clinical depression recognition using deep neural networks: a comparative analysis. Comput. Methods Prog. Biomed. **211**, 106433 (2021)

31. Othmani, A., Kadoch, D., Bentounes, K., Rejaibi, E., Alfred, R., Hadid, A.: Towards robust deep neural networks for affect and depression recognition from speech. In: Del Bimbo, A., et al. (eds.) ICPR 2021. LNCS, vol. 12662, pp. 5–19. Springer, Cham (2021). https://doi.org/10.1007/978-3-030-68790-8_1

32. Ozdas, A., Shiavi, R.G., Silverman, S.E., Silverman, M.K., Wilkes, D.M.: Investigation of vocal jitter and glottal flow spectrum as possible cues for depression and near-term suicidal risk. IEEE Trans. Biomed. Eng. **51**(9), 1530–1540 (2004)

33. Pareja, A., et al.: EvolveGCN: evolving graph convolutional networks for dynamic graphs. In: Proceedings of the AAAI Conference on Artificial Intelligence, vol. 34, no. 4, pp. 5363–5370 (2020)

34. Pham, V.T., et al.: Independent language modeling architecture for end-to-end ASR. arXiv preprint arXiv:1912.00863 (2019)

35. Prendergast, M.: Understanding Depression. Penguin Group Australia (2006)

36. Ringeval, F., et al.: AVEC 2017: real-life depression, and affect recognition workshop and challenge. In: Proceedings of the 7th Annual Workshop on Audio/Visual Emotion Challenge, pp. 3–9 (2017)

37. Rustem, M.K., Makarov, I., Zhukov, L.E.: Predicting psychology attributes of a social network user. In: Proceedings of the Fourth Workshop on Experimental Economics and Machine Learning (EEML 2017), Dresden, Germany, 17–18 September 2017, pp. 1–7. CEUR WP (2017)

38. Sainath, T.N., Vinyals, O., Senior, A., Sak, H.: Convolutional, long short-term memory, fully connected deep neural networks. In: 2015 IEEE International Conference on Acoustics, Speech and Signal Processing (ICASSP), pp. 4580–4584. IEEE (2015)

39. Satt, A., Rozenberg, S., Hoory, R.: Efficient emotion recognition from speech using deep learning on spectrograms. In: Interspeech, pp. 1089–1093 (2017)

40. Seo, Y., Defferrard, M., Vandergheynst, P., Bresson, X.: Structured sequence modeling with graph convolutional recurrent networks. In: Cheng, L., Leung, A.C.S., Ozawa, S. (eds.) ICONIP 2018, Part I. LNCS, vol. 11301, pp. 362–373. Springer, Cham (2018). https://doi.org/10.1007/978-3-030-04167-0_33

41. Shirian, A., Guha, T.: Compact graph architecture for speech emotion recognition. In: 2021 IEEE International Conference on Acoustics, Speech and Signal Processing (ICASSP), ICASSP 2021, pp. 6284–6288. IEEE (2021)

42. Tikhomirova, K., Makarov, I.: Community detection based on the nodes role in a network: the telegram platform case. In: van der Aalst, W.M.P., et al. (eds.) AIST 2020. LNCS, vol. 12602, pp. 294–302. Springer, Cham (2021). https://doi.org/10.1007/978-3-030-72610-2_22

43. Trigeorgis, G., et al.: Adieu features? End-to-end speech emotion recognition using a deep convolutional recurrent network. In: 2016 IEEE International Conference on Acoustics, Speech and Signal Processing (ICASSP), pp. 5200–5204. IEEE (2016)
44. Vaswani, A., et al.: Attention is all you need. In: Advances in Neural Information Processing Systems, pp. 5998–6008 (2017)
45. Wang, H., Liu, Y., Zhen, X., Tu, X.: Depression speech recognition with a three-dimensional convolutional network. Front. Hum. Neurosci. 15 (2021)
46. Wang, P.S., et al.: Use of mental health services for anxiety, mood, and substance disorders in 17 countries in the who world mental health surveys. Lancet 370(9590), 841–850 (2007)
47. Yang, L., Sahli, H., Xia, X., Pei, E., Oveneke, M.C., Jiang, D.: Hybrid depression classification and estimation from audio video and text information. In: Proceedings of the 7th Annual Workshop on Audio/Visual Emotion Challenge, pp. 45–51 (2017)
48. Zlochower, A.J., Cohn, J.F.: Vocal timing in face-to-face interaction of clinically depressed and nondepressed mothers and their 4-month-old infants. Infant Behav. Dev. 19(3), 371–374 (1996)

Social Network Analysis

Research Papers Recommendation

Olga Gerasimova[1]([✉]) [iD], Anna Lapidus[1], and Ilya Makarov[1,2,3] [iD]

[1] HSE University, Moscow, Russia
{ogerasimova,iamakarov}@hse.ru, anyalapidus@list.ru
[2] University of Ljubljana, Ljubljana, Slovenia
[3] Artificial Intelligence Research Institute (AIRI), Moscow, Russia

Abstract. The work is devoted to academic papers recommendation task considered as link prediction on a static citation network. We compare several graph embeddings, text-based and fusion models in the link prediction problem on academic papers citation dataset. We showed that fusion models of graph and text information outperform other approaches based on graph or text information alone. We prove this via an extensive set of experiments with different train/test splits that our fusion models are robust and retain superior performance even with a reduced train set.

Keywords: Citation network · Graph embedding · Recommendation system

1 Introduction

At the start of each research, it is important to search for relevant earlier studies, thus placing the foundation for further investigations. Having large citation data, one can formulate the problem in the form of recommending research papers based on citation network data. Recently the number of diverse information resources has noticeably increased and the amount of published papers also grows every year. As the result, the necessity to automate the process of searching relevant papers and to fill missing citations based on the current citation information is of great importance for writing a correct research paper.

Different algorithms were proposed to make this process easier and suggest relevant articles that match the scientific interests of the researchers. A general approach to papers recommendation is based on keyword search. There are various improvements for classical keyword search such as ranking candidate papers according to certain quality measures, for instance, citing rate, or extending a set of keywords with similar terms [4,44]. An alternative approach suggests taking the advantage of citation networks and uses different network statistics (centrality measures, PageRank, etc.) to calculate ranking measures for each paper in a network [4,34].

The article was prepared within the framework of the HSE University Basic Research Program.

More advanced network embedding techniques inspired by word vector representation algorithms in Natural Language Processing have been introduced to improve efficiency for the citation prediction task [19]. In addition, the usage of text information associated with network nodes improves the quality of network representations in different tasks [46].

This work aims to compare the performance of text-based methods and network embedding techniques, including fusion models via text-associated network embeddings, for research paper recommendations based on citation networks. In this work, the recommendation is considered as a link prediction on the pair of node or text embeddings or both in a citation network using binary classification. We showed that fusion models are robust and flexible for solving recommender system problems via various comparison experiments and ablation studies.

The structure of the paper is as follows. In Sect. 2, we consider works in related fields. Next, in Sect. 3, we formulate a link prediction task on the citation network. Section 4 contains a dataset description. In Sect. 5, we present a detailed description of the conducted experiments. Then, Sect. 6 explores the results of the experimental part. Finally, we make the conclusion and discuss future work.

2   Related Work

To conduct an efficient and thorough search for scientific papers, which are relevant for research, is highly important. For that reason, academic papers recommendation problem has been widely studied and various approaches to handle this issue were proposed.

2.1   Text-Based Methods

Main text-based methods combine techniques such as keywords search, text similarity [12] or content-based filtering with ranking algorithms based on global relevance measures [2,38,45].

Research papers recommendation in terms of text similarity assumes that two papers are to be relevant to each other if the texts are lexically and semantically similar. Jaccard similarity [12] takes into account only common words for considered texts and, therefore, it does not allow to capture similarity except for the case with the direct matching of words in two texts. Explicit Semantic Analysis [8] presents a method that applies Cosine similarity between TF-IDF (term frequency - inverse document frequency) vector representations of two documents to evaluate text similarity. Cosine similarity has an advantage compared to Euclidean distance because even if the distance of two texts is large due to the difference in lengths of the texts, the angle between text vectors could be small. Though TF-IDF allows using important words as text features, it provides a sparse representation of text and could not capture the semantic similarity between different words.

Basic frequency-based model Bag of Words (BOW) [15] was extended via Continuous BOW (CBOW) and Skip-gram [32] word embedding models presenting dense word vectors that incorporate information about the context of a word in a text. The most common word embedding models include Word2Vec [32], FastText [17], etc. Word Mover's Distance (WMD) [20] presents another approach to handle the problem of similarity evaluation for the texts with no direct matching of words between two texts. WMD measures the minimum distance between words from two texts based on word2vec vectors to derive text similarity value.

Traditional keywords-based methods used in information retrieval could be less efficient for scientific literature search because it requires iterative keywords extension to cover a wider range of related papers. For that reason, some approaches for papers recommendation suggest using citation information [4].

2.2 Structural Graph Embeddings

Network embedding algorithms provide another approach to handle the problem of research papers recommendations using citation information. A recommendation problem could be stated in terms of link prediction for a pair of nodes in a citation network, which are presented via network embedding techniques [23–25, 27, 29].

Network embeddings are low dimensional representations of graph elements such as nodes, edges, subgraphs, or the whole graph. In this paper, we would mention only node and edge embeddings. Following the survey presented in [13], graph embedding techniques could be divided into three major classes, including Matrix Factorization based (LLE, SPE, GraRep, HOPE [3, 5, 33, 37]), Random-Walk based (DeepWalk, HARP, node2vec [7, 14, 35]). and Deep Learning methods (GCN, GAE, SDNE, DGNR [6, 18, 19, 41]). Based on graph properties, such as heterophily and homophily, it may be important to efficiently fuse information from structure and attribute representations.

2.3 Text-Attributed Network Embedding

Extended network embedding methods were developed to handle even attributed networks. TADW [42] suggests learning node embeddings by joint minimization of two distances that are the distance between embedding matrix and structure matrix and the distance between embedding matrix and attribute information matrix. PCTADW [39] presents a neural-network-based method that extends the TADW algorithm considering the direction of edges in the graph. The core idea is to process structure matrices reflecting incoming edges and outgoing ones independently using textual features matrix as additional information.

Graph Convolutional Networks (GCN) provide another method that allows to incorporate text features in node representation. GCN model takes node features matrix and graph structure matrix as input, and produces output node representations trained by a nonlinear function [18].

An alternative approach to include text attributes in network embedding learning is proposed in [16], where the ABRW algorithm computes biased transition matrix defined as a weighted sum of network transition matrix and text features transition matrix. The features transition matrix is constructed based on certain text similarity measures between text attributes for corresponding node pairs.

2.4   Recommender Systems

As for a recommendation task, we consider a large group of methods, which apply recommender systems techniques. According to [2], content-based filtering presents the prevalent approach in scientific papers recommender systems and provides algorithm based on authors' and papers' features association. Relevant papers search presupposes to use dominant words, word n-grams, topics as features for evaluating similarity between an author and a paper [11]. Recommendations obtained with content-based filtering are very similar to previous items associated with a user, which causes limitations for the approach.

Collaborative filtering is another widely-used approach [31,43] that finds similar users if they rate items similarly. In terms of research papers recommendation, this method would recommend papers cited by similar authors, where the similarity of authors is defined based on common citations.

Due to the sparsity of research papers network, recommender system approaches could be less promising. The research papers citation network contains few users (authors) and a large number of items (papers). Therefore, a low number of authors with common citations could be found, and, as the result, it is hard to induce efficient recommendations based on co-citation similarity [2].

In our research, we consider papers recommendation as a link prediction problem on citation networks similar to our previous studies on co-authorship studies [9,10,21,22,26]. In fact, having extensive information on the current citations and texts, we are able to predict missing links from different representations of paper and their relations with others in the field. In the next section, we formulate the method in detail.

3   Link Prediction on Citation Network

A citation network is presented as a graph, where nodes correspond to research papers. Directed edges of the graph indicate citation links between papers. So, such a network provides information on the relations between papers.

Link prediction is formulated as binary classification on graph edges, where the label '1' corresponds to the existing citation link between two papers and label '0' corresponds to the absence of an associated link.

Most of the approaches suggest obtaining edge representation based on its source and target nodes vectors because the number of edges can be considerably larger and learning direct edge embeddings does not allow to obtain representations for new links. If edge representation is built as an aggregation of nodes vectors, one can choose aggregation function as averaging, Hadamard product, L_1 or L_2-norm. We used Hadamard product, which usually shows the best performance.

Directed edges representation can be constructed from source and target node representations, if the node embedding algorithm provides two vectors for each node [33] or with special asymmetric edge function learned under embedding construction process [1].

For methods, which need text vector representation to input, different embeddings of words in the text were used. Details on network embedding choice and fusion models for network and attribute information can be found in [28] and [30], respectively.

Citation networks are useful for papers similarity analysis because academic databases have recently allowed access to the full text of the articles. Based on these networks, estimation of pairwise similarity of papers can be applied for recommendations. Similarity measures based on co-occurrence typically take into account co-citation for a couple of papers (number of papers that cite both considered papers) or common references [40].

4   Dataset

To evaluate the performance of graph embedding and text embedding models for the task of scientific papers recommendation, we used AAN Anthology Network Corpus [36]. Now, the AAN dataset contains more than 20000 papers on computational linguistics and natural language processing domains from the ACL Anthology (Association for Computational Linguistics).

AAN Corpus (2014 release) presents 19 924 papers published from 1982 to 2014, where each paper is attributed with metadata, including author, title, venue, and year. Full text or abstract is also available, but only for part of these papers. Apart from the corpus of ACL papers, the project also provides a citation network composed of 124 842 citation links.

Metadata Preprocessing. The article's metadata was presented as a single text document and it was parsed in order to extract necessary attributes, in particular, abstract and year. Text data preprocessing included extracting only alphabetic characters. We removed papers without abstract from the dataset for further experiments, because our main goal is to investigate using text information in terms of attributed network embeddings or pure text vectorization for the link prediction problem. As a result of the preprocessing steps, the citation network contains 13 236 articles connected by 67 704 links.

The final version of the citation graph is illustrated via the Gephi[1] visualization platform, see the Fig. 1. The greatest connected graph component contains 12 323 ($\approx$93% of nodes).

---

[1] https://gephi.org.

Fig. 1. The paper citation graph

5   Experiments Settings

5.1   Negative Edges Generation

The binary classification model requires both positive and negative samples for training, which arises the negative edges generation problem. We use random choice to form a set of negative edges, but we choose only not connected nodes within the network. To keep the balance between classes for classification models, the number of both sets such as generated negative links and existing positive links should be the same.

While our network is a directed graph, the direction of the links should also be considered. Hence, a negative sample should include inverse positive edges and we can train the algorithm to distinguish between oppositely directed links. So, the negative samples consist of inverse positive edges and the same number of randomly chosen pairs of non-connected nodes. The size of negative samples is twice the size of positive samples.

It is worth mentioning that the direction of predicted links may matter for our recommendation task because the link should start from a recent paper and point to an earlier published paper. However, we also considered several undirected network embeddings, which could be applied for nodes representation to any network.

5.2   Train and Test Splitting

Link prediction aims at predicting edges for a network with a part of hidden edges used as a test set. The strategy of removing edges for the test set can be based on a random choice of edges or depends on time associated with nodes appearance in the network. We decided to consider static task on the citation network in terms of predicting missing citations. That is why train and test splitting was not based on time modality as for temporal tasks.

One of the questions addressed in this work is whether the fraction of removed links for the test set significantly affects the result of link prediction. We compared results of link prediction for 10%, 30%, and 50% of links removed from the whole graph for the test set. In all the experiments, we report quality metrics averaged over five random seeds for a train set with std below 3-point precision.

5.3   Models Parameters

To obtain network nodes embeddings, we used methods such as Node2vec [14], HOPE [33], and fusion models, such as TADW [42], ABRW [16], and GCN [18]. We describe parameters, which we set for these models in Table 1.

Table 1. Parameters for network embeddings models.

Model	Parameters
Node2Vec	$d = 128$ - embedding dimension,
	$l = 80$ - random walks length,
	$n = 10$ - number of random walks per node
HOPE	$d = 100$ - embedding dimension,
	$\beta = 0.01$ - decay factor for Katz proximity
TADW	$d = 128$ - embedding dimension,
	$lamb = 0.2$ - a hyperparameter that controls
	The weight of regularization terms
ABRW	$d = 128$ - embedding dimension,
	$\alpha = 0.8$ - weight of structure transition matrix,
	$t = 30$ - top-k value,
	$l = 80$ - random walks length,
	$n = 10$ - number of random walks per node
GCN	$d = 64$ - embedding dimension,
	$epochs = 200, dropout = 0.5, learning\ rate = 0.01$
	Adam optimizer was used

6   Results

6.1   Link Prediction Based on Network Embedding

First, we compare the performance of structural network embeddings, such as Node2vec [14] and HOPE [33], working with undirected and directed networks, respectively. The results on link prediction for AAN citation network are shown in Table 2. In the case of directed model HOPE, we apply negative sampling with the inverse relation as described in Sect. 5.1 to learn that citation should always be asymmetric relation. The classification was performed using a Random Forest classifier with 10 decision trees and the minimum number of samples for internal node splitting set to 2. All results were presented for the test set with train/test split 70%/30%. The evaluation was conducted using classification metrics including Precision, Recall, and Accuracy.

Table 2. Comparing network embeddings for link prediction using Random Forest classifier.

Method	Directed	Precision	Recall	Accuracy
Node2Vec	−	0.608	0.488	0.587
diNode2Vec	+	0.618	0.518	0.599
HOPE	+	**0.932**	**0.644**	**0.833**

Second, we study how the train/test split affects the result of link prediction. The classification algorithm was trained with 10%, 30%, and 50% of removed links, on which we test our algorithms. The results of link prediction in terms of Accuracy for different training settings are presented in Table 3.

Table 3. Comparing network embeddings for link prediction using Random Forest classifier for different train/test splittings.

Method	Directed	10%	30%	50%
Node2Vec	−	0.586	0.587	0.583
diNode2Vec	+	0.604	0.599	0.592
HOPE	+	0.852	0.833	0.799

As we can see from the experiments, undirected Node2vec shows the worse performance, while directed Node2vec shows better results but still loses in quality to high-order asymmetric proximity preserving HOPE model.

6.2   Link Prediction Based on Text-Based Methods

Second, we formulate research papers recommendation based on text information alone via the following text embedding models: BOW, TF-IDF, Word2Vec, and FastText with mean and weighted versions described below.

For BOW and TF-IDF vector representations, terms with document frequency value exceeding 0.7 within the collection and unique terms were ignored. As for Word2Vec, we tested three options: 300- or 64- dimensional embeddings, and pretrained 300-dimensional embeddings. Finally, we used pretrained 300-dimensional FastText words embeddings that were then aggregated with equal weights to obtain text embedding (Mean FastText) or weighted according to TF-IDF value corresponding to each term in the text (Weighted FastText).

For classification, we used the same Random Forest model. Results on the test set are presented in Table 4 for train/test split 70%/30% and Table 5 for different train/test splits.

Table 4. Comparing text embeddings for link prediction using Random Forest classifier.

Text embedding	Precision	Recall	Accuracy
BOW	**0.835**	0.694	0.778
TF-IDF	0.831	0.694	0.776
Mean FastText	0.659	0.535	0.629
Weighted FastText	0.712	0.580	0.673
Word2Vec (d = 64)	0.823	0.723	0.784
Word2Vec (d = 300)	0.827	**0.731**	**0.789**
Word2Vec (pretrained, d = 300)	0.722	0.594	0.683

Table 5. Comparing text embeddings for link prediction using Random Forest classifier for different train/test splittings in terms of Accuracy.

Method	10%	30%	50%
BOW	0.782	0.778	0.773
TF-IDF	0.785	0.776	0.775
Mean FastText	0.632	0.629	0.617
Weighted FastText	0.682	0.673	0.667
Word2Vec (d = 64)	0.784	0.784	0.777
Word2Vec (d = 300)	0.801	0.789	0.787
Word2Vec (pretrained, d = 300)	0.690	0.683	0.681

Interestingly, BOW and TF-IDF showed good performance in all the tasks, however, the best accuracy was shown by non-pretrained Word2vec with $d = 300$ and the ability to capture the context of keywords representing the topic modeling from small text data in an efficient and consistent manner. Pretrained model degrades quality due to domain shift in text corpora. Variations of FastText aggregation do not stand out due to weighting over words oversmooths predictions when applied for link prediction. In addition, many papers have a high intersection of keywords, but it does not give information on citation (order in years) when averaging word representations.

Text Similarity Approach. For text similarity-based recommendation, we used cosine similarity on different text embeddings. Similarity measure can be viewed as a probability that two papers are related to each other and one of them could be recommended as a relevant paper for another one. For this approach, the similarity measure threshold had to be determined. Thresholds were chosen for each type of text embeddings, independently, via maximizing accuracy on the train set. The results of classification based on text similarity for text embedding representations and corresponding threshold values are presented in Table 6.

Table 6. Text similarity-based link prediction on different text embeddings.

Text embedding	Threshold	Precision	Recall	Accuracy
BOW	0.555	0.774	0.690	0.745
TF-IDF	0.523	**0.833**	**0.725**	**0.790**
Mean FastText	0.998	0.587	0.573	0.584
Weighted FastText	0.872	0.663	0.510	0.620
Word2Vec (d = 64)	0.821	0.794	0.718	0.766
Word2Vec (d = 300)	0.822	0.793	0.718	0.765
Word2Vec (pretrained, d = 300)	0.893	0.707	0.624	0.682

Although TF-IDF with distributional features of used words showed the best result, one can see that text information alone works worse than structural network embedding HOPE.

6.3   Fusion Approach

Finally, we evaluate research papers recommendations based on models fusing text and graph information: GCN, TADW, and ARBW with parameters described in Table 1. All the models were compared with different initializations by the best text embeddings from the previous experiments: BOW, TF-IDF, and Word2Vec ($d = 64, 300$).

For classification, we used the same Random Forest model. Results on the test set are presented in Table 7 for train/test split 70%/30%, and Table 8 and Table 9 for different train/test splits based on Accuracy and Recall, respectively. For both metrics, there is a tendency that the larger the train set, the lower the metric value. It can be explained as follows: the test set contains links that are badly predicted and increasing the size of the train set leads to the situation

Table 7. Comparing fusion models for link prediction using Random Forest classifier.

Text embedding	Precision	Recall	Accuracy
GCN + BOW	0.847	0.680	0.778
GCN + TFIDF	0.812	0.621	0.738
GCN + Word2Vec (d = 64)	0.901	**0.786**	0.850
GCN + Word2Vec (d = 300)	**0.903**	**0.786**	**0.851**
TADW + BOW	0.673	0.530	0.636
TADW + TFIDF	0.690	0.547	0.651
TADW + Word2Vec (d = 64)	0.794	0.689	0.755
TADW + Word2Vec (d = 300)	**0.804**	**0.699**	**0.764**
ABRW + BOW	0.614	0.484	0.590
ABRW + TFIDF	**0.639**	**0.508**	**0.611**
ABRW + Word2Vec (d = 64)	0.622	0.487	0.595
ABRW + Word2Vec (d = 300)	0.632	0.499	0.605

Table 8. Comparing fusion models for link prediction using Random Forest classifier for different train/test splittings in terms of Accuracy.

Method	10%	30%	50%
GCN + BOW	0.789	0.778	0.750
GCN + TFIDF	0.759	0.739	0.704
GCN + Word2Vec (d = 64)	0.862	0.850	0.833
GCN + Word2Vec (d = 300)	0.861	0.851	0.833
TADW + BOW	0.632	0.636	0.631
TADW + TFIDF	0.655	0.651	0.649
TADW + Word2Vec (d = 64)	0.761	0.755	0.752
TADW + Word2Vec (d = 300)	0.775	0.764	0.765
ABRW + BOW	0.588	0.590	0.581
ABRW + TFIDF	0.613	0.611	0.605
ABRW + Word2Vec (d = 64)	0.599	0.595	0.592
ABRW + Word2Vec (d = 300)	0.600	0.605	0.595

when the part of these links becomes larger. Also, models ranking does not depend on metrics.

In all the cases, basic models based on the GCN model show the best quality while TADW and ARBW appeared to be non-robust to different text embeddings used for fusion models initialization. In fact, the combination of Word2vec text embedding with a simple GCN model allows one to efficiently fuse graph and text contexts for the link prediction problem.

Table 9. Comparing fusion models for link prediction using Random Forest classifier for different train/test splittings in terms of Recall.

Method	10%	30%	50%
GCN + BOW	0.692	0.680	0.633
GCN + TFIDF	0.650	0.621	0.568
GCN + Word2Vec (d = 64)	0.805	0.786	0.753
GCN + Word2Vec (d = 300)	0.800	0.786	0.759
TADW + BOW	0.528	0.530	0.532
TADW + TFIDF	0.554	0.547	0.550
TADW + Word2Vec (d = 64)	0.699	0.689	0.688
TADW + Word2Vec (d = 300)	0.713	0.699	0.701
ABRW + BOW	0.473	0.484	0.471
ABRW + TFIDF	0.509	0.508	0.496
ABRW + Word2Vec (d = 64)	0.493	0.487	0.486
ABRW + Word2Vec (d = 300)	0.495	0.499	0.489

7   Conclusion and Discussion

In our research, we compared graph-based, text-based, and fusion models for suggesting research paper recommendations by solving the link prediction problem. We showed that a combination of graph and text information enriches feature representation achieving the best results in the task, and these conclusions are robust across different train/test splits. Data and code are available at GitHub.[2]

In our future studies, we aim to study dynamic citation networks and extend our approaches for capturing time-dependent structures, where new nodes corresponding to recently published papers appear, therefore, temporal network models can be useful for network embedding representation learning and consequently might improve the performance on the papers recommendation task.

References

1. Abu-El-Haija, S., Perozzi, B., Al-Rfou, R.: Learning edge representations via low-rank asymmetric projections. In: Proceedings of the 2017 ACM on Conference on Information and Knowledge Management, pp. 1787–1796 (2017)
2. Beel, J., Gipp, B., Langer, S., Breitinger, C.: Paper recommender systems: a literature survey. Int. J. Digit. Libr. **17**(4), 305–338 (2016)
3. Belkin, M., Niyogi, P.: Laplacian eigenmaps and spectral techniques for embedding and clustering. In: NIPS, vol. 14, pp. 585–591 (2001)
4. Bethard, S., Jurafsky, D.: Who should I cite: learning literature search models from citation behavior. In: Proceedings of IC CIKM, pp. 609–618 (2010)

---

[2] https://github.com/Olga3993/Research-Papers-Recommendation.git.

5. Cao, S., Lu, W., Xu, Q.: GraRep: learning graph representations with global structural information. In: Proceedings of IC CIKM, pp. 891–900 (2015)
6. Cao, S., Lu, W., Xu, Q.: Deep neural networks for learning graph representations. In: Proceedings of the AAAI Conference on Artificial Intelligence, vol. 30 (2016)
7. Chen, H., Perozzi, B., Hu, Y., Skiena, S.: Harp: hierarchical representation learning for networks. In: Proceedings of AAAI, vol. 32 (2018)
8. Gabrilovich, E., Markovitch, S., et al.: Computing semantic relatedness using Wikipedia-based explicit semantic analysis. In: IJcAI, vol. 7, pp. 1606–1611 (2007)
9. Gerasimova, O., Makarov, I.: Higher school of economics co-authorship network study. In: 2019 2nd International Conference on Computer Applications & Information Security (ICCAIS), pp. 1–4. IEEE (2019)
10. Gerasimova, O., Syomochkina, V.: Linking friends in social networks using Hash-Tag attributes. In: van der Aalst, W.M.P., et al. (eds.) AIST 2020. LNCS, vol. 12602, pp. 269–281. Springer, Cham (2021). https://doi.org/10.1007/978-3-030-72610-2_20
11. Giles, C.L., Bollacker, K.D., Lawrence, S.: Citeseer: an automatic citation indexing system. In: Proceedings of JCDL, pp. 89–98 (1998)
12. Gomaa, W.H., Fahmy, A.A., et al.: A survey of text similarity approaches. Int. J. Comput. Appl. **68**(13), 13–18 (2013)
13. Goyal, P., Ferrara, E.: Graph embedding techniques, applications, and performance: a survey. Knowl.-Based Syst. **151**, 78–94 (2018)
14. Grover, A., Leskovec, J.: Node2vec: scalable feature learning for networks. In: Proceedings of ACM SIGKDD, pp. 855–864 (2016)
15. Harris, Z.: Distributional structure. Word **10**(2–3), 146–162 (1954)
16. Hou, C., He, S., Tang, K.: Attributed network embedding for incomplete attributed networks. arXiv:1811.11728 (2018)
17. Joulin, A., Grave, E., Bojanowski, P., Douze, M., Jégou, H., Mikolov, T.: FastText. zip: compressing text classification models. arXiv:1612.03651 (2016)
18. Kipf, T.N., Welling, M.: Semi-supervised classification with graph convolutional networks. arXiv:1609.02907 (2016)
19. Kipf, T.N., Welling, M.: Variational graph auto-encoders. arXiv:1611.07308 (2016)
20. Kusner, M., Sun, Y., Kolkin, N., Weinberger, K.: From word embeddings to document distances. In: ICML, pp. 957–966. PMLR (2015)
21. Makarov, I., Bulanov, O., Gerasimova, O., Meshcheryakova, N., Karpov, I., Zhukov, L.E.: Scientific matchmaker: collaborator recommender system. In: van der Aalst, W.M.P., et al. (eds.) AIST 2017. LNCS, vol. 10716, pp. 404–410. Springer, Cham (2018). https://doi.org/10.1007/978-3-319-73013-4_37
22. Makarov, I., Bulanov, O., Zhukov, L.E.: Co-author recommender system. In: Kalyagin, V.A., Nikolaev, A.I., Pardalos, P.M., Prokopyev, O.A. (eds.) NET 2016. SPMS, vol. 197, pp. 251–257. Springer, Cham (2017). https://doi.org/10.1007/978-3-319-56829-4_18
23. Makarov, I., Gerasimova, O.: Link prediction regression for weighted co-authorship networks. In: Rojas, I., Joya, G., Catala, A. (eds.) IWANN 2019. LNCS, vol. 11507, pp. 667–677. Springer, Cham (2019). https://doi.org/10.1007/978-3-030-20518-8_55
24. Makarov, I., Gerasimova, O.: Predicting collaborations in co-authorship network. In: 2019 14th International Workshop on Semantic and Social Media Adaptation and Personalization (SMAP), pp. 1–6. IEEE (2019)

25. Makarov, I., Gerasimova, O., Sulimov, P., Korovina, K., Zhukov, L.E.: Joint node-edge network embedding for link prediction. In: van der Aalst, W.M.P., et al. (eds.) AIST 2018. LNCS, vol. 11179, pp. 20–31. Springer, Cham (2018). https://doi.org/10.1007/978-3-030-11027-7_3

26. Makarov, I., Gerasimova, O., Sulimov, P., Zhukov, L.E.: Recommending co-authorship via network embeddings and feature engineering: the case of national research university higher school of economics. In: Proceedings of the 18th ACM/IEEE on Joint Conference on Digital Libraries, pp. 365–366. ACM (2018)

27. Makarov, I., Gerasimova, O., Sulimov, P., Zhukov, L.E.: Dual network embedding for representing research interests in the link prediction problem on co-authorship networks. PeerJ Comput. Sci. **5**, e172 (2019)

28. Makarov, I., Kiselev, D., Nikitinsky, N., Subelj, L.: Survey on graph embeddings and their applications to machine learning problems on graphs. PeerJ Comput. Sci. **7**, e357 (2021)

29. Makarov, I., Korovina, K., Kiselev, D.: JONNEE: joint network nodes and edges embedding. IEEE Access **9**, 1–14 (2021)

30. Makarov, I., Makarov, M., Kiselev, D.: Fusion of text and graph information for machine learning problems on networks. PeerJ Comput. Sci. **7**, e526 (2021)

31. McNee, S.M., et al.: On the recommending of citations for research papers. In: Proceedings of the 2002 ACM Conference on Computer Supported Cooperative Work, pp. 116–125 (2002)

32. Mikolov, T., Chen, K., Corrado, G., Dean, J.: Efficient estimation of word representations in vector space. arXiv:1301.3781 (2013)

33. Ou, M., Cui, P., Pei, J., Zhang, Z., Zhu, W.: Asymmetric transitivity preserving graph embedding. In: Proceedings of the 22nd ACM SIGKDD International Conference on Knowledge Discovery and Data Mining, pp. 1105–1114 (2016)

34. Page, L., Brin, S., Motwani, R., Winograd, T.: The PageRank citation ranking: bringing order to the web. Technical report, Stanford InfoLab (1999)

35. Perozzi, B., Al-Rfou, R., Skiena, S.: DeepWalk: online learning of social representations. In: Proceedings of ACM SIGKDD, pp. 701–710 (2014)

36. Radev, D.R., Muthukrishnan, P., Qazvinian, V., Abu-Jbara, A.: The ACL anthology network corpus. Lang. Resour. Eval. **47**(4), 919–944 (2013)

37. Shaw, B., Jebara, T.: Structure preserving embedding. In: Proceedings of the 26th Annual International Conference on Machine Learning, pp. 937–944 (2009)

38. Strohman, T., Croft, W.B., Jensen, D.: Recommending citations for academic papers. In: Proceedings of the 30th Annual International ACM SIGIR Conference on Research and Development in Information Retrieval, pp. 705–706 (2007)

39. Sun, K., Zhong, S., Xu, H.: Learning embeddings of directed networks with text-associated nodes-with application in software package dependency networks. In: 2020 IEEE International Conference on Big Data (Big Data), pp. 2995–3004. IEEE (2020)

40. Tian, H., Zhuo, H.H.: Paper2vec: citation-context based document distributed representation for scholar recommendation. arXiv:1703.06587 (2017)

41. Wang, D., Cui, P., Zhu, W.: Structural deep network embedding. In: Proceedings of the 22nd ACM SIGKDD International Conference on Knowledge Discovery and Data Mining, pp. 1225–1234 (2016)

42. Yang, C., Liu, Z., Zhao, D., Sun, M., Chang, E.: Network representation learning with rich text information. In: Twenty-Fourth International Joint Conference on Artificial Intelligence (2015)

43. Yang, C., Wei, B., Wu, J., Zhang, Y., Zhang, L.: Cares: a ranking-oriented CADAL recommender system. In: Proceedings of the 9th ACM/IEEE-CS Joint Conference on Digital Libraries, pp. 203–212 (2009)
44. Yu, X., Gu, Q., Zhou, M., Han, J.: Citation prediction in heterogeneous bibliographic networks. In: Proceedings of the 2012 SIAM International Conference on Data Mining, pp. 1119–1130. SIAM (2012)
45. Zarrinkalam, F., Kahani, M.: SemCiR. Program (2013)
46. Zhang, Z., Cui, P., Zhu, W.: Deep learning on graphs: a survey. IEEE Trans. Knowl. Data Eng. (2020)

Multimodal Space of Users' Interests and Preferences in Social Networks

Evgeniia Shchepina$^{(\boxtimes)}$ ⓘ, Evgeniia Egorova ⓘ, Pavel Fedotov ⓘ,
and Anatoliy Surikov ⓘ

ITMO University, St. Petersburg, Russia

jannyss@mail.ru

Abstract. Social networks users leave numerous digital footprints in cyberspace. A part of digital footprints can be represented as a topology of relationships, for example, group subscriptions and music subscriptions. In this paper we combine our clustering results obtained for different modalities into the single multimodal space of interests and preferences. This paper aims to build a model of users' interests in the multimodal space and obtain a comprehensive conclusion about users' interests. To do this, we build the graphs based on data of separate modalities, find communities in these graphs, consider given communities in a single space further highlighting communities of users' interests. As a result of applying various clustering algorithms, such as Leiden, Louvain, K-Means and Agglomerative Clustering, 5, 15 and 5 communities were identified and well interpreted for music, groups subscriptions and result multimodal interests, respectively. The constructed model showed better results for analysis of user similarity in comparison with the baseline model. Scientific novelty of our approach is in the proposed method of multimodal clustering heterogeneous data on the interests and preferences of social network users. The distinctive feature of this method from traditional approaches, such as biclustering, is the possibility of flexible scaling the number of initial modalities. As a result, the average performance of the model increased by 12% in accuracy and by 11% in F1-score compared to the baseline model. The model can be used independently both in social research and as an intermediate predictor (embeddings) to improve the output quality of other models, for example, models for obtaining the psychological type of users.

Keywords: Community detection · Multimodal space of interests · Social network analysis

1 Introduction

The online personality emerged with the expansion of social networks. Users spend 2 h 22 min on average per day social networking [1]. The user's profile provides information about their interests and preferences. We propose an

---

Supported by The Russian Science Foundation.

approach to model users' interests and preferences. The approach is to cluster two different modalities, which are music subscriptions and group subscriptions, and then combine the results into the single multimodal space of interests and preferences.

To develop the model, we collected open data from the social network. This social network contains detailed information about users: personal information, groups and music subscriptions, videos, personal photos and pictures, friends list. After that we processed the data and presented it as graphs. Different modalities can provide a representation of users' interests and preferences in unimodal structures, such as weighted graphs, where vertices are two types of points of interest: musicians and groups. The weights of edges are proportional to the mutual number of the user's subscriptions.

The rest of the paper is organized as follows. In Sect. 2, we overview the papers related. The proposed approach is described in Sect. 3. Section 4 contains all results of clustering different modalities and the multimodal space interests as well as validation of constructed model. The conclusion is in Sect. 5.

2   Related Work

Music data processing is becoming more relevant due to the development of the music industry, social networks, and music streaming services. Music clustering can help build recommendation systems that can provide users with potentially favored songs and artists individually. Work [2] describes classical music clustering based on song audio signal components like 'average volume' or 'modal note'. Utilizing such information hierarchical clustering in ternary trees is done. The results show good but not perfect clustering of different classical music genres and music pieces with a normalized tree benefit score less than 0.9 on medium-sized and big datasets, hence this approach needs some improvement. Authors of the study [3] group composers using only context-based data describing features that cannot be extracted from audio signals, for example, a composer's cultural background. The data combines the information about interpersonal musical influences between artists and different ecological categories, including country location and period of creativity. The application of PCA and K-means clustering to obtain medium-sized data successfully groups artists into well-interpreted clusters. The results show that such input information permits to build a model that roughly coincides with common musical knowledge; the results are promising for future publications.

An interesting approach of data clustering is represented in work [4]. At first, researchers transform datasets to probabilistic graph models that can explain data uncertainty. Then, the graphs are clustered by the graph algorithms, such as FDBSCAN [5] and PDBSCAN [6]. The results are evaluated by the Davies-Bouldin index [7], the Dunn index [8], and the Silhouette coefficient [9] and demonstrate an overall better performance of the approach proposed. Therefore, in some cases graph clustering can be more appropriate than the traditional one, which is proved in [10]. The authors convert several real datasets into graphs, cluster them by the Louvain graph clustering algorithm and compare the results with traditional clustering algorithms such as K-means and hierarchical clustering. On average the comparison shows the superiority of the approach over traditional clustering.

Community detection is a well-studied field. Community detection techniques are useful for social media algorithms to discover users' interests and cultural preferences.

Subramani Kumar and et al. [11] applied community detection to social networks using density-based clustering. They compared two well-known concepts for community detection that are implemented as distance functions in the algorithms SCAN [12] and DEN-GRAPH [13], the structural similarity of nodes and the number of interactions between nodes, respectively, to evaluate the advantages and limitations of these approaches. The hierarchical approach is used for clustering to avoid the problem of choosing an appropriate density threshold for community detection as well as a strict limitation for the applicability. All experiments are performed on Twitter data and Enron datasets with different characteristics.

Multimodal embeddings construction is used to analyze data from social networks. Moreover, different modalities can be ranked according to their relevance for a particular model and dataset [14]. Such models can be used in recommendation models [14,15] and predictive models [16]. In particular, recommendation algorithms often use collaborative filtering based on an items similarity matrix [17]. In addition, many works consider the interests of users from different social networks as multimodal data, compare the same person's data, build embeddings of their mutual interests to improve recommendation algorithms [16].

3   Methods

Building a model of users' interests in the multimodal space is done as follows. At the first step data containing the information about each modality is processed separately. Number of modalities can be any. For each of them we collect open data about users' subscriptions in a social network. The data is preprocessed, its entities are selected by corresponding thresholds and graphs in which nodes are dataset entities and edges are relationships between the entities are built. Communities are found in these graphs using community detection algorithms. On the basis of obtained communities, the probability of being assigned to any cluster i in any modality is calculated for every user j according to the following formula:

$$P(i,j) = \frac{N(i,j)}{S(j)} \tag{1}$$

where $N(i,j)$ is the number of unique entities from cluster i belonging to user j, $S(j) = \sum_{k=1}^{n} N(k,j)$ - total amount of unique entities belonging to user j, where n - total number of clusters. Further combine all the communities into a single network in the multimodal space. The vertices of this graph are found intramodal communities and the edges - connections between these communities whose weights reflect the probabilities of assigning users to both vertices. This graph is clustered and the communities of interest in a single multimodal space are obtained.

In our research two graph clustering algorithms are applied. The first one is the Louvain algorithm [18], which is based on optimizing a quality function (modularity, for example) in two phases: (1) local moving of nodes; (2) grouping them together. In the first phase nodes are moved to the communities in such a way that makes the biggest increase of the quality function. The second phase is an aggregation of the network using the obtained in the first step communities. Then, each community becomes a node in a new network and the process repeats until the quality function achieves its maximum and cannot increase anymore. The main disadvantage of this approach is that the algorithm may find arbitrary badly connected communities. This problem is solved by the second algorithm, the Leiden algorithm [19] that always finds connected communities. It involves three steps: local moving of nodes, refinement of the partition and creating the network based on the obtained partition using non-refined partition for the creation of the initial aggregated network. Then, the steps repeat until no improvements can be made. The Leiden algorithm stops using the idea of greedy nodes merging. Conversely, at the refinement stage a single node can be added to any community, the main idea is that the operation increases the quality function.

Figure 1 is illustrated this process.

In the proposed method, the absence or significant data noise in one of the modalities does not become an insurmountable obstacle in the construction of the output. The addition of new modalities can improve the accuracy and robustness of the final output. This feature of the proposed method makes it suitable for developing incremental and online learning models with a flexible approach to the number and form of input data and the unchanged form of final output.

4   Experiments

4.1   Group Subscriptions

Data. In this paper we use the dataset collected from the social network VK - Russian online social media and social networking service which allows users to message each other, create and join groups, listen to music, watch videos, play games and much more. 'Group subscription' from the user means that he joins the group. 'Music subscription' means that the user adds the audio to the playlist. The group subscriptions dataset contains 7598 users and 215487 group subscriptions, and information about the groups (name, status, number of subscribers). All the data is anonymized.

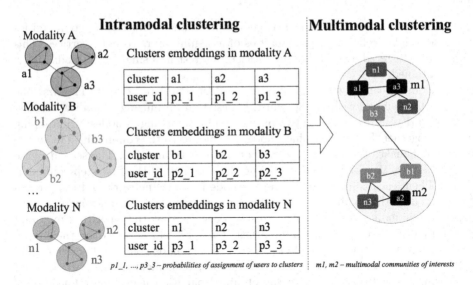

Intramodal clustering   **Multimodal clustering**

Clusters embeddings in modality A

cluster	a1	a2	a3
user_id	p1_1	p1_2	p1_3

Clusters embeddings in modality B

cluster	b1	b2	b3
user_id	p2_1	p2_2	p2_3

...

Clusters embeddings in modality N

cluster	n1	n2	n3
user_id	p3_1	p3_2	p3_3

p1_1, ..., p3_3 – probabilities of assignment of users to clusters   *m1, m2 – multimodal communities of interests*

Fig. 1. Illustration of the proposed method, which is in two parts: intramodal and multimodal clustering. Embeddings in intramodal clustering represent probabilities of users belonging to clusters (communities).

a)   b)   c)

Fig. 2. (a) The results of the detection community. (b) The results of the detection communities in musical subscriptions. (c) The results of the detection communities in multimodal analysis.

We process the data as follows:

1. Select groups by threshold;
2. Construct the adjacency matrix and the graph Fig. 2 (where nodes are groups, and edges are the relationship between groups by the number of user subscriptions).

Table 1. Description of groups clusters.

Community number	Top groups	Community size	Interpretation
0	Government of St. Petersburg; Accidents and Emergencies—St. Petersburg—Peter Online—SPb;	194	St. Petersburg news groups
4	Healthy lifestyle—I AM MY OWN COACH; Fitness and health. Diets and sports; SPORT - IS LIFE; WORKOUT;	194	Sports and healthy lifestyle groups
9	KudaGo: St. Petersburg; The Village Petersburg; St. Petersburg events; KASSIR.RU	171	Groups about activities and entertainment
14	Science and Technology; Roskosmos; High Technologies—High Tech; PostScience	168	Science and technology groups

The result is a graph divided into 15 communities. The interpretation of the communities is presented in Table 1. The best algorithm quality metrics are obtained using the Leiden algorithm, as seen in Table 4.

4.2 Music Subscriptions

Data. The dataset about users' musical subscriptions contains four columns: user id, user's track id, title of the track and song artist. The dataset contains 116155 unique artists from the music subscriptions of 2664 users, that is less than the number of users in the group subscriptions dataset despite the fact that the same group of users is analyzed in the study. The reason for this is that not all the users shows their musical subscriptions at their page, therefore they cannot be analyzed.

At first, the data is preprocessed. This part of the research is obligatory because Vkontakte music service allows users to download and rename audio by themselves that lead to the possibility of incorrectly written songs and song artists names. Rows containing null and incorrect values (for example, not numeric values in numeric columns) are dropped. Artists' names are converted to lowercase, all the symbols except letters, numbers and connection symbols ('_', '&') are deleted from them. Further, we identify the songs that are performed by several artists by the separators 'feat', '&' and '_' and split them in such a way that we obtain several rows instead of one, where each of them describes only one unique song artist.

Many artists' names are written incorrectly. To edit them we choose a threshold for the number of artists - 30 - and a threshold for an artist name length - 7, which were determined as the best during the testing. Bigger number of artists threshold leads to very superficial clustering results, smaller - to excessively detailed results, that do not detect common music interests. As for the artist name length, longer threshold length helps to correct very few artists names, shorter - leads to change of correctly written artists names, that are similar to other. If an artist's name is longer or equal to 7 characters and is in at least 30 different users' subscriptions, call it 'correct'. If it is longer than 7 characters but his songs are listened to by less than 30 users, it can be written incorrectly because of more rare writing. 'Correct' and 'potentially incorrect' artists' names are compared by the function SequenceMatcher from the difflib Python library which compares pairs of text sequences. If similarity ratio is bigger than 85%, a 'potentially incorrect' artist's name is changed into 'correct'.

Users that have at least 10 unique artists and artists that are at least in 30 unique users' music subscriptions are selected for further research. Selection of minimum number of occurrences in users' music for artists less than 30 leads to detection of weakly distinguishable groups. We build the adjacency matrix, with each cell describing the number of users listening to both artists from row and column indices. The number of the users after the preprocessing - 2263, the number of artists - 1065. This matrix is converted into an undirected weighted graph and clustered by the Leiden algorithm with the parameter of maximum number of nodes in the community equal to 296 due to the best interpretation of such clustering comparing with clustering with different or no maximum nodes limitation. The results of the clustering into 5 communities are presented in Fig. 2. Figure 2 shows that the clustering algorithm detects 4 big highly distinguishable communities (they are painted in blue, red, black and yellow colours) and one faintly discernible community - the smallest one. Table 2 represents the description of each community.

Interpretation of the results proves that the Leiden algorithm manages to divide the artists by genres and their audience using only the data on music affiliation to users.

Several clustering metrics are represented in Table 4, which shows that the Leiden algorithm can detect more detailed groups with the higher clustering results: bigger Silhouette and Calinski-Harabaz score, less Davies-Bouldin score. Obtained results demonstrate the detection of high-level users' music interests and are valid for the used dataset. For the larger and more representative audience of the social network data number of clusters and identified music genres and audiences will be probably bigger.

After the clustering we calculate the probability of user attribution to the cluster i by dividing the number of unique artists from the cluster i in the user's music by the total number of unique artists in the user's music.

Table 2. Description of music clusters.

Community number	Top artists	Community size	Interpretation
1	Coldplay, Red Hot Chili Peppers, Moby, Arctic Monkeys	296	Alternative and indie rock English artists
2	Basta, Kasta, Armin van Buuren, Madonna	296	Russian and English rap and pop artists - popular among Russian people under the age of 30
3	Zemfira, Splean, Noize MC, Leningrad	215	Russian and English rock artists - popular among Russian people over the age of 25
4	Lana Del Rey, Massive Attack, Eminem, Kanye West	138	English artists of all genres - popular among people of all ages
5	Linkin Park, Muse, The Prodigy, Metallica	125	English artists of different rock subgenres

4.3 Multimodal Analysis

People see objects, hear sounds, sense texture and taste. In our research, we propose an approach that explores several modalities. We used music information in the form of text and information about users' subscriptions. In the future, modalities such as video, audio and images may be added to our model.

For the analysis of multimodal data on users' interests, the communities obtained at stages group and music are considered in a single space for the further clustering purpose. Based on the probabilities embeddings of assigning users to music and groups communities, a graph is built. The vertices of this network are these communities, the edges are the connection between these communities. The weight of the edge between two vertices is proportional to the averaged sum of the probabilities of assigning users to both vertices.

Clustering is performed by two algorithms based on the estimation of the distance between vertices: the K-Means and Agglomerative Clustering algorithms, which have identified 5 and 6 communities respectively. All communities (music and groups) in the multimodal representation are equal, that is, they have equal weights equal to 1. Table 4 represents multimodal clustering metrics. We have calculated and compared the metrics and as a result have chosen the K-Means algorithm for further analysis. The number of communities found is based on metrics for assessing the quality of clustering. The table shows the results only for the best number of communities. The visualization of the best clustering is shown in Fig. 2.

In the visualization music communities are marked with black circles Table 3.

The users belonging to the selected communities have certain similarities in their interests. In addition, we can even make an assumption about other

Table 3. Description of multimodal communities.

Community number	Top items in communities	Community size	Interpretation
1	Groups: Interior design ideas, E.squire, Closing Studio, Healthy Lifestyle, Life Psychology, Joker's Psychology, Just do it, Recipes, Women's Magazine, Fashion and Style, Feminine Thoughts Music: Basta, Kasta, Lana Del Rey, Massive Attack	7	Interests including recipes, fashion, also some groups dedicated to an active lifestyle and healthy food, motivating quotes, popular female pop singers and life-affirming music
2	Groups: Government of St. Petersburg; Accidents and Emergencies—St. Petersburg—Petersburg Online—SPb	1	Community contains only one group cluster - St. Petersburg news groups
3	Groups: St. Petersburg events, Cinema, World of Discovery, Useful education, Books, Smart Money, Business magazine Music: Coldplay, Red Hot Chili Peppers, Zemfira, Splean, Noize MC	6	Groups about business, books and cinema. Music communities contain English and Russian rock artists for middle-aged people
4	Groups: English everyday, English School, Cheerful Sociopath, Optimist, Sarcasm Music: Linkin Park, Muse, The Prodigy, Metallica, Nickelback	3	Music artists of different more heavy rock subgenres, groups about studying English and groups with interesting posts and memes
5	Groups: Science and Facts, Science and Technology, Bookmania, Roskosmos, What if. . .?	3	Science and Technology groups

interests for users of each community. By examining statistics on community users, such as gender, age, city of residence. So, we found out that in community no.3 women are predominantly, and in community no.4 there are men. For all the users included in the selected three communities, the average probabilities of assignment to each community are determined and the following diagram in Fig. 3 is built on the basis of this data.

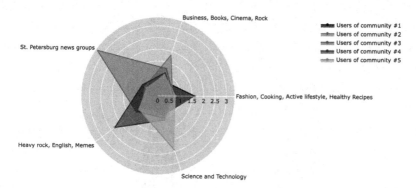

Fig. 3. Average probabilities of assignment of users to communities.

For the users of each community, the probability of being assigned to this community is the highest, and the probability of being assigned to other communities ranges from 0.1 to 0.3. However, the probabilities of being assigned to community with Business, Books and Cinema interests are high - they vary from 0.2 to 0.3. It can be assumed that such values are due to the fact that this cluster includes many common interests.

We have displayed the users of our sample in a single space using the convolution t-SNE method. The result of this visualization is shown in Fig. 4. You can see that communities of users with different interests are well separated.

Table 4. Clustering metrics.

Modalities	Algorithm	Communities	Modularity	Silhouette	Calinski-Harabaz	Davies-Bouldin
Groups	Leiden	15	0.191	0.234	181	3.524
	Louvain	6	0.11	−0.363	161.82	3.249
Music	Leiden	5	−0.028	−0.002	6.304	9.704
	Louvain	3	−0.025	−0.011	4.857	14.845
Multimodal	K-Means	5	−0.047	0.079	3.514	1.493
	Agglomerative Clustering	6	−0.049	0.09	3.281	1.326

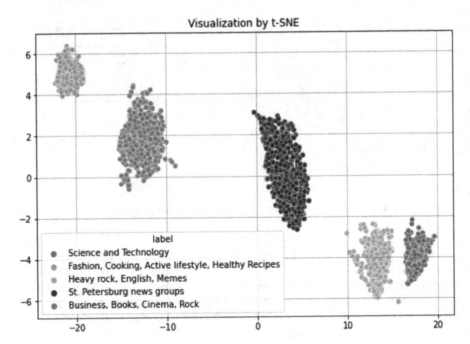

Fig. 4. Visualization of users in multimodal space using t-SNE.

4.4 Model Validation

To validate the model, we perform the following experiment: we select 100 random pairs from a sample of 2262 users, such that 50 of them lie at the minimum Euclidean distance from each other in the multimodal space of interests, and the other 50 - at the maximum Euclidean distance from each other. The resulting sample is marked by experts as follows: each expert is asked to familiarize himself with the personal pages of two users and conclude whether their interests are similar. The expert takes into account: general information, posts on the user's wall and public subscription data. The resulting markup was used to validate the model.

We build an alternative model as a baseline based on LDA analysis of topics in posts by users, published by them on their personal pages. The model was built with the number of topics parameter equal to 10, the parameter is based on the indicators of the coherence of the received topics. The thematic vectors obtained for all posts on the user's personal page are averaged, making up the final vector of interests of this user. Thus, the Euclidean distance between the vectors of their interests is the criterion for determining the proximity of the interests of two users.

The validation results in Table 5 show that the quality of work of both classifiers significantly exceeds the random classifier and is at a similar level, which indicates the consistency of both approaches to identify the interests of the user.

Table 5. Model validation metrics

	Accuracy	Precision	Recall	F1
Model on interests	78%	68%	85%	76%
Model on topics	66%	55%	78%	65%

It can be noted that the model based on the multimodal space of interests shows 12% better accuracy, 13% better precision, 7% better recall, and 11% better F1 score.

5 Conclusion and Future Work

We have built the model of multimodal embeddings of the user's description in the space of his interests. By analyzing detected communities at each step of clustering, similar interests were assigned to similar communities. All music and groups communities have a high-quality interpretation, and only part of communities in final clustering have qualitative interpretation. The scientific novelty lies in the proposed method of multimodal clustering data on the interests and preferences of social network users and the structure of our model. The constructed model has been validated in comparison with topics model.

In future works, we plan to consider not only the modalities of musical interests and subscriptions to groups, but also other available information on social networks as modalities to compile a complete portrait of the user and most accurately display the user in the space of their interests. Future research will focus on considering the dynamic change in the users' interests within a time period. For this, we will construct unique trajectories of interests. In the future, this model can significantly improve predictive algorithms and algorithms on topologies.

Acknowledgments. This research was financially supported by The Russian Science Foundation, Agreement no.17-71-30029 with co-financing of Bank Saint Petersburg.

References

1. Tankovska, H.: Daily time spent on social networking by internet users worldwide from 2012 to 2020 (2020). https://www.statista.com/statistics/433871/daily-social-media-usage-worldwide/. Accessed 29 Apr 2021
2. Vitányi, P., Cilibrasi, R., De Wolf, R.: Algorithmic clustering of music (2004)
3. Georges, P., Nguyen, N.: Visualizing music similarity: clustering and mapping 500 classical music composers. Scientometrics **120**(3), 975–1003 (2019). https://doi.org/10.1007/s11192-019-03166-0
4. Halim, Z., Khattak, J.H.: Density-based clustering of big probabilistic graphs. Evol. Syst. (2019)
5. Ducange, P., Marcelloni, F., Bechini, A., Criscione, M., Renda, A.: Fdbscan-apt: a fuzzy density-based clustering algorithm with automatic parameter tuning (2020)

6. Zhou, F., Xie, Y.H., Ma, Y.H., Liu, Y.A.: PDBSCAN: parallel DBSCAN for large-scale clustering applications (2012)
7. Davies, D.L., Bouldin, D.W.: A cluster separation measure. IEEE Trans. Pattern Anal. Mach. Intell. (1979)
8. Dunn, C.: Well-separated clusters and optimal fuzzy partitions (1974)
9. Rousseeuw, P.J.: Silhouettes: a graphical aid to the interpretation and validation of cluster analysis. J. Comput. Appl. Math. (1987)
10. Liu, Z., Barahona, M.: Graph-based data clustering via multiscale community detection. Appl. Netw. Sci. (2020)
11. Ntoutsi, I., Kroger, P., Subramani, K., Velkov A., Kriegel, H.P.: Density-based community detection in social networks (2011)
12. Feng, Z., Xu, X., Yuruk, N., Schweiger, T.A.J.: Scan: a structural clustering algorithm for networks (2007)
13. Barth, A., Falkowski, T., Spiliopoulou, M.: Dengraph (2007)
14. Wang, X., He, X., Huang, X., Tao, Z., Weiand, Y., Chua, T.S.: MGAT: multimodal graph attention network for recommendation. Inf. Process. Manag. 50(5), 102277 (2020)
15. Guo, J., Du, X., Lv, J., Song, B., Guizani, M.: Interest-related item similarity model based on multimodal data for top-n recommendation. IEEE Access 7, 12809–12821 (2019)
16. Pal, G., O'Halloran, K.L., Jin, M.: Multimodal approach to analysing big social and news media data. Discourse Context Media (2021)
17. Varlamis, I., Eirinaki, M., Gao, J., Tserpes, K.: Recommender systems for large-scale social networks: a review of challenges and solutions. Futur. Gener. Comput. Syst. 78, 413–418 (2018)
18. Lambiotte, R., Blondel, V.D., Guillaume, J.L., Lefebvre, E.: Fast unfolding of communities in large networks. J. Stat. Mech.: Theor. Exp. (2008)
19. Van Eck, N.J., Traag, V.A., Waltman, L.: From Louvain to Leiden: guaranteeing well-connected communities. Sci. Rep. (2018)

Citation Network Applications in a Scientific Co-authorship Recommender System

Vladislav Tishin[1,2], Artyom Sosedka[1,2], Peter Ibragimov[2],
and Vadim Porvatov[1,2(✉)]

[1] Sberbank, 117997 Moscow, Russia
eighonet@gmail.com
[2] National University of Science and Technology "MISIS", 119991 Moscow, Russia

Abstract. The problem of co-authors selection in the area of scientific collaborations might be a daunting one. In this paper, we propose a new pipeline that effectively utilizes citation data in the link prediction task on the co-authorship network. In particular, we explore the capabilities of a recommender system based on data aggregation strategies on different graphs. Since graph neural networks proved their efficiency on a wide range of tasks related to recommendation systems, we leverage them as a relevant method for the forecasting of potential collaborations in the scientific community.

Keywords: Graph machine learning · Neural networks · Recommender systems · Social graphs

1 Introduction

Since the advent of scientific communities, there has been a high demand in the area of collaboration recommendations. Due to the complex nature of interconnections between researchers, this domain has not reached a successful automation for a long time.

According to the underlying graph structure of collaboration networks, we propose to use recently emerged graph neural networks (GNN) to efficiently predict research cooperation between scientists. This branch of machine learning has readily proved its outstanding performance in a wide range of areas related to recommender systems [15]. Such algorithms as Node2Vec [4], Attri2Vec [16], and GraphSAGE [6] can be trained to capture structural features of co-authorship network. Embeddings produced by these methods can be effectively applied to the different forecasting tasks including prediction of network connections as well [8].

Graph neural networks allow us not only to boost performance in straightforward link prediction task on co-authorship network, but to improve the quality on such a forecasting challenge via aggregation of additional information from

E. Burnaev et al. (Eds.): AIST 2021, LNCS 13217, pp. 293–299, 2022.
https://doi.org/10.1007/978-3-031-16500-9_24

the citation graph. Future development of the discussed pipeline can lead to the simplification of the collaboration assessment process for the R&D team management.

2   Related Work

Recommender systems for scientific communities have a long history of development. Early approaches in this area [1,13] were based on deterministic network information, which lacked the ability to represent complex features of graph data.

Learning algorithms in the area of link prediction resolved a variety of issues related to capture of graph intricacies. The explicit examples of such models are local random walk [8] and local naive Bayes [9].

Various metrics such as content similarity LDAcosin [3] were also applied in this field. Another popular method [11] leverages linear regression on feature vectors of nodes and set of graph measures [10]. However, implementation of a learning feature extractor instead of deterministic metrics would significantly boost the performance of such pipelines. Applying this idea to the collaboration network structures, usage of graph neural networks becomes native [17].

Despite the promising results achieved by graph neural networks, different techniques could be implemented to further increase their efficiency. Those methods include the alteration of graph topology [14] and the usage of task-independent techniques like Node2Vec in order to create initial node representations which serve as better inputs for GNNs [5].

3   Data

We use classic HEP-TH dataset [7] as the subject for further development and extension. It consists of citation and co-authorship graphs obtained from the arXiv papers published between January 1993 and April 2003 in the area of high energy physics theory. Unfortunately, there is no connection between these two parts of the dataset (authors' IDs were not provided in the citation network) which makes its initial second part worthless for our purposes.

To our best knowledge, the potential of the HEP-TH citation network was never explicitly revealed. Due to the presence of highly useful paper metadata (such as author lists or abstracts), the range of the dataset usage can be significantly extended. In order to complete our current research, we perform processing of the citation graph metadata aiming to restore the corresponding co-authorship network. To preserve homogeneous nature of the reconstructing graph, we discarded from the citation graph all anonymous papers.

Along with the information about authors and abstracts, HEP-TH involves a journal reference field with the publisher output information. The presence of such data provides us an opportunity to extract ISSNs of indexed publications and further parse scientific metrics (quartile, h-index, and impact factor) from the "SCImago Journal & Country Rank" website.

4 Methods

Explored architecture consists of two subsequently applied graph neural networks. The first model generates vector representations of the publications according to their annotations and the structure of the citation network. As its input we leverage the directed citation graph $G(V, A, X)$, where $V = \{v_1, v_2, \ldots, v_n\}$ corresponds to the set of graph nodes (articles), $A : n \times n \rightarrow \{0, 1\}$ is the adjacency matrix (each edge encodes citation between two papers), and $X : n \times d \rightarrow \mathbf{R}$ denotes the matrix of node features (vectorized abstracts via pre-trained FastText [2]).

We perform a set of computational experiments using previously addressed unsupervised methods GraphSAGE, Node2Vec, and Attri2Vec. In the following, we briefly discuss each of them to clarify their usage as the feature extractors.

GraphSAGE. This method aggregates information about the set of neighbor nodes $N(v)$ in order to produce embedding of node v

$$h_v^{l+1} = \sigma(W^l \cdot \text{CONCAT}(h_v^l, h_{N(v)}^{l+1})), \tag{1}$$

where $l + 1$ is the current number of a convolutional layer, $h_v^0 = x_v$, W^l is the matrix of learning parameters, and $h_{N(v)}^{l+1}$ can be extracted by different aggregation functions like Max Pooling or Mean aggregator.

Node2Vec. This algorithm generates sequences of nodes via second-order random walks and utilizes them as the input data for a skip-gram model. The skip-gram generates pairs from input and context nodes in order to cast them to the feedforward neural network. Its weights can be used as the desired node embeddings as a result of the following function optimization:

$$\max_f \sum_{u \in V} \left[-\log Z_u + \sum_{n_i \in N(u)} f(n_i) \cdot f(u) \right], \tag{2}$$

where $Z_u = \sum_{v \in V} \exp(f(u) \cdot f(v))$ is per-node partition function and f($\cdot$) corresponds to the mapping function.

Attri2Vec. The last considered model uses the image $f(x_i)$ of node v_i with feature vector x_i in the new attribute subspace to predict its context nodes. In this method, the task is to solve the joint optimization problem

$$\min_{W^{in}, W^{out}} -\sum_{i=1}^{|V|} \sum_{j=1}^{|V|} n(v_i, v_j) \log \frac{\exp\left(f(v_i) \cdot w_j^{out}\right)}{\sum_{k=1}^{|V|} \exp\left(f(v_i) \cdot w_k^{out}\right)}, \tag{3}$$

where $n(v_i, v_j)$ is the number of times that v_j occurs in v_i context within t-window size in the generated set of random walks, W_{in} is the weight matrix from the input layer to hidden layer and W_{out} is the weight matrix from the hidden layer to the output layer.

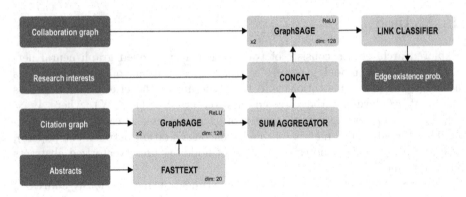

Fig. 1. Example configuration of the described architecture based on GraphSAGE convolutions.

Link prediction step follows after the publications embeddings generating. For this task, we consider the co-authorship graph $\hat{G}(\hat{V}, \hat{A}, \hat{X})$, where $\hat{V} = \{\hat{v}_1, \hat{v}_2, \ldots, \hat{v}_m\}$ is the set of graph vertexes (authors), $\hat{A} : m \times m \rightarrow \{0, 1\}$ is the adjacency matrix (each edge encodes collaboration between the two authors), and $\hat{X} : m \times k \rightarrow \mathbf{R}$ denotes the matrix of node features (one-hot encoded research interests of authors). In order to supply the predictive model by additional data about publications of the authors, we extend each element $\hat{x}_i$ of the matrix $\hat{X}$ to $\hat{x}_i'$ as follows:

$$\hat{x}_i' = \text{CONCAT}(x_i, \sum_{e \in e_i} e), \tag{4}$$

where e_i denotes the set of publications embeddings of ith author.

After the concatenation, the extended graph was translated as an input to the two-layer GraphSAGE with link classifier. It constructs the embedding of the potential links applying a binary operator to the pair of node embeddings (we consider L1, L2, Hadamard operator, average, and inner product [12]). Finally, these link embeddings are passed through the dense classification layer to obtain probabilities of links existence in the network. The whole pipeline is illustrated in Fig. 1.

The last model is trained by minimizing the binary cross-entropy loss

$$L = -\frac{1}{N} \sum_{i=1}^{N} y_i \cdot \log \hat{y}_i + (1 - y_i) \cdot \log (1 - \hat{y}_i), \tag{5}$$

where N is the output size, y_i is the true link labels, $\hat{y}_i$ is the predicted link existence probabilities.

5 Results

We performed training and evaluation on the PC with 1 GPU Tesla V100 and 96 GB of RAM. The weights of the neural networks were updated by Adam

Table 1. Results of different models on test sample

Article embedding	Author embedding	LP op.	Accuracy	AUC-ROC	F1-score
–	GraphSAGE (Mean)	L2	0.8793	0.9442	0.8817
FastText	GraphSAGE (Mean)	Had	0.8844	0.9486	0.8828
GraphSAGE (Mean)	GraphSAGE (Mean)	Had	0.8895	0.9568	**0.8911**
GraphSAGE (Mean)	GraphSAGE (Mean)	L2	**0.8928**	0.9531	0.8885
GraphSAGE (Mean)	GraphSAGE (MaxPool)	L1	0.8638	**0.9617**	0.8489

optimizer. We divided graph edges into train, validation and test samples in ratio 3:1:2.

We evaluated the set of models with various link embedding operators, representation learning models, and aggregation functions. As the main quality measurements, we chose binary accuracy, AUC-ROC, and F1-score. Received embeddings from the models were tested on the supervised GraphSAGE setup applied to the link prediction task.

We conducted experiments with more than 80 different configurations and represented key results in Table 1. The first two baselines were selected as the best among the approaches using either vectors of authors interests without citation network information or just embeddings of the abstracts. Comparison of these simpler architectures with full models which includes two graph neural networks could be interpreted as an ablation study.

As it is shown in the table, the proposed aggregation algorithm positively influences the quality of scientific collaboration forecasting. The model without any citation data significantly suffers from the lack of expressive input features as well as the pipeline including only the abstracts of the papers. Obtained result allow us to report that proposed aggregation strategy efficiently utilizes citation graph properties.

6   Conclusion and Outlook

In the present paper, we briefly introduced the two-stage pipeline for the collaboration prediction task. Performed computational experiments reveal the perspective of citation data utilization in sense of co-authorship network extension. The embeddings generated by GNNs effectively capture the network properties including its topology and vectorized abstracts represented as features of the corresponding graph nodes.

Our main contributions in the present work are the following:

1. We perform extraction of the co-authorship graph from the corresponding HEP-TH citation network.

2. Presence of useful metadata allows us to parse the scientific significance measures of the publications (e.g., impact factor).
3. We aggregate structural data from the citation graph and apply it to the co-authorship network in order to evaluate its influence on the link prediction quality.

Along with the future improvements of link prediction methods in the area of scientific collaboration, we intend to explore qualitative and quantitative assessment approaches of emerged links. As the probabilistic estimation of collaborations is not sufficient, it is important to extend it by less abstract metrics. In the following, we are going to leverage the average impact factor and the total number of publications for this task.

Acknowledgements. We acknowledge fruitful discussions with Natalia Semenova.

References

1. Alinani, K., Wang, G., Alinani, A., Narejo, D.H.: Who should be my co-author? recommender system to suggest a list of collaborators. In: 2017 IEEE International Symposium on Parallel and Distributed Processing with Applications and 2017 IEEE International Conference on Ubiquitous Computing and Communications (ISPA/IUCC), pp. 1427–1433. IEEE (2017)
2. Bojanowski, P., Grave, E., Joulin, A., Mikolov, T.: Enriching word vectors with subword information. Trans. Assoc. Comput. Linguist. **5**, 135–146 (2017)
3. Chuan, P.M., Son, L.H., Ali, M., Khang, T.D., Huong, L.T., Dey, N.: Link prediction in co-authorship networks based on hybrid content similarity metric. Appl. Intell. **48**(8), 2470–2486 (2017). https://doi.org/10.1007/s10489-017-1086-x
4. Grover, A., Leskovec, J.: node2vec: Scalable feature learning for networks. In: Proceedings of the 22nd ACM SIGKDD International Conference on Knowledge Discovery and Data Mining, pp. 855–864 (2016)
5. Gupta, C., Jain, Y., De, A., Chakrabarti, S.: Integrating transductive and inductive embeddings improves link prediction accuracy. arXiv preprint arXiv:2108.10108 (2021)
6. Hamilton, W.L., Ying, R., Leskovec, J.: Inductive representation learning on large graphs. In: Proceedings of the 31st International Conference on Neural Information Processing Systems, pp. 1025–1035 (2017)
7. Leskovec, J., Kleinberg, J., Faloutsos, C.: Graph evolution: densification and shrinking diameters. ACM Trans. Knowl. Discovery Data **1** (2006)
8. Liu, W., Lü, L.: Link prediction based on local random walk. EPL (Europhys. Lett.) **89**(5), 58007 (2010)
9. Liu, Z., Zhang, Q.M., Lü, L., Zhou, T.: Link prediction in complex networks: a local naïve bayes model. EPL (Europhys. Lett.) **96**(4), 48007 (2011)
10. Makarov, I., Bulanov, O., Gerasimova, O., Meshcheryakova, N., Karpov, I., Zhukov, L.E.: Scientific matchmaker: collaborator recommender system. In: van der Aalst, W.M.P., et al. (eds.) AIST 2017. LNCS, vol. 10716, pp. 404–410. Springer, Cham (2018). https://doi.org/10.1007/978-3-319-73013-4_37
11. Makarov, I., Bulanov, O., Zhukov, L.E.: Co-author recommender system. In: Kalyagin, V.A., Nikolaev, A.I., Pardalos, P.M., Prokopyev, O.A. (eds.) NET 2016. SPMS, vol. 197, pp. 251–257. Springer, Cham (2017). https://doi.org/10.1007/978-3-319-56829-4_18

12. Makarov, I., Gerasimova, O., Sulimov, P., Korovina, K., Zhukov, L.E.: Joint node-edge network embedding for link prediction. In: van der Aalst, W.M.P., et al. (eds.) Analysis of Images, Social Networks and Texts, pp. 20–31. Springer International Publishing, Cham (2018)
13. Sie, R.L., Drachsler, H., Bitter-Rijpkema, M., Sloep, P.: To whom and why should i connect? co-author recommendation based on powerful and similar peers. Int. J. Technol. Enhanced Learn. 4(1–2), 121–137 (2012)
14. Singh, A., et al.: Edge proposal sets for link prediction. arXiv preprint arXiv:2106.15810 (2021)
15. Ying, R., He, R., Chen, K., Eksombatchai, P., Hamilton, W.L., Leskovec, J.: Graph convolutional neural networks for web-scale recommender systems. In: Proceedings of the 24th ACM SIGKDD International Conference on Knowledge Discovery & Data Mining, pp. 974–983. KDD 2018, Association for Computing Machinery, New York, NY, USA (2018). https://doi.org/10.1145/3219819.3219890
16. Zhang, D., Yin, J., Zhu, X., Zhang, C.: Attributed network embedding via subspace discovery. Data Min. Knowl. Disc. 33(6), 1953–1980 (2019). https://doi.org/10.1007/s10618-019-00650-2
17. Zhang, M., Chen, Y.: Link prediction based on graph neural networks. Adv. Neural. Inf. Process. Syst. 31, 5165–5175 (2018)

Theoretical Machine Learning
and Optimization

How Fast Can the Uniform Capacitated Facility Location Problem Be Solved on Path Graphs

Alexander Ageev[1] , Edward Gimadi[1,2] , and Alexandr Shtepa[2(✉)]

[1] Sobolev Institute of Mathematics, pr. Koptyuga 4, 630090 Novosibirsk, Russia
{ageev,gimadi}@math.nsc.ru
[2] Department of Mechanics and Mathematics, Novosibirsk State University,
Pirogova 1, 630090 Novosibirsk, Russia
shoomath@gmail.com

Abstract. The Capacitated Facility Location Problem (CFLP) aims to open a subset of facilities, each of which has an opening cost and a capacity constraint, to serve all the given client demands such that the total transportation and facility opening costs are minimized. We consider the network CFLP in which the clients and the facilities are located at the vertices of a given edge-weighted transportation network graph. The network CFLP is NP-hard even when the network is a simple path. However, the Uniform CFLP (UCFLP) where the capacities of all facilities are identical is known to be polynomial-time solvable on path graphs. In this paper we revisit the dynamic programming algorithm by Ageev, Gimadi and Kurochkin [Diskret. Analys. Issl. Oper., 2009] for the UCFLP on path graphs and show how its time complexity can be improved to $\mathcal{O}(m^2 n^2)$, where m is the number of possible facility locations, n is the number of clients, which is the best possible time complexity within the framework of the considered algorithm. Also we carried out the comparison of the proposed strictly polynomial algorithm with the best pseudo-polynomial algorithms.

Keywords: Uniform Capacitated Facility Location Problem · Path graph · Exact algorithm · Dynamic programming · Time complexity · Polynomial algorithm · Pseudo-polynomial algorithm · Comparison analysis

1 Introduction

The Uncapacitated Facility Location Problem (UFLP) is a classic optimization problem for determining the optimal places of opening facilities or warehouses. It has many variations and many applications in logistics, telecommunications,

Supported by the program of fundamental scientific researches of the SB RAS (project No. 0314-2019-0014), and by the Russian Foundation for Basic Research, (project No. 20-31-90091).

health care, planning robust systems, clustering, computer vision, etc. For a detailed overview we refer the reader to book [11] and the papers [1,5,6,15,16].

In the UFLP we are given a set of n clients, each having a certain demand of a product, a set of m possible facility locations, each having a certain opening cost, and transportation costs of delivering of the product unit from the i-th facility to the j-th client. The problem is to determine which facilities to open, so that the total opening costs and the transportation costs of serving all the clients are minimized. The problem is strongly NP-hard since the SET COVER problem reduces to it [7].

The natural generalization of the classic UFLP is the Capacitated Facility Location Problem (CFLP), in which each facility i has a capacity a_i, which is the maximum amount of product it can produce. The CFLP aims to open a subset of facilities, each of which has an opening cost and a capacity constraint, to serve all the given client demand such that the total transportation and facility opening costs are minimized. We consider the network CFLP, in which the clients and the facilities are located at the vertices of a given edge-weighted transportation network graph. The network CFLP is NP-hard even when the network is a simple path.

There are two variants of the CFLP: the single allocation CFLP, where the demand of a client must be served by only one facility, and the multiple allocation CFLP, where a client can be served by multiple facilities simultaneously. The problem is NP-hard in both statements and it remains so even in the metric case [11], where an edge-weighted transportation network graph is given, the clients and the facilities are located at the vertices of this graph, and the unit product delivery cost from the i-th facility to the j-th client is defined according to the shortest path distance. In what follows, by CFLP we mean exactly its multiple allocation version. The metric CFLP is known to be constant-factor approximable (see, for example, [14]) with currently the best approximation factor of $(5 + \varepsilon)$ [4].

There are several types of CFLP according to the assumptions concerning facility capacities. If it is allowed to place several facilities at one site, the problem is called *soft CFLP*. Alternatively, if it is allowed to place only one facility at one site, the problem is called *hard CFLP* or *CFLP with hard capacities* [2], [12]. Soft capacities make problem easier [12]. Usually, if we consider CFLP we examine hard CFLP and in this paper we investigate hard CFLP, as well.

For the case of arbitrary capacities Mirchandani et al. [13] suggested a pseudo-polynomial algorithm for the multiple allocation CFLP on a path graph with running-time $\mathcal{O}\left(mB\min\{a_{\max}, B\}\right)$, where $B = \sum_j b_j$ is the total demand, and $a_{\max}$ is the maximum facility capacity. In [9], the pseudo-polynomial algorithm with improved running-time $\mathcal{O}(mB)$ is presented for the CFLP with hard capacities on a path graph.

If the capacities of all facilities are identical, such problem is called the Uniform CFLP (UCFLP). In 2004, Ageev [2] presented the first polynomial-time algorithm for the multiple allocation UCFLP on a path graph with time complexity $\mathcal{O}(m^5 n^2 + m^3 n^3)$. Later the time complexity of this algorithm was improved up to $\mathcal{O}(m^4n^2)$ in [3]. And another speeding up of the algorithm with running time $\mathcal{O}(m^3n^2)$ was proposed in paper [8].

In this paper, we revisit the dynamic programming algorithm by Ageev, Gimadi and Kurochkin [3] for the UCFLP on path graphs (UCFLP-path) and show how its time complexity can be improved to $\mathcal{O}(m^2n^2)$, which is the best possible time within the framework developed in [2,3].

The UCFLP-path has several real-world applications in optimization, for example, location of identical step-down transformers for a high-voltage power line in rural area and scheduling identical deliveries from a retailer to various households in the same neighborhood [13].

The rest of the paper is organized as follows. In the second section the UCFLP-path is formulated and we give some notation, which will be used further. The third section is devoted to some useful properties of solutions for the considered problem. In the fourth section we propose a revisited algorithm for UCFLP-path and prove the main result of the paper. The fifth section is about comparison analysis of the proposed algorithm with the best pseudo-polynomial algorithms. And in the last sixth section we describe some directions for future research.

2   Problem Statement and Basic Notation

In the UCFLP-path we are given an input edge-weighted graph $P = (V_P, E_P)$, where $|V_P| = n + m$ and edges of E_P form simple path on V_P. Let the vertices of P be enumerated from left to right in the order of their bypassing. Let $\mathcal{F} = \{v_1, \ldots, v_m\} \subseteq V_P$ be the given set of facility locations, where v_i is the i-th leftmost facility in $\mathcal{F}$, $1 \leq i \leq m$, and $\mathcal{C} = \{u_1, \ldots, u_n\} \subseteq V_P$ be the given set of clients with u_j being the j-th leftmost client in $\mathcal{C}$, $1 \leq j \leq n$.

Each facility $v_i \in \mathcal{F}$ has a nonnegative integral capacity a and a nonnegative opening cost f_i. Each client $u_j \in \mathcal{C}$ has a nonnegative integral demand b_j. Transportation of a unit of product from i-th facility to j-th client costs c_{ij}, which is defined as the weight of simple path from v_i to u_j. Without loss of generality we assume that $a \leq \sum_{j=1}^{n} b_j = B$.

The problem can be formulated as the following integer linear program:

$$\min\left(\sum_{v_i \in S} f_i + \sum_{v_i \in S} \sum_{u_j \in \mathcal{C}} c_{ij} x_{ij}\right), \tag{1}$$

subject to

$$\sum_{v_i \in S} x_{ij} = b_j, \ 1 \leq j \leq n, \tag{2}$$

$$\sum_{u_j \in \mathcal{C}} x_{ij} \leq a, \ 1 \leq i \leq m, \tag{3}$$

$$x_{ij} \in \mathbb{Z}_+, \ v_i \in \mathcal{F}, \ u_j \in \mathcal{C}, \tag{4}$$

where x_{ij} is equal to the amount of the demand of client u_j that is served by facility v_i. The constraints (2) guarantee that each client demand is satisfied while the constraints (3) ensure that the facility capacities are not exceeded. Finally, the restrictions (4) mean that the problem variables are in the set of nonnegative integers $\mathbb{Z}_+$.

By (S, x) we denote a feasible solution of the problem with the set S and the quantities $x = (x_{ij})$. For any $1 \leq i' \leq i'' \leq m$, by *facility segment* $[i', i'']$ we denote a subset $\{v_i \in \mathcal{F} \mid i' \leq i \leq i''\}$, and similarly for $1 \leq j' \leq j'' \leq n$ by *client segment* $[j', j'']$ we denote a subset $\{u_j \in \mathcal{C} \mid j' \leq j \leq j''\}$.

Let (S, x) be a feasible solution of (1)–(4). We say that the i-th facility serves the j-th client if $x_{ij} > 0$. We say that the i-th facility is *consumed*, if $\sum_{j=1}^{n} x_{ij} = a$, and *non-consumed*, if $0 < \sum_{j=1}^{n} x_{ij} < a$. Note that consumed and non-consumed facilities are necessarily open.

3   Properties of UCFLP-Path Solutions

In this section we discuss the key properties of UCFLP-path optimal solutions shown in [2,3] which allow to construct a polynomial-time dynamic programming algorithm for the problem.

In what follows, we consider only those problems for which inequality

$$\sum_{j=1}^{n} b_j \leq ma$$

is true, since otherwise, the total capacity of all facilities is not enough to meet the total demand of clients and the problem has no solution.

Each possible solution (S, x) of an input $\mathcal{I}$ now can be represented as a bipartite graph with vertex set $\mathcal{F} \cup \mathcal{C}$ and edge set $E = \{(i, j) \mid v_i \in \mathcal{F}, u_j \in \mathcal{C} : x_{ij} > 0\}$.

The first useful property is that there are optimal solutions for the UCFLP-path, in which the client set is split into continuous segments, and for each client segment $[j', j'']$ there is one open facility in the solution that fully serves the clients in this segment, except, possibly, the clients j' and j''.

Lemma 1 [2,3]. *An instance $\mathcal{I}$ of the UCFLP-path has an optimal solution (S, x) satisfying the following property:*
(1^0) *for any $v_{i'}, v_{i''} \in \mathcal{F}$ and $u_{j'}, u_{j''} \in \mathcal{C}$ such that $i' < i''$ and $j' < j''$ either $x_{i'j''} = 0$, or $x_{i''j'} = 0$.*

Note that property (1^0) also holds in the case of arbitrary capacities, and leads to a pseudo-polynomial time algorithm for the CFLP on path graphs [13]. In the graph terms the proposition asserts that instance $\mathcal{I}$ has an optimal solution (S, x) whose graph is a forest.

The following crucial property of the UCFLP-path optimal solutions further specifies the structure of an optimal solution and concerns the non-consumed facilities.

Lemma 2 [2,3]. *Let $\mathcal{I}$ be an instance of the UCFLP-path. Then it has an optimal solution (S, x) satisfying property (1^0) and the following property: (2^0) for any non-consumed facilities i' and i'', the facility segment $[i', i'']$ contains open facilities i^* and i^{**} such that they do not share a client and there are no open facilities in the segment $[i^* + 1, i^{**} - 1]$.*

In the graph terms this lemma states that there is an optimal solution (S, x) whose graph is a forest and no two non-consumed facilities of (S, x) lie in the same connected component of the graph.

4 Polynomial-Time Algorithm for the UCFLP-Path

In this section we discuss a polynomial-time dynamic programming algorithm [2,3] for the UCFLP-path based on the structural properties (1^0) and (2^0), and show how its time complexity can be improved to $\mathcal{O}(m^2 n^2)$.

For any $i \in \{0, 1, \ldots, m\}$ and $j \in \{0, 1, \ldots, n\}$, let $R(i, j)$ denote the cost of an optimal solution to the subproblem of $\mathcal{I}$ with the first i facilities and the first j clients, except auxiliary values $R(i, 0)$, $0 \le i \le m$ and $R(0, j)$, $1 \le j \le n$. Therefore the value of $R(m, n)$ is the cost of an optimal solution with respect to the original UCFLP-path. An optimal solution that satisfies properties (1^0) and (2^0) can be found by the following dynamic programming algorithm:

$$R(i, 0) := 0, \ 0 \le i \le m, \quad R(0, j) := +\infty, \ 1 \le j \le n. \tag{5}$$

And for all $1 \le i \le m$, $1 \le j \le n$

$$R(i, j) = \min_{1 \le i' \le i, 1 \le j' \le j} \left\{ R(i' - 1, j' - 1) + Q_{i', i}(j', j) \right\}, \tag{6}$$

where $Q_{i', i}(j', j)$ is the cost of an optimal solution to the instance with the set of facilities $[i', i]$, $1 \le i' \le i \le m$, and the set of clients $[j', j]$, $1 \le j' \le j \le n$ in which at most one open facility in $[i', i]$ is non-consumed.

On the basis of work [3] we can formulate the next statement.

Theorem 1.

$$Q_{i', i''}(j', j'') = \min_{1 \le t \le r, \ i' \le i \le i''} \left\{ \widetilde{G}^L_{i-1, t-1}(j') + S_{it}(j', j'') + \widetilde{G}^R_{i+1, r-t}(j'') \right\} \tag{7}$$

where
$\widetilde{G}^L_{i-1, t-1}(j')$ *is the optimal cost for opening $t - 1$ consumed facilities in $[i', i]$;*
$\widetilde{G}^R_{i+1, r-t}(j'')$ *is the optimal cost for opening $r - t$ consumed facilities in $[i, i'']$;*
$S_{it}(j', j'')$ *is the cost of choosing the i-th facility of the segment $[i, i'']$ to be the t-th open and the only non-consumed facility.*

Proof. Let us introduce some additional notation:

$B_j = \sum_{k=1}^{j} b_k$ is the total demand of the client segment $[1, j]$, $1 \le j \le n$;

$C_{ij} = \sum_{k=1}^{j} c_{ik} b_k$ is transportation costs associated with the satisfaction of the total client demand from segment $[1, j]$ by the i-th facility, $C_{i0} = 0$;

$r = \lceil (B_{j''} - B_{j'-1})/a \rceil$ is number of facilities to be opened in the segment $[i', i'']$, to satisfy the total demand of the segment $[j', j'']$, provided that $[i', i'']$ is a simple segment.

In fact, we need to determine which facility in segment $[i', i'']$ will be non-consumed. The remaining open facilities are consumed, their number is equal to $(r - 1)$, i.e., each of them produces the product of the maximum capacity a. For a visual understanding, it is convenient to consider a straight line segment of length $\sum_{j=j'}^{j''} b_j$ that matches the total demand of the segment $[j', j'']$. Our goal is to cover it by a set of segments consisting of the $(r - 1)$ segments of length a and one segment of length $\sum_{j=j'}^{j''} b_j - (r - 1)a$. Thus, "moving" along the segment of total demand a segment of length $\sum_{j=j'}^{j''} b_j - (r - 1)a$, corresponding perhaps to non-consumed facility, we can efficiently calculate the value $Q_{i', i''}(j', j'')$.

We denote by $g_{i,t,j'}$ the opening and transportation costs of consumed facility $i \in [i', i'']$, provided that among facilities from $[i', i - 1]$ are exactly $(t - 1)$ open and all of them are consumed (we assume that $t < r$).

Intersection of service area of facility i with neighboring open facilities occurs for clients j_{t-1}^* and j_t^*. The following two cases are possible:

1. $j_{t-1}^* = j_t^*$. In this case the consumed facility i serves only one client j_t^* and the required costs

$$g_{i,t,j'} = f_i + c_{ij_t^*} a.$$

2. $j_{t-1}^* < j_t^*$. In this case the i-th consumed facility serves all client demands from interval (j_{t-1}^*, j_t^*). The transportation costs of this client interval is

$$(C_{ij_t^* - 1} - C_{ij_{t-1}^*}).$$

Transportation costs for serving the leftmost client j_{t-1}^* are equal to

$$c_{ij_{t-1}^*} \left(B_{j_{t-1}^*} - B_{j'-1} - (t - 1)a \right).$$

For the rightmost client j_t^*, we similarly obtain

$$c_{ij_t^*} \left(ta - (B_{j_t^* - 1} - B_{j'-1}) \right).$$

Thus, we find the required costs

$$g_{i,t,j'} = f_i + \left(C_{ij_t^*-1} - C_{ij_{t-1}^*} \right) +$$

$$+ c_{ij_t^*} \left(ta - (B_{j_t^*-1} - B_{j'-1}) \right) +$$

$$+ c_{ij_{t-1}^*} \left(B_{j_{t-1}^*} - B_{j'-1} - (t-1)a \right).$$

Define an auxiliary set of clients $T = \{j_1^*, \ldots, j_{r-1}^*\}$ as follows:

$$B_{j_t^*-1} - B_{j'-1} < ta \le B_{j_t^*} - B_{j'-1}, \ 1 \le t < r.$$

If the first k open facilities $\{i_1, i_2, \ldots, i_k\}$, $k \le r$ are consumed, then by definition of T for any $1 \le s \le k - 1$, either the intersection of service areas of neighboring facilities i_s and i_{s+1} occurs on client j_s^*, or if the service areas do not overlap the facility i_s serves the client j_s^*, and the facility i_{s+1} serves the client $j_s^* + 1$ (following j_s^* among the clients in $[j', j'']$).

We denote by $G_{i,t}^L(j')$ the optimal opening and transportation costs of t consumed facilities in the segment $[i', i]$, and the facility i is necessarily open (we assume that $|\{i', \ldots, i\}| = i - i' + 1 \ge t$). We also introduce $\widetilde{G}_{i,t}^L(j')$, which is optimal opening and transportation costs of the first t consumed facilities in the segment $[i', i]$, when the i-th facility does not have to be open.

The above quantities for all parameters are calculated by means of the following recurrence relations

$$G_{i,t}^L(j') = \begin{cases} g_{i,1,j'}, & \text{if } t = 1, \\ g_{i,t,j'} + \widetilde{G}_{i-1,t-1}^L(j'), & \text{if } 2 \le t \le i - i' + 1. \end{cases}$$

$$\widetilde{G}_{i,t}^L(j') = \begin{cases} \min_{i' \le k \le i} G_{k,1}^L(j') = \min_{i' \le k \le i} g_{k,1,j'}, & \text{if } t = 1, \\ \min_{i' \le k \le i, \ (k-i'+1) > t} G_{k,t}^L(j') = \min\{G_{i,t}^L(j'), \widetilde{G}_{i-1,t}^L(j')\}, & \text{if } 2 \le t \le i - i' + 1. \end{cases}$$

Thus, we can efficiently compute the values $\widetilde{G}_{i,t}^L(j')$.

Using an identical scheme, we determine the parameters $\widetilde{G}_{i,t}^R(j'')$ and $G_{i,t}^R(j'')$, which denote the opening and transportation costs of t consumed facilities in segment $[i, i'']$ analogically to the previous definitions. The difference is that previously we looked at open consumed facilities ranging from left end of segment $[i', i'']$, and these facilities served some segment of clients $[j', j_s] \subseteq [j', j'']$. Now we consider open consumed facilities ranging from the right end of the segment $[i', i'']$. Performing this, they serve some segment of clients $[j_s, j''] \subseteq [j', j'']$. We denote by $g'_{i,t,j''}$ the opening and transportation costs of consumed facility $i \in [i', i'']$, provided that among facilities $[i + 1, i'']$ are exactly $(t - 1)$ open and they are all consumed. Parameters similar to B_j, $T = \{j_1^*, \ldots, j_{r-1}^*\}$ are denoted by B_j', $T' = \{j_1^{**}, \ldots, j_{r-1}^{**}\}$. It is easy to see that formulas for calculating them are built in the same way:

$$B_j' = \sum_{k=j}^n b_k, \ 1 \le j \le n,$$

$$r = \left\lceil (B_{j'}' - B_{j''+1}')/a \right\rceil.$$

We define an auxiliary set of clients T' as follows

$$T' = \{j_1^{**}, \ldots, j_{r-1}^{**}\},$$

$$B'_{j_t^{**}+1} - B'_{j''+1} < ta \le B'_{j_t^{**}} - B'_{j''+1}, \ t = 1, \ldots, r-1.$$

In a similar way we compute the costs $g'_{i,t,j''}$ by the following formula

$$g'_{i,t,j''} = f_i + \left(C_{ij_t^{**}+1} - C_{ij_{t-1}^{**}} \right)$$

$$+ c_{ij_{t-1}^{**}} \left(B'_{j_{t-1}^{**}} - B'_{j''+1} - (t-1)a \right) + c_{ij_t^{**}} \left(ta - (B'_{j_t^{**}+1} - B'_{j''+1}) \right).$$

And finally, quantities $\tilde{G}_{i,t}^R(j'')$ are calculated according to the equations

$$G_{i,t}^R(j'') = \begin{cases} g'_{i,1,j''}, & \text{if } t = 1, \\ g'_{i,t,j''} + \tilde{G}_{i+1,t-1}^R(j''), & \text{if } 2 \le t \le i'' - i + 1. \end{cases}$$

$$\tilde{G}_{i,t}^R(j'') = \begin{cases} \min\limits_{i \le k \le i''} G_{k,1}^R(j'') = \min\limits_{i \le k \le i''} g'_{k,1,j''}, & \text{if } t = 1, \\ \min\limits_{i \le k \le i'', \ (i''-k+1) > t} G_{k,t}^R(j'') = \min\left\{ G_{i,t}^R(j''), \tilde{G}_{i+1,t}^R(j'') \right\}, & \text{if } 2 \le t \le i'' - i + 1. \end{cases}$$

Note that it is optimal to open $r = r_{j'j''} = \lceil (b_{j'} + \ldots + b_{j''})/a \rceil$ facilities in the segment $[i', i'']$ for serving all the client demand in $[j', j'']$. So the idea of calculating $Q_{i',i}(j', j)$ is to determine r open facilities, and to find the sequential number t of the only non-consumed open facility.
And the last quantities are calculated by the following formula

$$S_{it}(j', j'') = f_i + (C_{ij_{r-t}^{**}-1} - C_{ij_{t-1}^*}) + c_{ij_{t-1}^*}(B_{j_{t-1}^*} - B_{j'-1} - (t-1)a)$$

$$+ c_{ij_{r-t}^{**}}(B'_{j_{r-t}^{**}} - B'_{j''+1} - (r-t)a), \ i - i' + 1 \ge t.$$

$\square$

Lemma 3. *The calculation of all values $\tilde{G}_{i,t}^L(j')$, $\tilde{G}_{i,t}^R(j'')$ and $S_{it}(j', j'')$ can be performed in $\mathcal{O}(m^2 n)$, $\mathcal{O}(m^2 n)$ and $\mathcal{O}(m^2 n^2)$ time, respectively.*

Proof. Let us estimate the time complexity of finding all the necessary quantities.

Computing of all values B_j, B'_j takes $\mathcal{O}(n)$, computing of all C_{ij} takes $\mathcal{O}(mn)$ operations.

Construction of sets T and T' for each fixed j' requires no more than $\mathcal{O}(m)$ actions, hence it takes $\mathcal{O}(mn)$ operations in total.

The quantities $g_{i,t,j'}$ and $g'_{i,t,j'}$ are calculated in time $\mathcal{O}(m^2 n)$.

Knowing all the previous parameters we compute all $\tilde{G}_{i,t}^L(j')$ and all $\tilde{G}_{i,t}^R(j')$ in $\mathcal{O}(m^2 n)$ time.

Calculating $S_{it}(j', j'')$ for fixed j', j'' requires $\mathcal{O}(m^2)$ operations. Thus computing these quantities for all parameter values requires no more than $\mathcal{O}(m^2 n^2)$ time.

$\square$

Now we show how to speed up the calculation of $Q_{i',i''}(j',j'')$ values to get a faster algorithm for the UCFLP-path.

Theorem 2. *The UCFLP-path can be solved in $\mathcal{O}(m^2n^2)$ time.*

Proof. By Lemma 3 all values $\widetilde{G}^L_{i,t}(j')$, $\widetilde{G}^R_{i,t}(j'')$ and $S_{it}(j',j'')$ can be computed in total $T_1 = \mathcal{O}(m^2n^2)$ time. Having these values calculated, we introduce the following notation for all $1 \leq i \leq m$, $1 \leq j' \leq j'' \leq n$:

$$\widetilde{Q}_i(j',j'') = \min_{1 \leq t \leq r} \left\{ \widetilde{G}^L_{i-1,t-1}(j') + S_{it}(j',j'') + \widetilde{G}^R_{i+1,r-t}(j'') \right\}. \tag{8}$$

Note that it now takes $T_2 = \mathcal{O}(m^2n^2)$ time to compute all values $\widetilde{Q}_i(j',j'')$. Taking into account the relations (7) and (8), we get the formula for computing $Q_{i',i''}(j',j'')$:

$$Q_{i',i''}(j',j'') = \min_{i' \leq i \leq i''} \widetilde{Q}_i(j',j'') = \min\{Q_{i',i''-1}(j',j''); \widetilde{Q}_i(j',j'')\}. \tag{9}$$

Thus, knowing all the values of $\widetilde{Q}_i(j',j'')$ and the previous value $Q_{i',i''-1}(j',j'')$, one can find the next value $Q_{i',i''}(j',j'')$ in constant time. Therefore calculating $Q_{i',i''}(j',j'')$ for all indices $1 \leq i' \leq i'' \leq m$, $1 \leq j' \leq j'' \leq n$, can be done in $T_3 = \mathcal{O}(m^2n^2)$ time.

Finally, from (5)–(6) it follows that knowing all values $Q_{i',i''}(j',j'')$, the calculation of all $R(i,j)$ takes $T_4 = \mathcal{O}(m^2n^2)$ time. Thus, the total time complexity of the algorithm is $T_1 + T_2 + T_3 + T_4 = \mathcal{O}(m^2n^2)$. So the time complexity of proposed algorithm is strictly polynomial.

The main result of the work is completely proved.   □

5   Comparison Analysis

We want to compare strictly polynomial algorithm presented in this paper with the algorithm MKT from [13] and two algorithms GSTHeap and GST from [9]. These algorithms are pseudo-polynomial and invented for non-unform case of the CFLP, we have different facility capacities a_i, $1 \leq i \leq m$, without loss of generality we assume that $a_i \leq B = \sum_{j=1}^{n} b_j$. These algorithms have time complexities $\mathcal{O}\left(mB \cdot \min\{a_{max}, B\}\right)$, $\mathcal{O}\left(mB \cdot \log\left(\min\{a_{max}, B\}\right)\right)$, and $\mathcal{O}(mB)$, respectively, where $B = \sum_{j=1}^{n} b_j$ is total demand, a_{max} is the maximum facility capacity. The second and the third algorithms are based on the Dynamic Programming scheme from the first one and have reduced time complexities. Since we assume that $a_i \leq B$ $i = 1, ..., m$, we have such time-complexities for these algorithms $\mathcal{O}(mB \cdot a_{max})$ and $\mathcal{O}(mB \cdot \log a_{max})$, and $\mathcal{O}(mB)$, respectively. Since we consider uniform case of the CFLP

$$O(mB \cdot a_{max}) = O(mB \cdot a) = O(mn\overline{b}a),$$

where $\bar{b} = \frac{1}{n} \sum_{j=1}^{n} b_j$ is average demand.

Similarly, for the second and the third algorithm

$$\mathcal{O}(mB \cdot \log a_{max}) = \mathcal{O}(mB \cdot \log a) = \mathcal{O}(mn\bar{b} \log a),$$

$$\mathcal{O}(mB) = \mathcal{O}(mn\bar{b}).$$

Let us compare algorithm AGS with time complexity $\mathcal{O}(m^2 n^2)$ from this work and the fastest algorithm GST with time complexity $\mathcal{O}(mn\bar{b})$. We define the parameter $N = mn$ and get such time complexities $\mathcal{O}(N^2)$ for AGS and $\mathcal{O}(N\bar{b})$ for GST. Let us consider the problems with $\bar{b} \leq N$. In Fig. 1 we can see gray line, which depicts the time complexity for algorithm AGS and the time complexity for GST with maximum $\bar{b} = N$ (these two graphics coincide) in logarithmic scales, solid black line, which shows the time complexity for algorithm GST with $\bar{b}$ decreased by 100 in logarithmic scales, doted black line, which pictures the time complexity for algorithm GST with $\bar{b}$ decreased by 1000 in logarithmic scales, and the dashed black line, which draws the time complexity for algorithm GST with $\bar{b}$ decreased by 10000 in logarithmic scales. $\lg T(N)$ is decimal logarithm of time complexity $T(N)$.

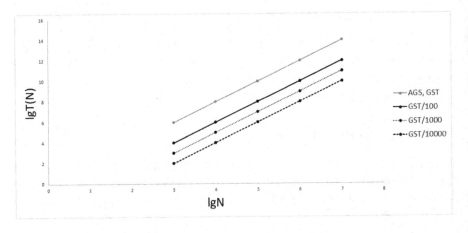

Fig. 1. Comparison of the time complexities for algorithms AGS and GST.

We can conclude that under the condition $\bar{b} \leq N = mn$ GST is competitive with AGS, despite GST is pseudo-polynomial. This result is quite interesting since in practical problems nearly always $\bar{b} \leq mn$.

Let us consider algorithm AGS, algorithm GSTHeap, and algorithm MKT. The second one with time complexity $\mathcal{O}(N\bar{b} \log a)$. We put $\bar{b} = N$, $a = 100$ and construct graphics for decimal logarithm of time complexity for these algorithms (see Fig. 2).

We can suggest, that GSTHeap outperforms MKT as expected and algorithm AGS from present paper is the best one.

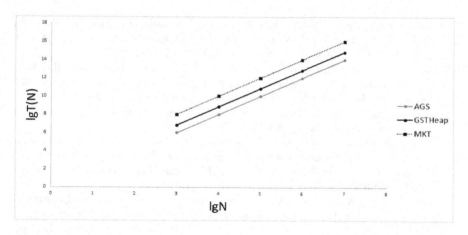

Fig. 2. Comparison of the time complexities for algorithms AGS, GSTHeap, and MKT.

6 Conclusion and Future Work

In this work we have shown how to improve the time complexity of the algorithm solving UCFLP-path from $\mathcal{O}(m^5 n^2 + m^3 n^3)$ in [2], $\mathcal{O}(m^4 n^2)$ in [3], $\mathcal{O}(m^3 n^2)$ in [8], to our final result $\mathcal{O}(m^2 n^2)$. Since the dynamic programming scheme (5)–(6) requires to precompute 4-dimensional matrix Q, the potential of the current variant of the algorithm is apparently exhausted, and to solve the problem faster one would need a different scheme.

Comparison analysis shows, that for the UCFLP-path pseudo-polynomial algorithm with time complexity $\mathcal{O}(mB)$ from [9] is competitive with strictly polynomial algorithm presented in current work, if $\bar{b} \leq mn$.

In [9,10] it was shown that one can get faster algorithms for facility location problems on path graphs, using the concavity of the transportation costs. This property was not used in the algorithm presented here, so for the further research it would be interesting to study the possible opportunities the concave costs can give for the UCFLP-path. It would be also interesting to study the (conditional) lower bounds on the time complexity for this problem.

References

1. Ageev, A.A.: A criterion of polynomial-time solvability for the network location problem. In: Integer Programming and Combinatorial Optimization, Proceedings of the IPCO II Conference, Campus Printing, pp. 237–245. Carnegi Mellon University, Pittsburg (1992)
2. Ageev, A.A.: A polynomial-time algorithm for the facility location problem with uniform hard capacities on path graph. In: Proceedings of the 2nd International Workshop Discrete Optimization Methods in Production and Logistics DOM 2004, Omsk, pp. 28–32 (2004)

3. Ageev, A.A., Gimadi, E.Kh., Kurochkin, A.A.: A polynomial-time algorithm for the facility location problem on the path with uniform hard capacities. Discret. Anal. Oper. Res. Novosibirsk: IM SO RAN **16**(5), 3–17 (2009)

4. Bansal, M., Garg, N., Gupta, N.: A 5-approximation for capacitated facility location. In: Epstein, L., Ferragina, P. (eds.) ESA 2012. LNCS, vol. 7501, pp. 133–144. Springer, Heidelberg (2012). https://doi.org/10.1007/978-3-642-33090-2_13

5. Billionet, A., Costa, M.-C.: Solving the uncapacitated plant location problem on trees. Discret. Appl. Math. **49**(1–3), 51–59 (1994)

6. Cornuéjols, G., Nemhauser, G.L., Wolsey, L.A.: The uncapacitated facility location problem. In: Discrete Location Theory, pp. 119–171. Wiley, New York (1990)

7. Garey, M.R., Johnson, D.S.: Computers and Intractability: A Guide to the Theory of NP-Completeness. Freeman, San Francisco (1979)

8. Gimadi, E.K., Kurochkina, A.A.: Time complexity of the Ageev's algorithm to solve the uniform hard capacities facility location problem. In: Evtushenko, Y., Jaćimović, M., Khachay, M., Kochetov, Y., Malkova, V., Posypkin, M. (eds.) OPTIMA 2018. CCIS, vol. 974, pp. 123–130. Springer, Cham (2019). https://doi.org/10.1007/978-3-030-10934-9_9

9. Gimadi, E., Shtepa, A., Tsidulko, O.: Improved exact algorithm for the capacitated facility location problem on a line graph. In: Proceedings of 15th International Asian School-Seminar Optimization Problems of Complex Systems (OPCS), pp. 53–57. IEEE (2019). https://doi.org/10.1109/OPCS.2019.8880248

10. Hassin, R., Tamir, A.: Improved complexity bounds for location problems on the real line. Oper. Res. Lett. **10**(7), 395–402 (1991)

11. Laporte, G., Nickel, S., Saldanha da Gama, F.: Location Science. Springer, Cham (2015)

12. Levi, R., Shmoys, D.B., Swamy, C.: LP-based approximation algorithms for capacitated facility location. Math. Program. **131**, 365–379 (2012). https://doi.org/10.1007/s10107-010-0380-8

13. Mirchandani, P., Kohli, R., Tamir, A.: Capacitated location problems on a line. Transp. Sci. **30**(1), 75–80 (1996). https://doi.org/10.1287/trsc.30.1.75

14. Pál, M., Tardos, E., Wexler, T.: Facility location with hard capacities. In: Proceedings of the 42nd IEEE Symposium on Foundations of Computer Science (FOCS), pp. 329–338 (2001). https://doi.org/10.1109/SFCS.2001.959907

15. Shah, D.: An unified limited column generation approach for facility location problem on trees. Ann. Oper. Res. **87**, 363–382 (1999)

16. Shah, R., Farach-Colton, V.: Undiscretized dynamic programming: faster algorithms for facility location and related problems on trees. In: Proceedings of the 13th Annual ACM-SIAM Symposium on Discrete Algorithms, San Francisco, California, SODA, pp. 108–115 (2002)

On a Weakly Supervised Classification Problem

Vladimir Berikov[1,2]([✉]) [iD], Alexander Litvinenko[3] [iD], Igor Pestunov[4] [iD],
and Yuriy Sinyavskiy[4] [iD]

[1] Sobolev Institute of Mathematics, Novosibirsk, Russia
[2] Novosibirsk State University, Novosibirsk, Russia
berikov@math.nsc.ru
[3] RWTH Aachen, Aachen, Germany
litvinenko@uq.rwth-aachen.de
[4] Federal Research Center for Information and Computational Technologies,
Novosibirsk, Russia
pestunov@ict.sbras.ru

Abstract. We consider a weakly supervised classification problem. It is a classification problem where the target variable can be unknown or uncertain for some subset of samples. This problem appears when the labeling is impossible, time-consuming, or expensive. Noisy measurements and lack of data may prevent accurate labeling. Our task is to build an optimal classification function. For this, we construct and minimize a specific objective function, which includes the fitting error on labeled data and a smoothness term. Next, we use covariance and radial basis functions to define the degree of similarity between points. The further process involves the repeated solution of an extensive linear system with the graph Laplacian operator. To speed up this solution process, we introduce low-rank approximation techniques. We call the resulting algorithm WSC-LR. Then we use the WSC-LR algorithm for analysis CT brain scans to recognize ischemic stroke disease. We also compare WSC-LR with other well-known machine learning algorithms.

Keywords: Weakly supervised classification · Uncertainty model · Manifold regularization · Low-rank approximation · Similarity matrix · Computed tomography

1 Introduction

Data is ubiquitous in modern science and technology. Unfortunately, very often, the data may be incomplete, noisy, or contain errors. This happens, for example, in manual annotation or labeling of data. In contrast to the automated one,

The study was carried out within the framework of the state contract of the Sobolev Institute of Mathematics (project no FWNF-2022-0015). The work was partly supported by RFBR grant 19-29-01175. A. Litvinenko was supported by funding from the Alexander von Humboldt Foundation.

manual data processing could be costly, slow, and subjected to errors. Usually, only a tiny part of data is accurately labeled; the rest is either unlabeled or labeled inaccurately. Additionally, due to the poor quality of the data, it could be hard to classify data, and its label becomes uncertain. We model this uncertain labeling by introducing probability distribution functions.

Weakly supervised classification is a branch of machine learning (ML), dealing with uncertainties in the data labeling. In the problem settings, it is assumed that some of the samples are labeled inaccurately. The model of inaccuracy can be formulated in different ways [33].

The contribution of this paper is a new scalable method for classifying weakly supervised data by using a combination of manifold regularization, low-rank matrix representation, and ensemble clustering. The manifold regularization technique considers the data geometry and later uses it in the regularization process. By approximating large covariance matrices, we speed up calculations and reduce memory requirements. Ensemble methods obtain a solution with high predictive ability based on several simpler algorithms with a high degree of diversity. We use ensemble clustering to find the structure of data in a robust way.

We consider a weakly supervised classification problem in the transductive learning setting: it is believed that the unlabeled test sample is known, and it can be used as additional information for the classifier.

To demonstrate the capability of the proposed method, we apply it for recognizing acute stroke (AS) using computed tomography (CT) scans. The process of annotation of CT brain images is rather complicated and time-consuming. By an extensive collection of data, not all images may contain indications of regions affected by the stroke. For example, a radiologist can only note that pathological signs are present in the given CT scan or point out some approximate area where a lesion has occurred. For analysis of such inexact information, special methods are required.

1.1   Related Works and Preliminary Results

Uncertainties can be classified in aleatory and epistemic uncertainties [17]. Aleatory (irreducible) uncertainty takes its origin in the stochastic nature of data: the measured outcomes can differ each time one runs similar experiments. Epistemic uncertainty comes from the lack of knowledge on the phenomenon under consideration. The uncertainty can be reduced by model updating using additional information on data and its properties.

Methods based on finding potentially erroneous labels and correcting them were introduced in [7,22]. Information about nearest neighbors of points is used for the correction. Another method is related to minimizing theoretical risk estimates taking into account the random labeling error [12]. Empirical risk functional is divided into two parts; the first component does not depend on the noise, and only the second part is affected by noisy labels. A survey on such a methodology (called noisy label learning) can be found in [28].

For a coarse-grained labeling model (also called multi-instance learning), class labels are indicated only for sets of instances [30]. For example, a collection of body cells is labeled as a group containing pathological examples. The goal is to predict class labels for new groups of observations. The authors of [13] propose an optimization algorithm using spherical separation to deal with multi-instance learning problems.

Crowdsourcing-based models [24,32] are used when the labeling is performed by many independent workers (groups of people, usually communicated via the Internet). The classification is performed with probabilistic or ensemble-based methods.

Another promising methodology related to the classification with incomplete labeling is so-called positive-unlabeled learning, where a learner only has access to positive examples and unlabeled data [2].

Weakly supervised learning models assume two basic assumptions: cluster assumption and manifold assumption. The former expects that objects from the same cluster often have the same label. The latter is based on the hypothesis that the manifold to which points with the same labels belong is smooth. Manifold regularization [3,29] relies on this assumption. Both the classification error and the regularizing component are minimized in the model fitting stage. In dense regions of the manifold, the prediction distribution should not change very fast. Graph Laplacian (GL) [3,31] is a convenient tool to estimate the classifier's smoothness. A serious issue for this type of methods is their non-scalability, i.e., the need to store in memory a non-sparse square GL matrix, and also a large cost of matrix operations.

In [4], a scalable method for weakly supervised regression based on GL regularization and low-rank matrix representation was introduced. In this method, the Wasserstein distance between distributions was used in the regularization term. As opposed to this work, we consider weakly supervised learning problem in the classification context. The same low-rank matrix technique is adopted to weakly supervised classification.

We apply the proposed method in the problem of medical data analysis. There are a large variety of works describing mathematical and engineering tools implemented in health care. The collaboration between computer scientists and health care professionals leads to the creation of innovative techniques such as predictive modeling of diseases or finding an optimal way of medical treatment [23].

The rest of the paper is organized as follows. Problem settings are defined in Sect. 2. Then in Sect. 3 we provide a short overview of existing weakly supervised learning methods and explain the details of our proposed method. Section 4 describes the task of ischemic stroke recognition using CT scans and reports on the results obtained with the proposed method and some other weakly supervised and fully supervised machine learning algorithms. The concluding part summarizes our achievements and presents some future plans.

2   Problem Settings and Notation

Let a dataset $\mathbf{X} = \{x_1, \ldots, x_n\}$ be given, where $x_i \in \mathbb{R}^d$ is a feature vector with dimensionality d, and n denotes the sample size. As usual, it is supposed that data instances are independently sampled from some statistical population and follow an identical distribution.

In a fully supervised classification, we assume a set $Y = \{y_1, \ldots, y_n\}$ be given, and each $y_i \in \{c_1, \ldots, c_K\}$, where K is the number of classes.

It is necessary to create a classifier which should forecast in some best way a class label for a new example from the same statistical population.

In the problem of semi-supervised transductive classification, the class labels are known only for a part of the data set $\mathbf{X}_1 \subset \mathbf{X}$ of comparatively small size. We assume that the labeled part is $\mathbf{X}_1 = \{x_1, \ldots, x_{n_1}\}$, and the rest is unlabeled part $\mathbf{X}_0 = \{x_{n_1+1}, \ldots, x_n\}$. The class labels for $\mathbf{X}_1$ are denoted by $\mathbf{Y}_1 = \{y_1, \ldots, y_{n_1}\}$. The problem is to predict the class labels for $\mathbf{X}_0$.

In the weakly supervised classification problem, we suppose that for some data points the class labels are uncertain. Let $Y_i = (Y_{i1}, \ldots, Y_{iK})$ denote a vector in which each element $Y_{ik} \geq 0$ equals to the known degree of belonging of a point x_i to class c_k, $\sum_{k=1}^{K} Y_{ik} = 1$. For accurately labeled instance $x_i \in X_1$ we have $Y_{ik} = \mathbb{I}[y_i = c_k]$, where $\mathbb{I}(\cdot)$ is the indicator function: $\mathbb{I}[true] = 1$, $\mathbb{I}[false] = 0$. For uncertainly labeled $x_i \in X_1$, two or more components of Y_i can take positive values. Vector Y_i equals zero for the unlabeled point $x_i \in X_0$. Denote by $Y_{1,0}$ the given class assignment matrix $Y_{1,0} = (Y_1, \ldots, Y_n)^T$.

We aim at predicting in some best way the likelihood of class labels for all instances from the unlabeled part of data: for each $x_i \in X_0$, find the decision vector $F_i = (F_{i1}, \ldots, F_{iK})$, where $\sum_{k=1}^{K} F_{ik} = 1$ and $F_{ik} \geq 0$. Let $F = (F_1, \ldots, F_n)^T$ denote the overall decision matrix.

The method proposed in this paper is based on a combination of manifold regularization and low-rank representation of similarity matrix. Let W be a similarity matrix with elements W_{ij} indicating the degree of similarity between points x_i and x_j. A function from the Matérn family [21] can be used for this purpose. It gained a widespread interest in recent years [15]. The Matérn covariance depends only on the distance $h := \|x - x'\|$:

$$W(\mathbf{h}; \boldsymbol{\theta}) = \frac{\sigma^2}{2^{\nu-1}\Gamma(\nu)} \left(\frac{h}{\ell}\right)^\nu K_\nu \left(\frac{h}{\ell}\right) + \tau^2 \mathbf{I}, \tag{1}$$

with parameters $\boldsymbol{\theta} = (\sigma, \ell, \nu, \tau)^\top$. Here σ^2 is the variance, τ^2 the nugget, I the identity matrix, $\nu > 0$ controls the smoothness of the random field, with larger values of ν corresponding to smoother fields, and $\ell > 0$ the spatial range parameter that measures how quickly the correlation of the random field decays with distance. A larger ℓ corresponds to a faster decay. $\mathcal{K}_\nu$ denotes the modified Bessel function of the second kind of order ν.

Let us introduce a diagonal matrix $D = diag(D_1, \ldots, D_n)$ with elements $D_i = \sum_j W_{ij}$. Matrix

$$L = I - D^{-1/2} W D^{-1/2}$$

of dimensionality $(n \times n)$ is called the normalized graph Laplacian. The elements of the matrix are: $L_{ij} = \delta_{ij} - \frac{W_{ij}}{\sqrt{D_i}\sqrt{D_j}}$, where $\delta_{ij} = \mathbb{I}(i = j)$, $i, j = 1, \ldots, n$.

3 Proposed Weakly Supervised Classification Scheme

In this Section, we define the optimization problem, and then solve it. The obtained solution gives us the optimal labeling.

3.1 Optimization Problem

Consider the following optimization problem:

$$\text{find } F^* = \arg \min_{F \in R^{n \times K}} Q(F)$$

$$= \frac{1}{2} \left(\sum_{x_i \in X_1} ||F_i - Y_i||^2 + \beta \sum_{x_i, x_j \in X} W_{ij} \left\| \frac{F_i}{\sqrt{D_i}} - \frac{F_j}{\sqrt{D_j}} \right\|^2 \right), \tag{2}$$

s.t. $\sum_k F_{ik} = 1$, $F \geq 0$, where $\beta > 0$ is a regularization parameter. The first sum on the right side of (2) is aimed to reduce the fitting error on labeled data; the second component plays the role of a smoothing function: its minimization means that if two points x_i, x_j (either labeled, weakly labeled or unlabeled) are similar, their decision vectors should not be very different. Note that a similar problem for the task of semi-supervised classification using clustering ensemble was considered in [5,6]. It is known that the optimized function is convex. To find the optimal solution, we use the method of Lagrange multipliers. We differentiate the Lagrange function $Q(F) + \sum_i \lambda_i (1 - \sum_k F_{ik})$ obtained from (2), and after transformations with normalized graph Laplacian get the equations (use Matlab notation):

$$F_{ik} - Y_{ik} + \beta F_{ik} - \beta L_{i,:} F_{:,k} - \lambda_i = 0, \quad i = 1, \ldots, n_1, \tag{3}$$

$$\beta F_{ik} - \beta L_{i,:} F_{:,k} - \lambda_i = 0, \quad i = n_1 + 1, \ldots, n. \tag{4}$$

Here λ_i, $i = 1, \ldots, n$, are the Lagrange multipliers; $L_{i,:}$ and $F_{:,k}^*$ are ith row of matrix L and kth column of matrix F^*, respectively, $k = 1, \ldots, K$.

Let us sum up (3), (4) over k, and from the conditions

$$\sum_k F_{ik} = 1, \quad \sum_k Y_{ik} = \begin{cases} 1, i = 1, \ldots, n_1 \\ 0, i = n_1 + 1, \ldots, n, \end{cases}$$

we obtain:

$$\beta - \beta \sum_k L_{i,:} F_{:,k} - K \lambda_i = 0, \quad i = 1, \ldots, n.$$

Because $\sum_k L_{i,:} F_{:,k} = \sum_k \sum_j L_{ij} F_{jk} = \sum_j L_{ij} \sum_k F_{jk} = \sum_j L_{ij} = D_i$ we have:

$$\lambda_i = \frac{\beta}{K}(1 - D_i), \quad i = 1, \ldots, n.$$

Denote

$$I_{10} := diag(I_{11}, \ldots, I_{nn})$$

where

$$I_{ii} := \begin{cases} 1, i = 1, \ldots, n_1 \\ 0, i = n_1 + 1, \ldots, n, \end{cases}$$

and let matrix R be of elements $R_{ik} = Y_{ik} + \frac{\beta}{K}(1 - D_i)$. Substituting λ_i and using the introduced matrix notation, Eqs. (3), (4) are represented in the form

$$(I_{10} + \beta L)F = R, \tag{5}$$

hence the optimal solution of (2) is

$$F = (I_{10} + \beta L)^{-1}R, \tag{6}$$

if the inverse matrix exists. One may note that it is always possible to choose the regularization parameter β to ensure the well-posedness of the problem.

3.2 Low-Rank Matrix Representation

Limitations of the above introduced method are 1) a high storage cost of the non-sparse graph Laplacian matrix of size $n \times n$, and 2) a high computational cost of matrix operations. A low-rank hierarchical matrix approximation [14,19, 20] is an elegant way for finding computationally efficient solution. Alternative options are to use Nyström method [10], low-rank tensor decomposition [11,18], or a low-rank representation of the averaged co-association matrix considered as similarity matrix [5].

Let us suppose that the similarity matrix W can be presented in the low-rank form

$$W = AA^\top, \tag{7}$$

where the matrix $A \in \mathbb{R}^{n \times m}$ has a small rank m, and $m \ll n$. Then from Eq. 6 we get

$$F = (G - \beta D^{-1/2}AA^\top D^{-1/2})^{-1}R,$$

where $G = I_{10} + \beta I$. It is possible to apply Woodbury matrix identity:

$$(S + UV)^{-1} = S^{-1} - S^{-1}U(I + VS^{-1}U)^{-1}VS^{-1},$$

where $S \in \mathbb{R}^{n \times n}$ is an invertible matrix, $U \in \mathbb{R}^{n \times m}$, $V \in \mathbb{R}^{m \times n}$, and get:

$$F = (G^{-1} + \beta G^{-1}U(I - \beta U^T G^{-1}U)^{-1}U^T G^{-1})R. \tag{8}$$

Here $U = D^{-1/2}A$. In the found solution, a matrix of significantly smaller dimensionality $(m \times m)$ should be processed. The cost of numerical inversion is of order $O(nm + m^3)$.

3.3   Algorithm

The scheme of weakly supervised classification algorithm based on graph Laplacian and low-rank decomposition of a similarity matrix (WSC-LR) is as follows.

Algorithm WSC-LR
Input:
Dataset $\mathbf{X}$ containing labeled, inaccurately labeled (denoted by $\mathbf{X}_1$), and unlabeled instances (denoted by $\mathbf{X}_0$);
Y_{10} : given class assignment matrix;
Output:
F: decision matrix for objects from $\mathbf{X}_0$.
Steps:
1. Determine the similarity matrix W_{ij}.
2. Find the normalized graph Laplacian in the low-rank representation (7);
3. Determine the predicted decision matrix according to (8);
end.

4   Practical Application: Recognition of Acute Stroke from CT Scans

In this Section, we apply the proposed method for the recognition the ischemic stroke disease by using computed tomography (CT) brain scans. The early diagnosis of acute stroke (AS) is crucial for successful further medical treatments, thus the automated analysis of CT images is of greatest significance. An anonymized sample of patients was obtained from the database of the International Tomography Center SB RAS, Novosibirsk, Russia. Forty-six patients (in total 6000 image slices) with diagnosed ischemic or hemorrhagic acute strokes were used in the study. In each image slice, a radiology specialist performs manual identification of the pathological density area in the substance of the brain. An example of such pathology is shown in Fig. 1 (selected by the green contour).

Non-contrast CT diagnosis of acute ischemic or hemorrhagic stroke is effective. However, the interpretation of non-contrast CT images is rather challenging. This is because the identification of early changes in AS on a CT image depends on both technical (image contrast, room shadow at the time of image analysis, monitor brightness) and human factors (specialist's experience, fatigue, way of interpretation of incoming clinical data). As a result, reliable differentiation of ischemic signs by a CT diagnostic specialist is rather difficult. Some uncertainties in the forecast are possible. It is known that even highly qualified specialists can make erroneous predictions with sufficiently large probability (up to 10%).

Deep convolutional neural networks and texture-based ML methods are widely used for the analysis of CT images [1, 26]. In contrast to deep neural networks (sometimes called "black box" for their non-transparent behaviour), texture-based models are easy to interpret, and the obtained results are more reliable in a certain sense. In many applications, the loss cost is very high (e.g., in medicine),

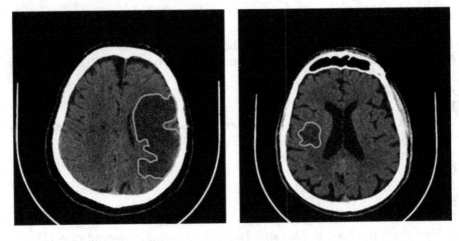

Fig. 1. Example of CT image.

and the transparency factor becomes of prime importance. Therefore, texture-based and other "transparent" methods are of high interest.

The uncertainty in the labeling of CT images takes its origin from the nature of the considered feature extraction technique: the extracted features describe properties of fragments which consist of sets of pixels of the digital image. Some of the fragments, especially those located near the border of the pathological area, can not be labeled accurately because they include pixels associated with either affected or unaffected by stroke (see the details of the technique in the following subsection).

4.1   Data Preprocessing and Feature Extraction

The processing of source data (CT image slices with lesion masks) begins with the construction of "healthy" brain tissue masks. Each mask covers tissues surrounded by the scull bone except the lesion areas. The constructed masks are used for feature extraction. Each source image within the scull bone is contrasted and cut into square fragments of $FS \times FS$ size, shifted by S, where FS and S are the preset parameters. Typically, the affected area size is much smaller than the healthy one, so the shift parameter should vary to get the balanced sample (we set $S = FS$ for healthy and $S = FS/4$ for affected tissues). Examples of masks and fragments emplacement for $FS = 16$ and 30 are shown in Fig. 2.

All fragments completely contained in a healthy tissue mask or having their centers inside an affected tissue mask are used in the feature extraction procedure for "healthy" and "affected" classes, respectively. Because the class labels assignments can be uncertain, we should use some uncertainty models. For each fragment, we define the degree ("probability") of its belonging to the affected class as a proportion of the fragment's area contained in the affected tissue mask.

The features used in the classification, listed in Table 1, include 20 texture features [9,16,27] and 10 statistical features [25].

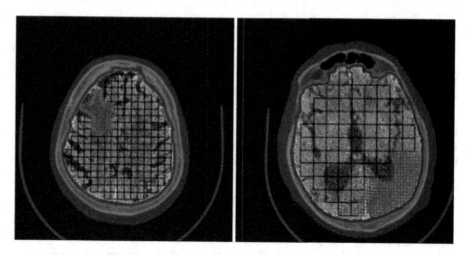

Fig. 2. Masks of healthy (purple contours) and affected (yellow contours) brain tissues and fragments emplacement for healthy and affected tissues (green and red squares respectively) for $FS = 16$ (left) and 30 (right). Red color saturation indicates the probability of assigning the fragment to the affected tissue. (Color figure online)

In this study, the gray-level co-occurrence matrix (GLCM) is calculated as a mean of four matrices corresponding to four different directions ($\theta = 0$, 45, 90, and 135°) and one distance ($d = 1$ pixel). Let N_g denote the number of gray levels for GLCM calculation; $P = P^{(I)}$ is normalized GLCM for fragment (rectangular part of image) $I = \{I(x,y),\ 1 \leq x,y \leq 2FS + 1\}$, $P(i,j) \in [0;1]$. This matrix is defined as follows:

$$P^{(I)} = \left\{ P^{(I)}(i,j),\ 0 \leq i,j < N_g \,\middle|\, \begin{array}{l} P^{(I)}(i,j): I(x_1,y_1) = i,\ I(x_2,y_2) = j, \\ |x_1 - x_2| \leq 1,\ |y_1 - y_2| \leq 1 \end{array} \right\}.$$

To determine the features, let us first introduce the following notation. Let

$$P_x(i) = \sum_{j=0}^{N_g-1} P(i,j),\quad \mu(P_x) = \sum_{i=0}^{N_g-1} iP_x(i),$$

$$\sigma(P_x) = \sqrt{\sum_{i=0}^{N_g-1} (i - \mu(P_x))^2 P_x(i)};$$

$$P_{x+y}(k) = \sum_{i=0}^{N_g-1}\sum_{j=0}^{N_g-1} P(i,j),\quad i+j = k,\quad 0 \leq k \leq 2N_g - 2\ ;$$

$$P_{x-y}(k) = \sum_{i=0}^{N_g-1}\sum_{j=0}^{N_g-1} P(i,j),\quad |i-j| = k,\quad 0 \leq k \leq N_g - 1;$$

$$HXY1 = -\sum_{i=0}^{N_g-1}\sum_{j=0}^{N_g-1} P(i,j)\log_2\left(P_x(i)\,P_x(j)\right);$$

$$HXY2 = -\sum_{i=0}^{N_g-1}\sum_{j=0}^{N_g-1} P_x(i)\,P_x(j)\log_2\left(P_x(i)\,P_x(j)\right).$$

Further, let H be the normalized histogram for the fragment, $H(i) \in [0,1]$, $i = 0,\ldots,N_g$, $\sum_i H(i) = 1$. Now all these notations can be used to define 30 features, which are summarized in Table 1.

4.2   Classification with Fully Supervised ML Algorithms

The classifiers for the acquired dataset were tested by the cross-validation technique. The dataset was partitioned by images to avoid correlation between test and train samples. In each run, the fragments of one image were considered as testing samples and fragments which belonged to other images as the training ones. The classification results were evaluated using the following metrics: sensitivity ($SENS$), specificity ($SPEC$), and balanced accuracy (BA)

$$SENS = \frac{TP}{TP+FN}, \quad SPEC = \frac{TN}{TN+FP}, \quad BA = \frac{SENS+SPEC}{2}, \quad (9)$$

where TP denotes the number of affected by stroke fragments correctly classified as affected (True Positive); TN is the number of healthy fragments correctly classified as healthy (True Negative); FP denotes the number of healthy fragments incorrectly classified as affected (False Positive); FN is the number of affected fragments incorrectly classified as healthy (False Negative). The obtained quality metrics were averaged over cross-validation runs.

For a fully supervised classification, we use only definitely labeled fragments in the learning step, i.e., completely contained in healthy (for healthy class) or affected (for affected class) tissue mask.

Software. The following machine learning algorithms from OpenCV 4.5.0 open-source software package [8] were used: k-NN with $k \in \{2, \ldots, 14\}$, Random Forest (regression and classification trees), Normal Bayes classifier and SVM with different kernels. The parameters of the algorithms were optimized by OpenCV built-in procedure. The classification results (including 3 best results for k-NN classifier) obtained for $FS = 30$ and $N_g = 9$ are presented in Table 2. The seven most relevant features used in the Random Forest and their importance estimates are shown in Table 3.

Interestingly, the set of most important features include basic intensity characteristics rather than more complicated attributes describing qualitative terms such as roughness, smoothness, or bumpiness.

4.3   Weakly Supervised Classification

We use the same data on acute stroke recognition and the extracted features as in the previous experiments. To verify our WSC-LR algorithm, we apply the same

Table 1. Components of the feature vector for classification of CT scans.

No	Feature description	Formula for calculation
1	Energy	$f_1 = \sum_{i=0}^{N_g-1} \sum_{j=0}^{N_g-1} (P(i,j))^2$
2	Contrast	$f_2 = \sum_{k=0}^{N_g-1} k^2 P_{x-y}(k)$
3	Correlation	$f_3 = \left(\sum_{i=0}^{N_g-1} \sum_{j=0}^{N_g-1} (i - \mu(P_x))(j - \mu(P_x)) P(i,j) \right) / \sigma(P_x)$
4	Variance	$f_4 = (\sigma(P_x))^2$
5	Homogeneity	$f_5 = \sum_{i=0}^{N_g-1} \sum_{j=0}^{N_g-1} \frac{P(i,j)}{1+(i-j)^2}$
6	Sum average	$f_6 = \sum_{k=0}^{2N_g-2} k P_{x+y}(k)$
7	Sum variance	$f_7 = \sum_{k=0}^{2N_g-2} \left(k - \sum_{k=0}^{2N_g-2} k P_{x+y}(k) \right)^2 P_{x+y}(k)$
8	Sum entropy	$f_8 = -\sum_{k=0}^{2N_g-2} P_{x+y}(k) \log_2(P_{x+y}(k))$
9	Entropy	$f_9 = -\sum_{i=0}^{N_g-1} \sum_{j=0}^{N_g-1} P(i,j) \log_2(P(i,j))$
10	Difference variance	$f_{10} = \sum_{k=0}^{N_g-1} k^2 P_{x-y}(k) - \left(\sum_{k=0}^{N_g-1} k P_{x-y}(k) \right)^2$
11	Difference entropy	$f_{11} = -\sum_{k=0}^{N_g-1} P_{x-y}(k) \log_2(P_{x-y}(k))$
12	Information measure of correlation-1	$f_{12} = \frac{f_9 - HXY1}{-\sum_{i=0}^{N_g-1} P_x(i) \log_2(P_x(i))}$
13	Information measure of correlation-2	$f_{13} = 1 - \exp(-2(HXY2 - f_9))$
14	Auto-correlation	$f_{14} = \sum_{i=0}^{N_g-1} \sum_{j=0}^{N_g-1} ij P(i,j)$
15	Dissimilarity	$f_{15} = \sum_{i=0}^{N_g-1} \sum_{j=0}^{N_g-1} |i-j| P(i,j)$
16	Cluster shade	$f_{16} = \sum_{i=0}^{N_g-1} \sum_{j=0}^{N_g-1} (i+j-2\mu(P_x))^3 P(i,j)$
17	Cluster prominence	$f_{17} = \sum_{i=0}^{N_g-1} \sum_{j=0}^{N_g-1} (i+j-2\mu(P_x))^4 P(i,j)$
18	Maximum probability	$f_{18} = \max_{0 \le i,j \le N_g-1} P(i,j)$
19	Inverse difference normalized	$f_{19} = \sum_{i=0}^{N_g-1} \sum_{j=0}^{N_g-1} \frac{P(i,j)}{1+|i-j|/N_g^2}$
20	Homogeneity normalized	$f_{20} = \sum_{i=0}^{N_g-1} \sum_{j=0}^{N_g-1} \frac{P(i,j)}{1+((i-j)/N_g)^2}$
21	Skewness	$f_{21} = \frac{\sum_{k=0}^{N_g-1} \left(H(k) - \sum_{k=0}^{N_g-1} k H(k) \right)^3}{(\sigma(H(k)))^3}$
22	Kurtosis	$f_{22} = \frac{\sum_{k=0}^{N_g-1} \left(H(k) - \sum_{k=0}^{N_g-1} k H(k) \right)^4}{(\sigma(H(k)))^4}$
23	Energy	$f_{23} = \sum_{k=0}^{N_g-1} (H(k))^2$
24	Entropy	$f_{24} = -\sum_{k=0}^{N_g-1} H(k) \log_2(H(k))$
25	Average	$f_{25} = \sum_{k=0}^{N_g-1} k H(k)$
26	Median value	$f_{26} = Med(H)$
27	Maximum value	$f_{27} = \max_{0 \le i \le N_g-1} H(i)$
28	Minimum value	$f_{28} = \min_{0 \le i \le N_g-1} H(i)$
29	Values range	$f_{29} = \max_{0 \le i \le N_g-1} H(i) - \min_{0 \le i \le N_g-1} H(i)$
30	Standard deviation	$f_{30} = \sigma(H(k))$

cross-validation technique. Both uncertainly and accurately labeled fragments participate in the learning process and the validation. The best parameters were found with a grid-search procedure.

Table 2. Classification results obtained by fully supervised ML algorithms.

Classification algorithm	Performance metrics		
	SENS, %	SPEC, %	BA, %
Random Forest (regression trees)	77.0	77.4	77.2
Random Forest (classification trees)	29.4	82.5	55.95
Normal Bayes classifier	36.2	60.9	48.55
SVM (histogram intersection kernel)	68.8	34.3	51.55
SVM (linear kernel)	48.9	52.4	50.65
SVM (polynomial kernel)	31.9	67.3	49.6
SVM (sigmoid kernel)	47.7	53.3	50.5
k-NN ($k = 2$)	50.3	56.7	53.5
k-NN ($k = 4$)	42.8	61.1	51.95
k-NN ($k = 13$)	28.7	74.8	51.75

Table 3. Most relevant features determined by feature importance ranking using Random Forest.

Features	Importance value
Median value	0.081
Maximum value	0.062
Minimum value	0.046
Entropy	0.041
Standard deviation	0.040
Information measure of correlation-1	0.038

The classification results are shown in Table 4. In WSC-LR, we use the similarity matrix found with a clustering ensemble of size 30; each ensemble variant is obtained using k-means with the number of clusters taken from the interval $[201, \ldots, 230]$. The low-rank representation of the similarity matrix is found in the same manner as in [5]. We also used Nyström approximation with Matérn kernel and parameters $\sigma = 1$, $\ell = 1.25$, $\nu \to \infty$, $\tau = 0$ (which define a well-known Gaussian or RBF kernel). The parameter β in the optimization function Eq. 2 is equal to 0.001.

The last row in Table 4 demonstrates the performance metrics for a semi-supervised classification algorithm SSC-LR introduced in [5]. In this algorithm, the uncertain examples are considered as though they are not labeled at all. The aforementioned clustering ensemble technique is used for deriving the similarity matrix. The algorithms are run on a dual-core Intel Core i5 processor with a clock frequency of 2.4 GHz and 8 GB RAM. The average time for one run of WSC-LR is 6.5 s.

Table 4. Classification results for weakly supervised WSC-LR and semi-supervised SSC-LR. The values of $SENS$, $SPEC$ and BA are defined in Eq. 9.

Algorithm	Performance metrics		
	$SENS$, %	$SPEC$, %	BA, %
WSC-LR (Nyström)	82.7	77.3	80
WSC-LR (cluster ensemble)	82.3	74.3	78.3
SSC-LR	80.5	72.4	76.45

5 Conclusion

We posed a weakly supervised classification problem. To solve this problem, we developed a new algorithm (WSC-LR), which considers uncertain information about labels. To overcome computational bottlenecks of the existing algorithms, we computed a low-rank (LR) approximation of the similarity matrix and then used low-rank matrix arithmetic to speed up calculations. One of the essential ingredients of the WSC-LR method is the similarity matrix. This matrix can either be estimated from the ensemble of methods or assumed analytically (e.g., taken from the Matérn class). In the latest scenario, we apply the Nyström approximation, which showed slightly better results (first row in Table 4).

In the numerical section, the problem of classification of CT scans for acute stroke recognition was considered. Different classification texture-based methods were tested. Our WSC-LR algorithm demonstrated better classification results than the semi-supervised SSC-LR algorithm, which does not consider uncertain labels. From the experiments, one may conclude that additional information on uncertain labeling helps WSC-LR achieve the best classification performance regarding balanced accuracy.

The proposed weakly supervised classification scheme based on low-rank similarity matrix approximation allows us to decrease the memory requirement of the manifold regularization-based algorithms from quadratic to linear (with respect to the dimension of the input data). The computing time is also reduced from cubic (matrix inversion in (6)) to linear. This improvement is especially vital for permanently growing datasets.

In the future, we plan to combine texture-based and deep convolutional neural nets methodologies to obtain easy-to-interpret and accurate classifications in the weakly supervised learning context.

References

1. Armi, L., Fekri-Ershad, S.: Texture image analysis and texture classification methods-a review. arXiv preprint arXiv:1904.06554 (2019)
2. Bekker, J., Davis, J.: Learning from positive and unlabeled data: a survey. Mach. Learn. **109**(4), 719–760 (2020)

3. Belkin, M., Niyogi, P., Sindhwani, V.: Manifold regularization: a geometric framework for learning from labeled and unlabeled examples. J. Mach. Learn. Res. **7**(85), 2399–2434 (2006). http://jmlr.org/papers/v7/belkin06a.html

4. Berikov, V., Litvinenko, A.: Weakly supervised regression using manifold regularization and low-rank matrix representation. In: Pardalos, P., Khachay, M., Kazakov, A. (eds.) MOTOR 2021. LNCS, vol. 12755, pp. 447–461. Springer, Cham (2021). https://doi.org/10.1007/978-3-030-77876-7_30

5. Berikov, V.: Semi-supervised classification using multiple clustering and low-rank matrix operations. In: Khachay, M., Kochetov, Y., Pardalos, P. (eds.) MOTOR 2019. LNCS, vol. 11548, pp. 529–540. Springer, Cham (2019). https://doi.org/10.1007/978-3-030-22629-9_37

6. Berikov, V., Litvinenko, A.: Semi-supervised regression using cluster ensemble and low-rank co-association matrix decomposition under uncertainties. In: Proceedings of 3rd International Conference on Uncertainty Quantification in CSE, pp. 229–242 (2020). https://doi.org/10.7712/120219.6338.18377. https://files.eccomasproceedia.org/papers/e-books/uncecomp_2019.pdf

7. Borisova, I.A., Zagoruiko, N.G.: Algorithm FRiS-TDR for generalized classification of the labeled, semi-labeled and unlabeled datasets. In: Aleskerov, F., Goldengorin, B., Pardalos, P.M. (eds.) Clusters, Orders, and Trees: Methods and Applications. SOIA, vol. 92, pp. 151–165. Springer, New York (2014). https://doi.org/10.1007/978-1-4939-0742-7_9

8. Bradski, G.: The OpenCV library. Dr. Dobb's J. Softw. Tools (2000)

9. Clausi, D.A.: An analysis of co-occurrence texture statistics as a function of grey level quantization. Can. J. Remote. Sens. **28**(1), 45–62 (2002)

10. Drineas, P., Mahoney, M.W., Cristianini, N.: On the Nyström method for approximating a gram matrix for improved kernel-based learning. J. Mach. Learn. Res. **6**, 2153–2175 (2005)

11. Espig, M., Hackbusch, W., Litvinenko, A., Matthies, H.G., Zander, E.: Iterative algorithms for the post-processing of high-dimensional data. J. Computat. Phys. **410**, 109396 (2020). https://doi.org/10.1016/j.jcp.2020.109396. https://www.sciencedirect.com/science/article/pii/S0021999120301704

12. Gao, W., Wang, L., Li, Y.F., Zhou, Z.H.: Risk minimization in the presence of label noise. In: Proceedings of the AAAI Conference on Artificial Intelligence, vol. 30, no. 1 (2016). https://ojs.aaai.org/index.php/AAAI/article/view/10293

13. Gaudioso, M., Giallombardo, G., Miglionico, G., Vocaturo, E.: Classification in the multiple instance learning framework via spherical separation. Soft. Comput. **24**(7), 5071–5077 (2019). https://doi.org/10.1007/s00500-019-04255-1

14. Grasedyck, L., Hackbusch, W.: Construction and arithmetics of $\mathcal{H}$-matrices. Computing **70**(4), 295–334 (2003)

15. Guttorp, P., Gneiting, T.: Studies in the history of probability and statistics XLIX: on the Matérn correlation family. Biometrika **93**, 989–995 (2006). https://doi.org/10.1093/biomet/93.4.989

16. Haralick, R.M., Shanmugam, K., Dinstein, I.H.: Textural features for image classification. IEEE Trans. Syst. Man Cybern. **6**, 610–621 (1973)

17. Hüllermeier, E., Waegeman, W.: Aleatoric and epistemic uncertainty in machine learning: an introduction to concepts and methods. Mach. Learn. **110**(3), 457–506 (2021)

18. Litvinenko, A., Keyes, D., Khoromskaia, V., Khoromskij, B.N., Matthies, H.G.: Tucker Tensor analysis of Matern functions in spatial statistics. Comput. Methods Appl. Math. **19**(1), 101–122 (2019). https://doi.org/10.1515/cmam-2018-0022

19. Litvinenko, A., Kriemann, R., Genton, M.G., Sun, Y., Keyes, D.E.: HLIBCov: parallel hierarchical matrix approximation of large covariance matrices and likelihoods with applications in parameter identification. MethodsX **7**, 100600 (2020). https://doi.org/10.1016/j.mex.2019.07.001. https://github.com/litvinen/HLIBCov.git

20. Litvinenko, A., Sun, Y., Genton, M.G., Keyes, D.E.: Likelihood approximation with hierarchical matrices for large spatial datasets. Comput. Stat. Data Anal. **137**, 115–132 (2019). https://doi.org/10.1016/j.csda.2019.02.002. https://github.com/litvinen/large_random_fields.git

21. Matérn, B.: Spatial Variation. Lecture Notes in Statistics, vol. 36, 2nd edn. Springer, Berlin (1986)

22. Muhlenbach, F., Lallich, S., Zighed, D.A.: Identifying and handling mislabelled instances. J. Intell. Inf. Syst. **22**(1), 89–109 (2004). https://doi.org/10.1023/A:1025832930864

23. Pardalos, P.M., Georgiev, P.G., Papajorgji, P., Neugaard, B.: Systems Analysis Tools for Better Health Care Delivery, vol. 74. Springer, Heidelberg (2013)

24. Raykar, V.C., et al.: Learning from crowds. J. Mach. Learn. Res. **11**(43), 1297–1322 (2010). http://jmlr.org/papers/v11/raykar10a.html

25. Saber, E.S., Tekalp, A.M.: Integration of color, edge, shape, and texture features for automatic region-based image annotation and retrieval. J. Electron. Imaging **7**(3), 684–700 (1998)

26. Skourt, B.A., El Hassani, A., Majda, A.: Lung CT image segmentation using deep neural networks. Procedia Comput. Sci. **127**, 109–113 (2018)

27. Soh, L.K., Tsatsoulis, C.: Texture analysis of SAR sea ice imagery using gray level co-occurrence matrices. IEEE Trans. Geosci. Remote Sens. **37**(2), 780–795 (1999)

28. Song, H., Kim, M., Park, D., Shin, Y., Lee, J.G.: Learning from noisy labels with deep neural networks: a survey. arXiv preprint arXiv:2007.08199 (2020)

29. van Engelen, J.E., Hoos, H.H.: A survey on semi-supervised learning. Mach. Learn. **109**(2), 373–440 (2019). https://doi.org/10.1007/s10994-019-05855-6

30. Xiao, Y., Yin, Z., Liu, B.: A similarity-based two-view multiple instance learning method for classification. Knowl.-Based Syst. **201–202**, 105661 (2020). https://doi.org/10.1016/j.knosys.2020.105661

31. Zhou, D., Bousquet, O., Lal, T.N., Weston, J., Schölkopf, B.: Learning with local and global consistency. In: Proceedings of the 16th International Conference on Neural Information Processing Systems, NIPS 2003, pp. 321–328. MIT Press, Cambridge (2003)

32. Zhou, Z.H.: Ensemble Methods: Foundations and Algorithms. CRCPress, Boca Raton (2012)

33. Zhou, Z.H.: A brief introduction to weakly supervised learning. Natl. Sci. Rev. **5**(1), 44–53 (2017). https://doi.org/10.1093/nsr/nwx106. https://academic.oup.com/nsr/article-pdf/5/1/44/31567770/nwx106.pdf

A Local Search Algorithm
for the Biclustering Problem

Tatyana Levanova[1,2]([✉])[ⓘ] and Ivan Khmara[1][ⓘ]

[1] Sobolev Institute of Mathematics, Omsk Branch,
Pevtsova str. 13, 644043 Omsk, Russia
levanovatv@omsu.ru, ivan-hmara@mail.ru
[2] Dostoevsky Omsk State University,
Prospekt Mira 55A, 644077 Omsk, Russia

Abstract. Biclustering is an approach to solving data mining problems, which consists in simultaneously grouping rows and columns of a matrix. In this paper, we solve the problem of finding a bicluster of the maximum size, the elements of which should differ from each other by no more than a given value. To solve it, a new local search algorithm has been developed, representing an iterative greedy search. For its implementation, problem-oriented neighborhoods are constructed, different rules for determining the difference of bicluster elements are used. The constructed algorithm is tested on various types of data, the results are compared with the well-known algorithm of Cheng and Church. In all the examples considered, the sizes of the found biclusters are not less than the biclusters of the Cheng and Church algorithm. At the same time, the difference between the elements of bicluster and their average value in most cases is smaller than for the Cheng and Church biclusters.

Keywords: Data mining · Biclustering · Heuristics · Local search

1 Introduction

In the tasks of identifying groups of genes that have common properties, searching for groups of Internet users with similar interests, analyzing social networks, and so on, there is a need to group data so that objects and their features are clustered simultaneously. The approach to solving this problem is called biclusterization.

This approach firstly was proposed by Hartigan in [12]. Its *Block clustering* uses variance to form clusters: the smaller the variance, the higher the quality of the bicluster. So that biclusters do not consist of only one element, the number of clusters is fixed. Hartigan notes that to obtain biclusters with other properties, various objective functions can be used in the algorithm.

This technique was actively developed almost 30 years later, when it became necessary to analyze large volumes of data. Cheng and Church [6] proposed a

---

Supported by the program of fundamental scientific researches of the SB RAS.

E. Burnaev et al. (Eds.): AIST 2021, LNCS 13217, pp. 330–344, 2022.
https://doi.org/10.1007/978-3-031-16500-9_27

biclusterization algorithm (*Node-deletion algorithm*) and applied it to biological data on gene expression. To assess the quality of biclusters, the mean squared residue MSR is used. They are looking for a bicluster for which $MSR < \delta$, here δ is the specified threshold value. Node-deletion algorithm is a greedy heuristic, the main idea of which is to delete and add rows and columns. First, several rows and/or columns are deleted simultaneously in the algorithm (Multiple node deletion phase). After that, there is a stage of multiple deletion of one row or column (Simple node deletion phase). Further, at the end of the algorithm, a row and/or column are repeatedly added to obtain the maximal bicluster (Addition phase). In one iteration, the algorithm builds a single bicluster with the given δ conditions. To construct a set of L biclusters, it is necessary to make L iterations. At the end of such iteration, the elements of the found bicluster are replaced randomly (Substitution phase).

The ideas of the CC algorithm were later used in the construction of new approaches to solving biclusterization problems, below we describe some examples of them.

Bleuler, Relic and Zitzler [4] combined the algorithm of Cheng and Church and the evolutionary algorithm. They were the first to apply the scheme of the evolutionary algorithm to the biclustering problem. Individuals are encoded with binary strings. The initial population is formed randomly with a uniform distribution according to the size of the solutions. Bit-wise mutation and tournament selection were used. To increase the size of biclusters the CC algorithm is applied as local search. MSR is used to calculate the fitness function. It is shown that combination of the evolutionary algorithm and the greedy heuristics of Cheng and Church works better than these algorithms separately.

Gallo et al. [10] also proposed a hybrid approach combining the evolutionary algorithm and the Cheng&Church algorithm. Bleuler's evolutionary algorithm and Gallo's hybrid algorithm (BiHEA) have similar ideas. In particular, in both algorithms, local search is performed using the CC algorithm. The difference lies in the crossover operator and the fitness function. BiHEA uses a two-point crossover, and the fitness function includes genetic variation. The novelty consists in working with the best biclusters found. Firstly, at the current iteration, a given number of the best biclusters (the so-called elite) are selected and included in the new population without changes. Secondly, the best biclusters found are stored throughout the algorithm.

Mitra and Bank [16] (M&B) were the first to propose a multi-criteria evolutionary algorithm for the biclusterization problem. Their work is based on a Non-Dominated Sorting Genetic algorithm NDGA-II [7]. Similarly to the CC algorithm, the search for biclusters with the maximum size is performed under the condition $MSR < \delta$. To improve the solution, Multiple nodes deletion, Single node deletion, Multiple nodes addition phases are used as in the Cheng&Church algorithm.

Some algorithms are based on a traditional one-dimensional clustering algorithm using an additional strategy that provides analysis of the second dimension. Yang et al. [23] used traditional clustering together with Single Value

Decomposition. At the final stage, these authors combine and filter clusters. In more detail, the idea behind this approach is as follows. In the first step, two different matrices for genes and conditions are obtained by using Single Value Decomposition. Secondly, after preprocessing, clustering is applied to both of two matrices by the Mixed Clustering algorithm, based on agglomerative hierarchical clustering. As a measure of the difference between the submatrices, the correlation score of the submatrix are used. As a result of these steps, from two matrices, two sets are obtained, consisting of m and n clusters. Of these, $m \cdot n$ biclusters are obtained, but some of them can not be a δ-corBicluster. A δ-corBicluster is defined as a bicluster with a submatrix correlation score lower than δ. To build δ-corBiclusters of the maximal size, at the final stage the Lift algorithm is done using Cheng and Church's idea of removing and adding nodes.

The article by Cheng and Church is still relevant and is used to search for biclusters, comparative analysis, and evaluation of the quality of algorithms. For example, the CC algorithm is used in a software platform Biclustering Analysis Toolbox (BicAT) [1]. The BICAT system contains algorithms for biclusterization and clusterization of genetic data. In addition to the Cheng&Church algorithm, it includes BiMax [20], ISA [3], xMotifs [18], OPSM [2] methods.

Binary inclusion-Maximal algorithm (BiMax) [20] applies graph theory to bicluster detection. The results of the experiment are presented in the form of a binary matrix E, the elements of which are equal to 1 if the condition affects the gene and 0 otherwise. A bicluster is defined as a submatrix of a matrix E for which all elements are equal to 1. Taking into account the values of the elements, the set of columns of the matrix E is divided into two subsets, forming two submatrices. Then the rows of the matrix are divided into three subsets in a special way. After rearranging the rows, three types of submatrices are formed. Submatrices consisting of 1 form a biclusters. Submatrices consisting only of zero elements are excluded from further consideration. For matrices consisting of 0 and 1, the process continues. From the point of view of graph theory, this algorithm performs a search for all maximal bicliques in a corresponding graph-based matrix representation.

Biclusterization algorithms were further developed in many works. Due to the NP-complexity of many biclusterization problems, the algorithms are usually heuristic in nature. Nowadays in genetics, it has become insufficient to consider the simultaneous dependence of genes and conditions and vice versa. In modern algorithms for a more adequate explanation of the results, it became necessary to take into account the relationships between genes [15]. Recommendation systems are an important part of many online services, as they make it easier to work with large datasets. The biclusterization technique is actively used in recommendation systems [17].

In addition to algorithms using metrics, some of which were described above, there are algorithms that have other mechanisms. These are the algorithms that use graphs [20,24], probabilistic models [18,21], linear algebra [3,13], and so on (see, for example [8,9,14]). More detailed information is provided in the reviews [5,19,22].

In this paper, to solve the biclusterization problem, a new local search algorithm is proposed, which is an iterative greedy search. It employs the idea of deleting and adding rows and columns in a similar way to Cheng and Church's algorithm, differing in local search in the neighborhoods and metric. The constructed algorithm is tested on various types of data, the results are compared with the well-known algorithm of Cheng and Church.

The paper is organized as follows. In Sect. 2, the notation, statement and mathematical model for the biclusterization problem are presented. A new local search algorithm and well known CC algorithm are introduced in Sect. 3. The results of the experimental studies are described in Sect. 4. Finally, some conclusions are given.

2   The Biclusterization Problem

Biclusterization problems can have various formulations. In addition, the definition of a bicluster can be different. In this section, we will formulate the biclusterization problem under consideration. Before that, we will provide short information and give the necessary definitions and notation.

2.1   Some Information and Definitions

The general definition of a bicluster is as follows. Suppose that we are given a matrix of data $A = (a_{ij})$, with a set of rows I and a set of columns J; $i \in I, j \in J$. Let $\overline{I} \subseteq I$ and $\overline{J} \subseteq J$ be subsets of rows and columns. A bicluster B is a submatrix $\overline{A} = (\overline{I}, \overline{J})$ of a matrix A, such that its rows exhibit similar behavior on columns, and columns on features. The biclustering problem is to find a set $B = \{A_k\}$ consisting of k biclusters, $k \geq 1$. Thus, as a result, one and several biclusters can be obtained.

Biclusters can have different structures. The elements of matrix A can be only included in one bicluster, no more than one bicluster or in several biclusters. Depending on this, the biclusters may overlap or have no common elements. To determine the difference of bicluster elements, distance, deviation from the average value, and other functions can be used [6].

The above definition of a bicluster is not the only one. There are other definitions of it in the literature. It depends on which biclusterization problem needs to be solved. For example, in the definition of a bicluster, you can consider restrictions on the content of a bicluster or its size. It is possible to take into account the structure of the bicluster in this definition etc. [17]. In our research, we will use the general definition of a bicluster.

2.2   Problem Statement

In this paper, we consider the biclustering problem, which consists of the following: for a given matrix, it is necessary to find a bicluster of the maximum size such that the difference of its elements does not exceed the given value.

Definition 1 (Solution). The solution will be called a bicluster (submatrix) $\overline{A} \subseteq A$, $\overline{A} = (\overline{I}, \overline{J})$, $\overline{I}$ is a set of rows, $\overline{J}$ is a set of columns.

Definition 2 (Size). The size of the bicluster will be called the total number of elements included in the bicluster.

For a matrix A with m rows and n columns, the standard deviation is calculated by the formula:

$$S = \sqrt{\frac{\sum_{i=1}^{n} \sum_{j=1}^{n} \left(a_{ij} - \sum_{i=1}^{n} \sum_{j=1}^{n} \frac{a_{ij}}{mn} \right)^2}{mn}}.$$

Let us modify this formula for a bicluster. We will introduce the variables: x_i is equal to 1 if the row with the number i is included in the bicluster; $y_j = 1$ if the column with the number j is included in the bicluster; otherwise $x_i = 0$, $y_j = 0, i \in I, j \in J$. Define the number of rows in the bicluster as

$$m' = \sum_{i=1}^{m} x_i,$$

and the number of columns in the bicluster as

$$n' = \sum_{j=1}^{n} x_j.$$

Then the size of the bicluster is equal to

$$m' \cdot n' = \sum_{i=1}^{m} x_i \sum_{j=1}^{n} x_j.$$

An element is included in a bicluster if $x_i \cdot y_j = 1$. Using these notations, the formula for the standard deviation for a bicluster $\overline{A}$ has the form:

$$S(\overline{A}) = \sqrt{\frac{\sum_{i \in I} \sum_{j \in j} \left(a_{ij} x_i y_j - \sum_{i \in I} \sum_{j \in J} \frac{a_{ij} x_i y_j}{\sum_{i \in I} x_i \sum_{j \in J} y_j} \right)^2}{\sum_{i \in I} x_i \sum_{j \in J} y_j}}. \tag{1}$$

Then the mathematical model of the biclusterization problem under consideration looks like:

$$\max \sum_{i \in I} x_i \sum_{j \in J} y_j, \tag{2}$$

$$\sqrt{\frac{\sum_{i \in I} \sum_{j \in j} \left(a_{ij} x_i y_j - \sum_{i \in I} \sum_{j \in J} \frac{a_{ij} x_i y_j}{\sum_{i \in I} x_i \sum_{j \in J} y_j} \right)^2}{\sum_{i \in I} x_i \sum_{j \in J} y_j}} \leq \delta, \tag{3}$$

$$x_i, y_j \in \{0; 1\}; i \in I, j \in J. \tag{4}$$

The expression (2) reflects the goal to find a bicluster of the maximum size. The constraint (3) guarantees that the elements should differ by no more than the given value of threshold δ.

3 Solution Algorithms

This article presents a new local search algorithm for biclusterization problem. It is a variant of the greedy heuristic, which is a local search for a submatrix of the maximum possible size. This submatrix has a characteristic of the difference of elements, which should not exceed the given value. The algorithm consists of a special purposeful deletion and addition of rows and columns. Two variants of the new algorithm are presented, which differ in the way of determining the difference between the elements.

3.1 Local Search Algorithms for Biclusterization Problem

An important stage in the development of local search algorithms is the construction of problem-oriented neighborhoods. To solve the biclusterization problem, two types of neighborhoods were proposed in this paper.

Definition 3 (Neighborhood 1). The external neighborhood $N_{out}(\overline{A})$ a solution $\overline{A}$ we will call the set of solutions $\hat{A}$ obtained from $\overline{A}$ by adding a row $\hat{i} \in I \setminus \overline{I}$ and (or) a column $\hat{j} \in J \setminus \overline{J}$.

Definition 4 (Neighborhood 2). By the internal neighborhood $N_{in}(\overline{A})$ a solution $\overline{A}$ we will call the set of solutions $\tilde{A}$ obtained from $\overline{A}$ by deleting a row $\tilde{i} \in \overline{I}$ or a column $\tilde{j} \in \overline{J}$.

The following notation is used in the algorithm:
$A = (I, J)$ is a matrix, I is a set of rows, J is a set of columns;
$\overline{A} \subseteq A$ is a submatrix, $\overline{A} = (\overline{I}, \overline{J})$, $\overline{I}$ is a set of rows, $\overline{J}$ is a set of columns;
$\overline{A}_i$ is a vector-row of the matrix $\overline{A}$, $i \in \overline{I}$;
$\overline{A}_j$ is a vector-column of the matrix $\overline{A}$, $j \in \overline{J}$;
$\hat{I} = I \setminus \overline{I}$ is a set of deleted rows;
$\hat{J} = J \setminus \overline{J}$ is a set of deleted columns;
$R(\overline{A}_i) = (r_i)$ is a vector of string characteristics, $i \in \overline{I}$;
$C(\overline{A}_j) = (c_j)$ is a vector of column characteristics, $j \in \overline{J}$;
$S(\overline{A})$ is a characteristic of the matrix $\overline{A}$ calculated by the formula (1);
δ is the threshold value, $\delta = 0.25 \cdot A_{IJ}$, where A_{IJ}, is a average value for all elements of the matrix; $(.)^{(t)}$ denotes the value on the iteration number t.

Describe the scheme of the algorithm.

Scheme of the bicluster search algorithm

Step 0. The matrix A is given; calculate the threshold δ; current solution $\overline{A} := A; t := 1, flag := 0$.

Iteration $t, t := t + 1$.

Step 1. Calculate the characteristic $S(\overline{A})$ for the submatrix $\overline{A}$.

Step 2. Choose a new solution in the internal neighborhood of $N_{in}(\overline{A})$. Perform the following actions.

2.1. Calculate the characteristics of vector-rows $R = (r_i)$ and the vector-column $C = (c_j), i \in \overline{I}, j \in \overline{J}$ as follows:

$$r^{(t)} = \max_i r_i, i \in \overline{I}, \overline{i} = argmax \ r^{(t)};$$

$$c^{(t)} = \max_j c_j, j \in \overline{J}, \overline{j} = argmax \ c^{(t)}.$$

2.2. Determine the largest deviation from the characteristic $S(\overline{A})$ for rows and columns:

$$\Delta r^{(t)} = r^{(t)} - S^{(t)}(\overline{A});$$

$$\Delta c^{(t)} = c^{(t)} - S^{(t)}(\overline{A}).$$

2.3. Delete a row or column the following way.
If $\Delta r^{(t)} \geq \Delta c^{(t)}$ & $\Delta r^{(t)} > 0$, then $\overline{I} := \overline{I} \backslash \overline{i}, flag := 1$;
Otherwise, if $\Delta r^{(t)} < \Delta c^{(t)}$ & $\Delta c^{(t)} > 0$, then $\overline{J} := \overline{J} \backslash \overline{j}, flag := 1$.
Recalculate the characteristic $S(\overline{A})$.

Step 3. Choose a new solution in the external neighborhood of $N_{out}(\overline{A})$.

3.1. Calculate the characteristics of the vector-rows $R = (r_i)$ and the vector-columns $C = (c_j), i \in \overline{I}, j \in \overline{J}$;

$$r^{(t)} = \min_i r_i, i \in \hat{I}, \hat{i} = argmin \ r^{(t)};$$

$$c^{(t)} = \min_j c_j, j \in \hat{J}, \hat{j} = argmin \ c^{(t)}.$$

3.2. Recalculate deviations from the characteristic $S(\overline{A})$ for rows and columns:

$$\Delta r^{(t)} = r^{(t)} - S^{(t)}(\overline{A});$$

$$\Delta c^{(t)} = c^{(t)} - S^{(t)}(\overline{A}).$$

3.3. Add a row and/or column.
If $\Delta r^{(t)} \leq 0$, then add the row $\overline{I} := \overline{I} \bigcup \{\hat{i}\}, flag := 1$.
If $\Delta c^{(t)} \leq 0$, then add the column $\overline{J} := \overline{J} \bigcup \{\hat{j}\}, flag := 1$.

Step 4. If at least one change has occurred (a row and/or column has been added and/or deleted) or $S(\overline{A}) > \delta$ then we move on to the next iteration. Otherwise, the algorithm stops.

Based on this scheme, two variants of local search algorithms for the biclustering problem are constructed. In Variant 1, the formula (5) is used to determine the difference between the elements (see below), and in Variant 2, this characteristic is calculated by the formula (1).

If you need to get several biclusters, then all the elements included in the solution are replaced with random values with a uniform distribution in the original matrix. Then the algorithm resumes, this is a generally accepted method. Such actions occur until the given number of biclusters is received.

To determine the quality of the proposed algorithm, we compare their work with the Cheng&Church algorithm (CC). These authors were the first to apply biclusterization to the problem of gene expression data analysis To determine the difference between the bicluster elements, Cheng and Church use Mean Squared Residue (MSR), which is calculated as follows:

$$MSR(B) = \frac{1}{|I| \cdot |J|} \cdot \sum_{i=1}^{|I|} \sum_{j=1}^{|J|} (b_{ij} - b_{iJ} - b_{Ij} + b_{IJ})^2, \tag{5}$$

where b_{ij} is an element of the matrix B; b_{iJ}; b_{Ij} and b_{IJ} represent the average values in the row i, column j and the average value for all elements of the matrix B, respectively. They are looking for a bicluster for which $\alpha < MSR < \delta$, here δ is the specified threshold value, α is given part of MSR. A scheme of the Cheng&Church algorithm is shown below [6].

Scheme of the Cheng&Church algorithm

Input: Expression Matrix EM; Threshold δ.
Output: List of Biclusters L.
Step 1. Preprocess the missing values of EM.
Step 2. List $L = 0$.
Step 3. Bicluster B.
Step 4. $B = EM$.
Step 5. B_δ = multiple node deletion phase (B, δ).
Step 6. B'_δ = simple node deletion phase (B_δ, δ).
Step 7. B''_δ = addition phase (B'_δ).
Step 8. $L = L + B''_\delta$.
Step 9. Substitution phase B''_δ, EM.
Step 10. End_repeat.
Step 11. Return L.

Note the differences in the steps of the algorithm proposed in this paper and the Cheng&Church algorithm. The differences are shown in the Table 1. The main difference is indicated in point 3. It consists in the fact that when the CC algorithm finishes working, the new algorithm can continue the search.

Table 1. Difference between algorithms.

No.	Cheng&Church algorithm	Proposed algorithms
1	Simultaneous deletion of multiple rows and/or columns	Simultaneous deletion of multiple rows and/or columns does not occur
2	Multiple deletion of one row or column	Single deletion of one row or column
3	Multiple addition of one row and/or column at the end of the iteration	Multiple addition of one row and/or column inside the iteration
4	Characteristics of the difference: MSR (5)	Characteristics of the difference: Variant 1: MSR (5) Variant 1: $S(\overline{A})$ (1)

4　Experimental Studies

Two variants of the new algorithm and the Cheng&Church algorithm were implemented in Java. The experiment was conducted on a computer with an Apple M1 processor with 8.0 GB of RAM. Since the algorithms are deterministic, each algorithm was run once. Two datasets were used to conduct the computational experiment. The first datasets is based on the school electronic diary. The well-known library Jester Databases for Recommender Systems and Collaborative Filtering Research [11] was used as the second datasets. Let's describe a datasets of test tasks.

4.1　School Datasets

In this experiment, datasets were taken from the electronic diary of students. In the first set, 99 students became objects, their final grades in 15 subjects for the first quarter became signs, i.e. the input matrix A has 99 rows and 15 columns. The purpose of this example is to get 1 bicluster (Table 2). Figure 1 is an example of first school datasets with visualized as the heatmap. In the black and white diagram, darker rectangles correspond to higher grades. It can be seen that the heatmap (c) has a more uniform color.

In the second set, the objects were 26 students and the signs were their final grades in 15 subjects, the input matrix A has 26 rows and 15 columns. The purpose of this example is to get 2 biclusters.

Thus, students with similar success in some subjects mostly got into biclusters. Let's pay attention to the solution that was obtained by Variant 1. Here, the trend of the variability of elements along the lines is observed. Note that the bicluster obtained during the operation of Variant 2 has the maximum dimension and the minimum characteristic of the difference of elements among the presented biclusters. The operation time of each of the three algorithms did not exceed 2 s. In the Table 3 the first bicluster for each algorithm for second datasets is presented.

Table 2. Bicluster for First School Datasets.

	Algorithm CC	Variant 1	Variant 2
Bicluster $\overline{A}$	444545554	4454554	4554555
	555555555	4454554	5555555
	444545554	4454554	5555555
	454545554	4454554	4555545
	443445544	4454554	4554555
	555555555	3343443	5555555
	444545544	3343443	5554555
	444545544	4454554	4555555
	333434433	4454554	4545545
	444545544	4454554	4545545
	444544544	4454554	4545555
	444445544	4454554	5545455
	444545544	4454554	5445455
	454555544	4454554	4545545
	454555544	4454554	5555545
	454545544	4454554	5555545
	444545544	4454554	5545545
	454545544	4454554	4544555
	444445544	4454554	5555555
	444545544	4454554	4555545
	554555554		4555545
	554555554		4555545
	444545544		4545555
	444545544		4555545
	554555554		4555555
	555555554		4555555
	454545554		5555555
	555555555		5554555
	444545554		5555555
	444545544		5544555
			5545555
			4554555
			4555555
			5554545
			5554555
			5554555
			5545545
			5555555
			4555555
			4554554
			4545555
			5555555
			5554555
			5555555
Dimension A	$99 \cdot 15 = 1485$		
Dimension $\overline{A}$	$30 \cdot 9 = 270$	$20 \cdot 7 = 140$	$44 \cdot 7 = 308$
$H(\overline{A})$	0,692	0,578	0,403

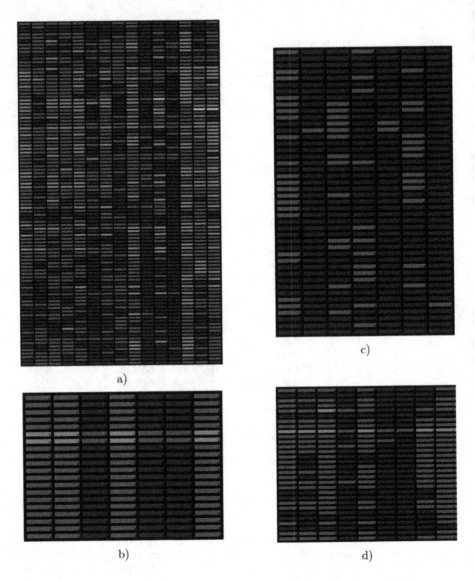

Fig. 1. Heatmap of the first school datasets: a) Full matrix, b) Bicluster using Variant 1, c) Bicluster using Variant 2, d) Bicluster using Cheng&Church algorithm.

Thus, students with similar success in some subjects got into biclusters. Let's pay attention to the solution that was obtained by Variant 1. Here, as in the previous examples, the trend of the variability of elements along the lines is preserved. Note that among the presented biclusters, the maximum size is observed for the bicluster obtained during the operation of Variant 2, and the minimum difference characteristic is for the bicluster obtained by Variant 1, but the size

Table 3. First Bicluster for Second School Datasets.

	Algorithm CC	Variant 1	Variant 2
Bicluster $\overline{A}$	44545455544	55545554	555455555555
	55555455545	55545554	555454555555
	55455455545	55545554	544555555555
	55555555555	55545554	544445455555
	54545455545	55545554	544555455555
	55545455545	55545554	544545455555
	54545455545	55545554	544455455555
	55545455545	55545554	545445455554
	54545455544	55545554	554545555554
	54555455544	55545554	554555555555
	55555555545	55545554	555445454555
	54545445545		545555555555
	55555455545		445455445444
	55555455545		545455555555
	54555455545		555445555555
	44545455445		444445455545
	55555455545		454455455554
	45555455545		544555555555
Dimension A	$26 \cdot 15 = 390$		
Dimension $\overline{A_1}$	$18 \cdot 11 = 198$	$11 \cdot 8 = 88$	$18 \cdot 12 = 216$
$H(\overline{A})$	0,597	0,433	0,447

of this bicluster is less than 2 times smaller than the bicluster obtained during the operation of Variant 2.

In the Table 4 the second bicluster for each of the three algorithms is presented. Recall that before getting the second biclusters, the elements that include into the first biclusters were replaced in the original matrix with random numbers with a uniform distribution in the range from 0 to 20. As a result, the second bicluster obtained by the CC algorithm contains elements that do not correspond to school grades, which means that these are elements that were replaced with random numbers, i.e., elements that included into the first bicluster. As for the result of Variant 1, the resulting bicluster also preserves the trend of variability in rows. During the work of the Variant 2, a bicluster was obtained, in which there is a group of children who mainly have a good rating in some subjects, in the first bicluster the excellent rating prevailed. The maximum size of the bicluster and the minimum characteristic of the difference of elements among the presented biclusters is observed in the bicluster found by Variant 2. The operation time of each of the three algorithms did not exceed 2 s.

Table 4. Second Bicluster for Second School Datasets.

	Algorithm CC	Variant 1	Variant 2
Bicluster $\overline{A}$	9 5	44444443	54444444
	13 10	44444443	44444444
	12 9	55555554	45445554
		55555554	44554545
		55555554	44454544
Dimension A	$26 \cdot 15 = 390$		
Dimension $\overline{A}_2$	$3 \cdot 2 = 6$	$5 \cdot 8 = 40$	$6 \cdot 8 = 48$
$H(\overline{A})$	5,183	0,591	0,446

4.2 Jester Datasets Library

To conduct this experiment, the datasets was taken from the Jester Datasets library for Recommendation Systems and Collaborative Filtering Studies [11]. The subjects were offered 100 jokes, which they had to evaluate. The rating range is from -1000 to 1000. For the experiment, we took the first 300 people who gave an assessment of each of the proposed jokes. Due to the fact that the data volume is quite large, the input matrix is A, we will not give the results of biclusters. We will only give the dimension and the obtained characteristics of the found biclusters (see Table 5).

Table 5. Bicluster for Jester Datasets.

	Algorithm CC	Variant 1	Variant 2
Dimension A	$300 \cdot 100 = 30000$		
Dimension $\overline{A}$	$2 \cdot 2 = 4$	$2 \cdot 4 = 8$	$2 \cdot 5 = 10$
$H(\overline{A})$	434,588	204,0	17,432

Note that the size of the obtained biclusters is extremely small compared to the matrix supplied to the input. This means that there is a very different sense of humor among these people. Nevertheless, the maximum dimension and the minimum characteristic of the difference of elements are observed in the bicluster found by Variant 2. The operation time of each of the three algorithms did not exceed 4 s.

5 Conclusions

We have considered the problem of biclusterization, which is relevant today. The new local search algorithm has been developed for the biclusterization problem. To assess the quality of the proposed algorithms, computational studies were conducted, a comparison with the Cheng&Church algorithm was performed. The analysis of the results showed that many objects with similar features fall into biclusters, which requires solving the biclusterization problem. Comparing the results of the work it was noticed that Variant 2 of the new algorithm in all the considered examples finds a bicluster of size no less than the sizes of the biclusters obtained as a result of the work of the other two algorithms. In addition, the lowest value of the characteristic of the difference of elements is almost always achieved on the same version of the new algorithm. Thus, the new algorithm in this implementation shows an advantage over the Cheng&Church algorithm.

References

1. Barkow, S., Bleuler, S., Prelic, A., Zimmermann, P., Zitzler, E.: BicAT: a biclustering analysis toolbox. Bioinformatics $22(10)$, 1282–1283 (2006)
2. Ben-Dor, A., Chor, B., Karp, R., Yakhini, Z.: Discovering local structure in gene expression data: the order-preserving submatrix problem. J. Comput. Biol. 10(3–4), 373–384 (2003)
3. Bergmann, S., Ihmels, J., Barkai, N.: Iterative signature algorithm for the analysis of large-scale gene expression data. Phys. Rev. E 67, 031902 (2003)
4. Bleuler, S., Prelic, A., Zitzler, E.: An EA framework for biclustering of gene expression data. In: Congress on Evolutionary Computation, pp. 166–173 (2004)
5. Busygin, S., Prokopyev, O., Pardalos, P.: Biclustering in data mining. Comput. Oper. Res. 35, 2964–2987 (2008)
6. Cheng, Y., Church, G.: Biclustering of expression data. In: 8th International Conference on Intelligent System for Molecular Biology, pp. 93–103 (2000)
7. Deb, K., Agrawal, S., Pratap, A., Meyarivan, T.: A fast elitist non-dominated sorting genetic algorithm for multi-objective optimization: NSGA-II. In: Schoenauer, M., et al. (eds.) PPSN 2000. LNCS, vol. 1917, pp. 849–858. Springer, Heidelberg (2000). https://doi.org/10.1007/3-540-45356-3_83
8. Dharan, S., Nair, A.: Biclustering of gene expression data using reactive greedy randomized adaptive search procedure. BMC Bioinform. 10(Suppl. 1, S27), 2964–2987 (2009)
9. Fan, N., Pardalos, P.: Multi-way clustering and biclustering by the ratio cut and normalized cut in graphs. J. Comb. Optim. 23, 224–251 (2012)
10. Gallo, C.A., Carballido, J.A., Ponzoni, I.: BiHEA: a hybrid evolutionary approach for microarray biclustering. In: Guimarães, K.S., Panchenko, A., Przytycka, T.M. (eds.) BSB 2009. LNCS, vol. 5676, pp. 36–47. Springer, Heidelberg (2009). https://doi.org/10.1007/978-3-642-03223-3_4
11. Goldberg, K.: Jester datasets for recommender systems and collaborative filtering research. http://eigentaste.berkeley.edu/dataset
12. Hartigan, J.A.: Direct clustering of a data matrix. J. Am. Stat. Assoc. 67(337), 123–129 (1972)
13. Henriques, R., Madeira, S.: BicPAM: pattern-based biclustering for biomedical data analysis. Algorithms Mol. Biol. 9(27), 224–251 (2014)
14. Ignatov, D.I., Kuznetsov, S.O., Poelmans, J.: Concept-based biclustering for internet advertisement. In: 2012 IEEE 12th International Conference on Data Mining Workshops, pp. 123–130 (2012)

15. Lazareva, O., et al.: BiCoN: network-constrained biclustering of patients and omics data. Bioinformatics **37**(16), 2398–2404 (2020)
16. Mitra, S., Banka, H.: Multi-objective evolutionary biclustering of gene expression data. Pattern Recogn. **39**(12), 2464–2477 (2006)
17. Moor, F.: A biclustering approach to recommender systems. In: Machine Learning (2019)
18. Murali, T.M., Kasif, S.: Extracting conserved gene expression motifs from gene expression data. In: Pacific Symposium on Biocomputing, pp. 77–88 (2003)
19. Pontes, B., Giráldez, R., Aguilar-Ruiz, J.: Biclustering on expression data: a review. J. Biomed. Inform. **57**, 163–180 (2004)
20. Prelić, A., et al.: A systematic comparison and evaluation of biclustering methods for gene expression data. Bioinformatics **22**(9), 1122–1129 (2006)
21. Segal, E., Taskar, B., Gasch, A., Friedman, N., Koller, D.: Rich probabilistic models for gene expression (2001)
22. Tanay, A., Sharan, R., Shamir, R.: Biclustering algorithms: a survey. Handb. Comput. Mol. Biol. **9** (2005)
23. Yang, W.H., Dai, D.Q., Yan, H.: Finding correlated biclusters from gene expression data. IEEE Trans. Knowl. Data Eng. **23**(4), 568–584 (2011)
24. Zhao, L., Zaki, M.: MicroCluster: efficient deterministic biclustering of microarray data. EEE Intell. Syst. **20**, 40–49 (2005)

Author Index

Printed in the United States
by Baker & Taylor Publisher Services